WORLD REVIEW

Environmental and Sustainability Education in the Context of the Sustainable Development Goals

Editors

Marco Rieckmann

Faculty I
Department of Education
University of Vechta
Germany

Rosalba Thomas Muñoz

University Center for Environmental Management
University of Colima
Mexico

CRC Press
Taylor & Francis Group
Boca Raton London New York

CRC Press is an imprint of the
Taylor & Francis Group, an **informa** business

A SCIENCE PUBLISHERS BOOK

First edition published 2024
by CRC Press
2385 NW Executive Center Drive, Suite 320, Boca Raton FL 33431

and by CRC Press
4 Park Square, Milton Park, Abingdon, Oxon, OX14 4RN

Library of Congress Cataloging-in-Publication Data (applied for)

ISBN: 978-0-367-70242-7 (hbk)
ISBN: 978-0-367-70243-4 (pbk)
ISBN: 978-1-003-14520-2 (ebk)

DOI: 10.1201/9781003145202

Typeset in Times New Roman
by Radiant Productions

Preface

The primary aim of this book is to provide a comprehensive overview of how countries across five continents have grappled with the integration of environmental and sustainability education (ESE) into their educational systems, all within the framework of the UN Sustainable Development Goals. We are deeply honored to have attracted leading scholars and experts in the field, who have generously contributed their unique insights and analyses on the pertinent issues. All of them share a steadfast commitment to ensuring that education serves as an effective cornerstone of sustainability. We extend our heartfelt appreciation to them, as well as to all those who played a role in the creation of this book.

The chapters within this volume serve as windows into the global landscape of environmental and sustainability education. Each country's distinct context, encompassing economic, political, environmental, cultural, and social factors, is meticulously examined. Historical references are provided to illuminate the unique journeys of each nation towards environmental and sustainability education, while also addressing the challenges, achievements, and limitations of these endeavors.

These chapters explore how each country interprets environmental and sustainability education within its own unique context, elucidating their approaches to monitoring progress in alignment with the Global Agenda. Emerging issues, driven by the specific needs of each country, are thoughtfully explored. This diversity raises important questions about whether a uniform discourse can provide tailored solutions to address the distinct global challenges faced by individual countries.

Our aim extends beyond the mere presentation of information; we aspire to stimulate critical reflection. We encourage readers to engage in introspection, mirroring the discerning assessments made by our esteemed contributing authors. This introspection empowers readers to actively participate in the processes within their own countries, with a particular emphasis on implementing concrete and pragmatic solutions that extend beyond rhetoric.

This volume is intended for scholars experienced in the field of environmental and sustainability education, who wish to reflect on their own findings against the backdrop of other countries' ESE histories. It is also intended for doctoral students who may need contextual knowledge about environmental and sustainability education in specific countries for international comparative studies, as well as for students interested in studying the global development of environmental and sustainability education.

In assembling this book, our aspiration is to provide readers with valuable insights, ignite meaningful discussions, and inspire proactive engagement in the realm of environmental and sustainability education. We invite you to delve into the diverse perspectives presented within these pages and contemplate the pivotal role of education in addressing global challenges. Together, let us endeavor to make effective and meaningful contributions to a sustainable future.

Marco Rieckmann
Rosalba Thomas Muñoz

Contents

Oceania

Introduction

Marco Rieckmann[a] and *Rosalba Thomas Muñoz*[b,*]

The relevance of environmental and sustainability education in times of global crisis

There is no doubt that life in educational institutions has been transformed by the Covid-19 pandemic. The challenges it presented for education were adapting to virtual delivery, following up processes that had initially been face-to-face and then migrating them to new environments. These challenges were encountered throughout the world, but were not experienced with the same intensity in all countries. The differential vulnerability of different locations arose from varying levels of marginalization, poverty, economic crises, academic disadvantage, cultural homogenization, inadequate legal and political frameworks and, of course, the environmental issues that are resulting from people's way of life around the world.

Environmental and sustainability education (ESE)[1] has been reflecting on these causes for several decades. With regard to the ecological challenges, it has continually argued that our habitats impose limits on us and that we must take account of those if we want to continue to enjoy the comforts that our surroundings offer. In some areas, the environment is still so rich that people are not yet even noticing the (global) crises; in others, the comforts have been gone for several decades and the inhabitants of those regions have long been asking us to acknowledge their experience. In actual fact, countries all over the world are in the same position. We have reached the limits of growth.

The sustainability discourse that emerged as a result of a profound reflection on these limits has provided much evidence. Countries with severe environmental

---

[a] University of Vechta, Germany.
[b] University of Colima, Mexico.
Email: marco.rieckmann@uni-vechta.de
* Corresponding author: rosthomas@ucol.mx

[1] In different parts of the world, different terms are used to refer to educational concepts that relate to the sustainable transformation of societies, including environmental education, global education, education for sustainable development, education for sustainability, and sustainability education. In order not to commit ourselves to one single term here and thus also to reflect the diversity of terms used in the chapters of this book, we are deliberately using the collective term "environmental and sustainability education".

damage, high rates of poverty, inequality and high social marginalization are just some of the examples. The concept of sustainable development, with its transformative mindset, challenges the model of civilization that has been the norm for several centuries (cf. Hopwood et al. 2005).

ESE sets out serious needs and urgent aspirations. The needs include adapting international institutional frameworks, such as the Agenda 2030 and the Sustainable Development Goals (SDGs), Climate Empowerment Actions, among others, to local contexts. It is important we learn how to tailor these policies to specific local needs, conserving energy and resources and addressing each country's risks at local level. And this has to happen in circumstances that are still characterized by pedagogical emergencies, the return to face-to-face teaching, the need for flexibility in the face of this change, historical lags in learning, bureaucratic evaluations, etc. Meanwhile, the search is also now on for new pedagogical models that explain the transformation of the world's different spheres, scales, sectors, values and perspectives, in order to ensure education remains relevant to students' experiences.

The power of education is unquestionable, as is the fact that the education sector is not the only one that facilitates learning and that, in many cases, it is not even the most effective. All those who consider themselves environmental and sustainability educators face the challenge of dealing with the particularities of each institution, sector or community that wishes to be involved with this process of development.

ESE has been delivered for decades all over the world. The UN Decade of Education for Sustainable Development (ESD), the subsequent Global Action Program on ESD and the current ESD for 2030 program (UNESCO 2020) have provided many stimuli to promote and deliver ESE at all levels and in all areas of education. SDG 4 calls for ESE to be integrated into all education systems worldwide, and at the same time, ESE contributes to progress towards the SDGs (UNESCO 2017). Although international consensus is increasingly emerging on the importance of ESE and its and goals, there are also many regional differences when it comes to understanding and delivering ESE.

The purpose of this book

This book aims to provide an overview of work with and delivery of ESE worldwide. It deals with developments in different countries as well as the differential understandings of ESE and progress with its delivery in the context of Agenda 2030 and particularly the SDGs. It provides readers with an international overview of the latest developments and trends in ESE, specific country case studies and the history and likely future needs with regard to ESE in these countries.

We consider this analysis to be of great importance, because it highlights the characteristics of different national discourses on ESE and the flexibility that must be exercised by those who seek to deliver these types of global discussion programs and activities; the success of ESE is largely dependent on such flexibility.

The target readership for the book are environmental and sustainability educators and researchers from all over the world, who wish to engage with the global context of ESE in order to inspire their own educational practice, their work in education policy or their research projects.

The structure and contents of this book

The book presents the latest developments with ESE, in the context of the Agenda 2030 and the SDGs, in selected countries—from Africa: Kenya, South Africa, and Uganda; from Asia: India and Japan; from Europe: Austria, the Czech Republic, Germany, Sweden, and the UK; from Latin America: Brazil, Chile, Ecuador, and Mexico; from North America: Canada and USA; and from Oceania: Australia and New Zealand. These countries were selected to ensure that all world regions were represented and to illuminate different traditions and developments with regard to ESE. Of course, it is not possible to fully represent the global development of ESE, but the book does provide a diverse overview and insight that will hopefully be stimulating for many readers.

Each chapter gives some background information about the education system and other key data with regard to the country in question, including the historical background to ESE. Each country's understanding and delivery of ESE are described, and emerging issues and trends, and current and future needs are discussed. This brings us to the contributions themselves.

Halimu Suleiman Shauri, Maarifa Mwakumanya, and Sellah Lusweti show that **Kenya** has mainstreamed ESD in ministries and agencies and created a legal and policy framework to achieve the goals of the UN ESD Decade. Kenya's National Education Sector Plan (NESP), 2013–2018, provided an education and training strategy to promote ESD, culminating in the ESD Policy for the Education Sector in 2017. The delivery of ESD in Kenya is currently guided by seven strategies: advocacy and vision building, consultation and ownership, partnership and networks, capacity building and training, research and innovation, use of ICT, and monitoring and evaluation.

Rob O'Donoghue and Eureta Rosenberg describe ESE in **South Africa** through the lens of the staging of risk. They explain the shift from ESE approaches focusing on behaviour change to more engaging, open-ended processes. The chapter illustrates its analysis with a case study that looks at the Fundisa for Change teacher training program and related Hand-Prints of Care materials that address Agenda 2030 and the SDGs.

James Musana and Ronald Bisaso characterize **Uganda** as a key partner in the Global Action Program (GAP) on ESD on the African continent and as one of the first countries in the world to align its national planning processes with Agenda 2030 and the SDGs. The chapter also describes the efforts that have been made in Uganda to integrate ESE into educational curricula and teaching methods. Challenges for the further implementation of ESE are also presented, including: lack of policies focusing on whole-institution approaches, awareness-raising, and capacity-building.

For **India**, Sudeshna Lahiri outlines the different ESE programs and approaches in higher education and school education, and shows how the SDGs have provided an incentive to develop policies to integrate ESE into academic studies. However, the author considers that environmental education in India will need to take the SDGs into account as it evolves towards ESD/ESE.

Yoko Mochizuki recalls that the UN Decade of ESD was proposed by **Japan**. The chapter describes efforts by diverse Japanese stakeholders to mainstream ESD in a range of fields, and also traces the transition from environmental education and development education to ESD in the country. In addition, the author addresses ESD needs in the context of the 'triple disaster' (earthquake, tsunami, and the Fukushima nuclear disaster) and the pandemic.

Franz Rauch, Christiana Glettler, Regina Steiner, and Mira Dulle describe environmental education and global citizenship education as ESE focal points in **Austria**. Their chapter also addresses the link between ESE and action research and outlines new Austrian-wide projects, collaborations and platforms (e.g., Platform Education 2030) that are emerging in the context of Agenda 2030 and particularly SDG 4, building on established initiatives, structures and networks.

Jan Činčera characterizes the **Czech Republic** as a Central European country that has transitioned from a totalitarian regime to democracy and, against this background, illustrates the development of ESE in the country. He also addresses the prominent role of environmental education centres in shaping national ESE discourse in the initial decades after the revolution in 1989. In addition, the chapter describes the interaction between theory and practice, and stresses the necessity of building bridges between different types of stakeholders in order to meet the sustainability challenges of the 21st century.

For **Germany**, Marco Rieckmann and Mandy Singer-Brodowski trace the transition from environmental education and development education to ESD and global learning. Their chapter also describes the emancipatory, competence-oriented understanding of ESD that prevails in Germany. In addition, they illustrate how the ESD Decade and the follow-up programs have gradually brought about a structural implementation of ESD in the German education system. However, the chapter also points out the significant remaining differences between the education sectors and the German federal states.

Stefan Bengtsson and Paul Plummer provide an overview of **Sweden**'s approach to ESD. The chapter summarizes key developments that have been formative for overall trends and trajectories in the national engagement with global ESD policy. The authors describe the position of ESD in education policy and syllabi in Sweden, and set out how ESD is understood and translated into teaching practice. They also clarify that ESD in Sweden involves a notion of citizenship education that is closely aligned with the Swedish notion of democracy.

For the **United Kingdom** (UK), Alun Morgan and Paul Warwick describe the different developments with regard to ESE in Scotland, Wales, Northern Ireland, and England. In addition, they address the special role of the third sector (charities, voluntary and community organizations, social enterprises and cooperatives) in the development of ESE. Their chapter outlines the learning dimensions of ESE as follows: solutions-focused approaches to the SDGs, social learning through dialogue, experiential learning through place-based activation and reflexive learning through contemplative engagement.

For **Brazil**, Luiz Marcelo de Carvalho and Marcelo Gules Borges illustrate on the one hand the consequences of the policies of President Bolsonaro's extreme right-wing government (2019–2022), which have resulted in the dismantling of many of the country's achievements in terms of social, educational and environmental policy and created serious difficulties situation for the further implementation of ESE. On the other hand, the chapter also highlights the questions that are being raised and the critical, anti-hegemonic approaches proposed by Brazilian environmental educators, who have also offered resistance to the socio-political context created by the Bolsonaro government.

Karl Böhmer Muñoz, Viviana Andrea Contreras Cabezas, and Magali Loreto Aceiton Perea show that ESE in **Chile** is predominantly an awareness-raising exercise with close links to public environmental management. UNESCO's ESD programs are not taken up in Chile. The chapter identifies a need for higher education institutions to undertake further research, develop new methodologies and reflection more deeply on environmental education and ESD, and for a firm commitment by all parties to take the necessary steps to overcome the challenges of climate change.

Silvia Mera, Patricia M. Aguirre, and Freddy H. Villota describe the historical process of the incorporation of environmental education into **Ecuador**'s national educational system, identifying national experiences, advances, achievements, and limitations. They highlight efforts that are currently being made to introduce programs and projects in order to increase knowledge, promote values and develop practical competences that will enable learners to address environmental problems and raise awareness of sustainable development.

For **Mexico**, Rosalba Thomas Muñoz, Helio García-Campos, and Teresita Maldonado Salazar describe the resistance of environmental education stakeholders and researchers to UNESCO's intention to replace the term "environmental education" by ESD, and explain how they ultimately adopted the term "environmental education for sustainability" (EES). The chapter recognizes the diversity of conceptual approaches and demonstrates openness towards worldwide educational trends; however, it also seeks to vindicate the theoretical contributions from both Mexican and Latin America, which have been critical of the notion of development that is restricted to the idea of economic growth.

Kristen Hargis, Marcia McKenzie, and Nicola Chopin provide an overview of the state of ESE in formal education across **Canada**. The chapter focuses on sustainability engagement in policy and practice in ministries of education, school divisions, primary and secondary schools, and higher education institutions across the country. Among other findings, the chapter shows that over half of ministries of education, school divisions, and higher education institutions have sustainability-specific policies, and that current responses to climate change in formal education in Canada are insufficient to meet Canada's Paris Agreement commitments.

Michaela Zint, Jessica Ostrow Michel, Sarah R. Collins, Erin Gallay, Emma C. Sloan, Maite Elizondo Piñeiro, Lauren Balotin, Veronica Correa, María Isabel Dabrowski, Bita Davoodi, Joseph Dierdorf, Daniela Fernández Méndez Jiménez, Jessica Miller, Isabel Nakisher, Connor Roessler, Paige Schurr, Peter Siciliano, and Joshua Thompson outline that, although the **United States of America** (USA) was an early leader in (inter)national environmental education and there are places where it receives more attention, the current status of ESE in the country, and thus its overall impact, is limited. However, the chapter also describes encouraging shifts in focus from environmental education to sustainability education (including the SDGs) and toward whole-institution sustainability initiatives, with partnerships and networks providing critical support.

Annette Gough, Alan Reid, and Robert B. Stevenson explain that environmental (and sustainability) education has now been on the formal education agenda in **Australia** for 50 years. However, the past decade has not been fruitful, and there has been neither a strong platform nor momentum at government level to achieve Agenda 2030 in Australia. Nonetheless, there are now strong indicators that the situation could be turned around. The change of government in 2022 has raised hopes that ESE will be formally recognized, administratively supported, and funded in a more substantial way.

Kerry Shephard, Sally Birdsall, Chris Eames, and Jenny Ritchie point out that, although **New Zealand** has some sophisticated policy, strategy and outcome-oriented sustainability educational frameworks, there is a need for organized and committed leadership on ESE, research and evaluation of educational change and impact, and educational commitment to the SDGs.

In the **concluding chapter**, Rosalba Thomas Muñoz and Marco Rieckmann trace general, cross-national trends and developments, but also highlight differences between terminologies and national approaches to ESE. Among other things, they note that some countries have a strong focus on the UNESCO programs, while others are going their own way—and sometimes criticising the UNESCO concepts.

We hope this book will act as a stimulus for further development of international discourse on ESE, but also for reflection on specific national or local contexts. We are very grateful to all authors for their contribution to this volume.

References

Hopwood, B., M. Mellor and G. O'Brien. 2005. Sustainable development: mapping different approaches. Sustainable Development 13(1): 38–52. https://doi.org/10.1002/sd.244.

UNESCO. 2017. Education for Sustainable Development Goals. Learning Objectives. UNESCO. http://unesdoc.unesco.org/images/0024/002474/247444e.pdf.

UNESCO. 2020. Education for sustainable development: a roadmap. ESD for 2030. UNESCO. https://unesdoc.unesco.org/ark:/48223/pf0000374802.

Africa

Chapter 1

Education for Sustainable Development in Kenya

Halimu Shauri, Maarifa Mwakumanya* and Sellah Lusweti

Introduction

In 2005 the United Nations launched the Decade of Education for Sustainable Development (UNDESD). This was the result of a long process of international deliberation on the sustainability of development models, which began in 1972 at the United Nations Conference on Human Development held in Stockholm (ESD Kenya Country Report 2015). In March 2006, African Ministers of Education made a commitment to implement the UNDESD in the context of the Second Decade of Education for Africa. Their statement of commitment emphasized the need to situate UNDESD activities within key policy initiatives such as the Millennium Development Goals (MDGs), the United Nations Declaration on the New Partnership for African Development (NEPAD), the African Union's Second Decade of Education Action Plan, and the Dakar Framework for Action aimed at achieving the Education for All goals.

Kenya's national strategy, Vision 2030, taps into the above frameworks and aims to make Kenya a middle level income country by 2030 (GoK 2008). The growth target for gross domestic product (GDP) of 10% over the next three decades will inevitably result in heavy demand for energy, water, solid waste management and the manufacture and use of hazardous and toxic substances in the country (Nyangena 2009). According to Nyangena (2009), this will be compounded by population growth, particularly in urban areas, due to rural-urban migration, growing urbanization and rising living standards.

Pwani University, P.O. Box 195-80108, Kilifi, Kenya.
* Corresponding author: m.mwakumanya@pu.ac.ke

Although a key goal of Kenya's Vision 2030 is the attainment of the status of a 'nation living in a clean, secure and sustainable environment' driven by the principles of sustainable development, exponential growth in the economy and the population may stretch the government's ability to meet the infrastructure and service needs of the Kenyan people. The supporting infrastructure for the collection and treatment and disposal of sewerage and solid wastes is often unable to cope with the amounts generated. In fact, according to the UN, the dilemma of expanding economic activities equitably while attempting to stabilize the rate of resource use and reduce environmental impacts poses an unprecedented opportunity and challenge to society (UNEP 2011). Thus, Kenya's vision, if not well forecasted and strategized, is likely to lead to water and air pollution, public health problems and urban environmental degradation.

Cognizant of the foregoing, Kenya's 2013–2018 National Education Sector Plan (NESP) provided an education and training strategy to promote Education for Sustainable Development (ESD) in the context of the United Nations Global Action Programme (GAP) on ESD (Republic of Kenya 2014). This led to the development of ESD Policy for the Education Sector in 2017. Kenya's ESD strategy places the focus on four strategic thrusts, namely,

1) Conservation of natural resources
2) Pollution and waste management
3) Arid and semi-arid lands (ASALs) and high-risk disaster zones
4) Environmental planning and governance.

Since Kenya is a resource-based economy, education needs to empower Kenyan citizens to make informed decisions about sustainable development. The education sector should provide citizens with the skills to ensure that their country meets the economic and social goals of Vision 2030, while ensuring sufficient resources to support future generations. Kenya's Vision 2030 provides for linkage between education and other sectors including health, water and sanitation, the environment, gender, housing, and youth (JKUAT 2010). It thus provides a clear pathway for embedding ESD in the country's development agenda.

ESD is a broad and evolving concept that can be largely interpreted as holistic and transformational education, addressing learning content and outcomes, pedagogy and the learning environment with a view to achieving societal transformation (UNESCO 2014). The ESD implementation strategy aimed to provide the appropriate environment and sufficient capacity to enable all sectors and stakeholders to make an effective contribution towards the achievement of sustainable development. It was therefore a major step towards sustainable development in Kenya, highlighting appropriate learning, capacity-building programmes and the development of skills in sustainable use of resources at all levels. The hope is that it will inculcate sustainability values among the country's citizenry and lead to informed decision making on a range of sustainability-related issues.

Political, economic, cultural and social contexts and conditions, and the education system in Kenya

Sustainable development issues in Kenya are complex and interlinked (ESD Kenya Country Report 2015). They can be classified as societal, economic and environmental. Societal issues impacting sustainable development include poor governance, corruption, bigotry with regard to cultural diversity, ethnic animosity and gender inequality.

The economic issues revolve around systems of production, consumption, investment and service delivery, all of which aim at enhancing the country's GDP. However, economic performance is impeded by a number of challenges including high levels of poverty and related issues. Current projections indicate that 56% of the Kenyan population lives below the poverty line, earning less than US $1.00 per day (GoK 2015). Further, the gap between the rich and the poor has continued to widen, with a per capita income of about KES 1,239 per month in the rural areas and KES 2,648 in urban areas; as such, a large proportion of Kenyan society is poor (*ibid*).

Other challenges affecting economic growth and performance include inadequate investment infrastructure leading to rising levels of unemployment; rural-urban migration; corporate irresponsibility and lack of accountability; and corruption. Inefficient and wasteful production systems lead to the unsustainable utilization of natural resources, resulting in their degradation. Furthermore, the poor enforcement of policies and regulations governing production and marketing hinder economic growth and the attainment of optimal performance (ESD Kenya Country Report 2015).

The environment sector faces a number of challenges too. These include frequent droughts especially in Northern and Coastal Kenya, natural disasters (especially flooding and landslides), acute water shortages, climate change and variability, loss of biodiversity, and poor waste management systems. This has resulted in degradation of land and loss of forest cover, which is currently estimated at 7% of the total land area (Toot and Vlosky 2019), falling far below the globally recommended 10% minimum. Moreover, about 88% of the country's total surface area is comprised of ASALs while desertification is on the rise as a result of fragile ecosystems. The robust industrial development experienced in the country over the last four decades has negatively impacted the environment. This has resulted in increased waste generation leading to unsustainable waste management practices.

Political analysis

Throughout the period of its political independence, Kenya has had fairly limited experience of adversarial multi-party politics. The first was the short-lived experiment with political pluralism from independence in 1963 to the 'Little General Election' in 1966. The second began in the early 1990s and included the multiparty elections of 1992 and 1997. During 1991–92, Kenya established the multi-party system, which put the then President Daniel Arap Moi in an uncertain position. The power brokering played into the environmental sector as politicians focused their energies on land ownership, using it as a political bargaining chip. One case in point was the triggering of violence and ethnic clashes to clear the Rift Valley area where the

Mau Forest, the largest forest complex in Kenya, is located. The politicians used local militia groups, who fiercely attacked the villages and burned houses, destroyed property, and brutally killed and raped hundreds of women. As a result, more than three hundred thousand people were displaced (Chaudhry 2019). Following these incidents, the South-West Mau Forest was opened up for new settlements, especially around Kericho and Bomet area. Ever since, Kenya's political elites have targeted the Rift Valley and the surrounding areas in electoral campaigns. Consequently, before and after every election period, there have been widespread and recurring clashes in the area, in particular in the 1991–92, 1997, 2002, and 2007–2008 electoral periods.

In 2003, due to growing publicity about the forestland excisions and severe degradation of the Kenyan forests, a land commission was set up to investigate cases of land grabbing. The commission reported that huge chunks of protected parts of the Mau Forest Complex had been cleared over the last two decades and awarded to the political elites, including some prominent multi-national companies. The Ndung'u Report identifies serious incidents where political factions, public officers and provincial administrators used various illegal means to gain personal benefits. The report also highlights various accounts of widespread corruption, which gave rise to human rights violations, land grabbing and abuse of power (The Ndung'u Report 2003). In these political wars, resource access and ownership was intertwined with ethnicity, both of which emerged as persistent features of political contests. Political activity since the renewal of competitive politics in 1992 has seen the reconstruction of ethnicity, ethnic mobilization and ethnic conflict as the main instruments of political contestation (Ajulu 2010).

Despite the political upheavals, until late 2007, Kenya was considered one of the most stable countries in Africa. It had functioned as East Africa's financial and communications hub, hosted the headquarters of many international non-governmental organizations, and been a magnet for tourism. Disputed elections in late December 2007 triggered outbreaks of violence across the country that killed more than six hundred people. That prompted fears that Kenya would split along tribal lines and descend into prolonged unrest (Council on Foreign Relations 2008).

The Council on Foreign Relations (2008) provides a diversity of perspectives on politics and environmental sustainability in Kenya. Most institutions—including the judiciary, parliament, and electoral commission—are subordinate to the president. The president appoints high court judges and electoral commissioners, has the power to dissolve parliament, and controls the federal budget. The extent of presidential power is a legacy of the colonial period, and has changed little since independence in 1963. Such powers have been considered the source of weak political and institutional systems for ensuring checks and balances. The consolidation of power in the executive branch means that elections are dominated by a winner-takes-all mentality.

Although members of parliament are elected by the general population, parliament has little power to address public grievances, as parliamentarians gain power through political alliances to which they have to pledge allegiance. Soon after elections, voters realize that their elected officials are unable to address their concerns about social and economic inequality, and this leads them to distrust institutions, producing a 'sense of disempowerment and disillusionment.' Further, according to the Council on Foreign Relations, the electoral commission's inability to resolve disputes over the legitimacy of vote tabulation following the December 2007 elections served as further evidence that Kenya's political institutions could not be considered independent. Widespread corruption has eroded public trust in political institutions. Kenya ranked 150 out of 180 countries on Transparency International's Corruption Perceptions Index, alongside the Democratic Republic of Congo and Liberia, both of which recently emerged from civil war.

The approval of Kenya's new constitution in 2010 and relatively peaceful elections in 2013 and 2018 represented milestones in the country's transition from political crisis. The new constitution introduced an expanded Bill of Rights that includes social, economic and cultural rights (with strong focus on the needs and entitlements of children and women); it reduces the president's powers; provides for better separation of powers between the three arms of government; circumscribes the power of security agencies; reforms the electoral system; devolves power to the counties, and introduces changes to the budgetary process.

With regard to environmental sustainability, the principle of sustainable development is one of the national values and principles of governance set out in Article 10 2(d) of the Constitution. The Constitution guarantees the right to a clean and healthy environment and further guarantees the right to have the environment protected for the benefit of present and future generations. Among other functions, the state is required to: ensure the sustainable exploitation, utilization, management and conservation of the environment and natural resources, and ensure the equitable sharing of the benefits that accrue; work to achieve and maintain tree cover of at least ten % of the land area of Kenya; and encourage public participation in the management, protection and conservation of the environment.

Economic analysis

There is wide agreement that it is challenging to reconcile economic growth with the principles of sustainable development (OECD 2019). Ever-accelerating production and consumption deplete natural resources, produce unmanageable amounts of waste, and lead to a rise in global temperatures. The OECD acknowledges the existence of many laudable and credible initiatives to promote sustainable production and consumption, but points out that their impact has been limited. The unprecedented growth of the 20th century has had both positive and negative effects on the complex interactions between people, the natural environment and economic systems. Economic growth has created immense wealth in some areas of the globe, but left others behind. In Kenya there are economic inequalities between some ethnic groups, and bitter, long-standing disputes over land, particularly in the Rift Valley. The 2020 UN Human Development Index ranks Kenya 143 of 189 countries with regard to income inequality (UNDP 2020).

Nevertheless, Kenya has the largest and most diverse economy in East Africa, with an average annual growth rate of over 5% for nearly a decade; and with the highest Human Development Index in the region. Historically, Kenya has experienced rapid economic growth since independence: during the 1960s the economy grew at an average of 6.5% per annum (Legovini 2002). This growth declined in the late 1970s to around 4%. Between 1990 and 2002 economic performance was more or less constant; however, it improved between 2003 and 2007, reaching to a high of 7.1% economic growth rate in 2007. Due to the aforementioned political upheavals and post-election violence, the growth rate dropped to 1.7% in 2008 (KNBS 2009). The economy is now picking up; Kenya registered average growth of 5.6% in the period 2014 to 2019, indicating a stable economy on its way to achieving the objectives of the Kenya Vision 2030 (KIPPRA 2020).

Kenya's entrepreneurial spirit and human capital give it huge potential for further growth and job creation, and for reducing poverty (Institute of Development Studies – ISD 2021). The recent discovery of oil and other mineral resources creates great potential for the Kenyan economy; however, for the extraction to be sustainable, the Kenyan Government must assiduously follow guidelines such as those provided by The International Resource Panel on how the extractive sector can establish modern governance structures that addresses resource security and efficiency. Interestingly, despite the decline in the country's absolute poverty rate, wealth has not been distributed equally. Kenya remains a highly unequal society by income, gender, and geographical location. Poverty is highest in the arid and semi-arid areas that represent about 80% of the land area and are inhabited by about 20% of the population. The ISD further notes that poverty also affects the coastal area, which receives fewer resources. Rapid population growth is another major challenge, further complicated by high unemployment rates, especially for younger people. More than 70% of Kenya's population is below the age of 30 and the population aged under 14 alone amounts to 43% (Institute of Development Studies 2021).

From time to time the Kenyan Government has come up with strategies to improve the country's economic performance. Firstly, through the Economic Recovery Strategy for wealth and employment creation (ERS), for the period 2003 to 2007, the implementation of which improved the country's economic performance significantly; and secondly, through Vision 2030. The Vision is the government's long term plan for 2008 to 2030, which aims to transform Kenya into a newly industrializing 'middle-income country providing a high quality life to its citizens by the year 2030.' Kenya is aiming to achieve a sustainable economic growth rate of 10% per annum, equitable social development and a democratic political process that is issues-based, people-centred, results-oriented, and accountable. The implementation of Vision 2030 is expected to deliver higher economic growth and trade is likely to increase as a result of regional integration (East African Community). The Government has also made significant effort to address inequalities between counties through differential budgetary allocation. The poorest counties have received the largest proportion of equitable transfers, driven mainly by the poverty factor in the Commission on Revenue Allocation (CRA) formula, and accounting for 18% of revenue allocations through equitable transfers (KIPPRA 2020). Turkana and Mandera, for instance, received 3.9% and 3.4% respectively between 2013/14 and 2018/19. It is also notable that most of the poor counties allocated a significant proportion of their spending to development.

As was noted earlier, income in Kenya is heavily skewed in favour of the rich and against the poor; the richest 10% of households in Kenya control more than 42% of the income, while the poorest 10% control 0.76% of the income (Society for International Development 2004). This means that while the richest Kenyans earn about 56 shillings, the poorest earn 1 shilling per day. According to the Economist (June 1 2013), a 1% increase in incomes in the most unequal countries reduces poverty by merely 0.6%. In the most equal countries, the same 1% growth yields a 4.3% reduction in poverty. Poverty and inequality are thus part of the same problem, and there is a strong case to be made for both economic growth and redistributive policies. From this perspective, Kenya's Vision 2030 quest of growing by 10% per annum must also ensure that inequality is reduced and all people benefit equitably from development initiatives and the resources that are allocated (KNBS 2014). Other significant downside risks that, according to KIPPRA (2020), could result in a slowdown in economic activity include: rising fiscal pressures with increased debt servicing costs and fiscal measures to cushion the economy from the effects of Covid-19; adverse weather conditions; desert locust invasion; and the impact of Covid-19 on the economy.

Social analysis

Upward social mobility, namely the movement of individuals, families, households and other categories of people within or between social strata, is important for sustainable development and inclusive growth (KIPPRA 2020). Upward social mobility can enhance social cohesion, create feelings of inclusion among disadvantaged groups and diffuse extremism. In a progressive society, access to education, health, social protection and employment should not hinge on parental income, health, education and employment. Under the Third Medium-Term Plan (MTP III), Kenya is aiming to grow the economy by 7.0% by 2022 (GoK 2018). The notion of social exclusion can contribute to the understanding of the nature of poverty, as well as help to identify causes of poverty that may otherwise be neglected. UN DESA (2009) adds that, even though there is no direct causal relationship between poverty and violent conflict, the poverty associated with high levels of inequality and exclusion/marginalization can be a major contributory factor for higher crime rates and increased risks of social tensions, social disintegration, and ultimately violent conflict. KIPPRA (2020) agrees, noting that when growth and development policies are not inclusive, they are likely to trigger social conflict and derail the trajectory of development. The critical challenge then remains to attain high and sustainable levels of growth and development whilst ensuring that such growth is socially and economically inclusive. It is therefore of paramount importance to create a conducive environment that ensures productive employment, low poverty levels, reduced inequality, and environmental sustainability.

The Medium-Term Plan 2018–2022, emphasized the importance of the Four-point strategic development agenda, dubbed the "Big Four" agenda, for improving the living standards of Kenyans and growing the economy (GoK 2018). In the medium term plan 2018–2022, the Kenya Government aims to achieve inclusive growth by: ensuring food security; expanding the manufacturing sector to create jobs; providing universal health care to enhance human capital; and providing affordable housing that can be accessed by low-income earners. The Government has also put in place policies to enhance youth empowerment, gender equality and equal opportunities for individuals with disabilities. Despite progressive policy and institutional reforms, the ability of the social protection sector to enhance social mobility and ensure more inclusive growth processes is being curtailed by a number of challenges. These include weak policies, lack of adequate coordination, low programme coverage, and duplication of benefits (KIPPRA 2020), which have lowered the expected impact of social protection programmes. Other challenges include potential ghost beneficiaries, and the lack of an integrated system linking all social assistance programmes across Ministries, Departments and Agencies (MDAs) in one easily accessible online portal.

Historical background and development of environmental and sustainability education in Kenya

The United Nations Conference on the Human Environment held in Stockholm in 1972 helped to focus attention on environmental concerns, with the global community acknowledging the need for interrelationships between the environmental and socio-economic issues of poverty and underdevelopment (UNESCO 2008). At its 57th session in December 2002 following the Johannesburg Summit in South Africa, the United Nations General Assembly 2002 declared the ten year period from 2005 to 2014 the Decade of Education for Sustainable Development (DESD), calling on governments to integrate the principles of sustainability into their educational strategies and action plans. Prior to the DESD in Kenya, the country had been implementing education for sustainable development through its education and development policies. In response to the DESD declaration, Kenya developed a national ESD strategy in 2008, supported by the National Environment Management Authority (NEMA), outlining the implementation of, and vision for, ESD in the Kenyan context, presenting ways to engage in change to promote sustainable development, and proposing action-based strategies to guide stakeholders in their journey towards sustainable development (Republic of Kenya 2008). The strategy describes ESD as 'education that enhances sustainable development in Kenya' and whose mission it is 'to provide an enabling environment and capacity for all sectors and stakeholders to contribute effectively towards the achievement of sustainable development'.

ESD is implemented and coordinated through seven strategies: (1) Advocacy and vision-building; (2) Consultation and ownership; (3) Partnership and networks; (4) Capacity-building and training; (5) Research and innovation; (6) The use of information and communication technologies (ICTs); and (7) Monitoring and evaluation (Republic of Kenya 2008). Several Regional Centres of Expertise (RCEs) have also been established to enhance this process (Republic of Kenya 2008). RCEs are a global initiative established by the United Nations University in 2005 to achieve the goals of the DESD by translating its global objectives into the context of the local communities in which they operate (UNESCO 2011). This has implications for the role of the universities involved in RCEs, as will be discussed below in more detail. In 2011, in line with the DESD principles, the Kenyan government's Ministry of Environment and Mineral Resources (MEMR) also published the National Education for Sustainable Development Policy. The goal of the policy is 'education that enhances sustainable development in Kenya'. This emphasis on ESD in Kenyan education includes higher education institutions (Kariaga et al. 2013).

In 2014, the DESD came to an end and, as a follow up, UNESCO launched the Global Action Programme (GAP) on Education for Sustainable Development for an initial phase of five years (2015–2019), aiming to build on the advocacy and awareness-raising of the DESD. The GAP, focused on scaling up: driving policy forward, transforming learning and training environments, capacity-building for educators and trainers, mobilizing youth, and accelerating sustainable solutions at local level. The GAP's activities and interventions directly addressed ESD while at the same time working to meet the 17 Sustainable Development Goals (SDGs), which were launched in 2015 and address issues relating to poverty, hunger, health, education, energy, work, industry, inequalities, cities, consumption, climate, ocean life, ecosystems, peace and partnership. In the context of the GAP implementation, Kenya developed its National Education Sector Plan and ESD Policy in 2017 to promote ESD. These strategies enabled the country to build education strategies, climate change plans and the Green Economy strategy into its implementation of the GAP. GAP implementation in Kenya included the promotion of whole-institution and Green Campus initiatives, including the establishment of Eco-schools and the Green Campus network (Kenya Green University Network) to integrate sustainability into the mainstream of all learning environments. For Kenya, ESD is the core of SDG 4 Target 4.7, which aims to ensure that all learners acquire the knowledge and skills needed to promote sustainable development, including through education for sustainable development and sustainable lifestyles, human rights, gender equality, promotion of a culture of peace and non-violence, global citizenship and appreciation of cultural diversity and of culture's contribution to sustainable development (UNESCO 2019).

Understanding of environmental and sustainability education in Kenya

In acknowledgement of the importance of ESD, the Republic of Kenya's Ministry of Education asked the State of Israel's Agency for International Development Cooperation in the Ministry of Foreign Affairs for its assistance to set up a joint project aimed at improving the Kenyan education system. "From Given towards ESD Driven" is the spiral process model that was adopted to provide mutually reinforcing principles and tools for integrating ESD into learning institutions' curricula and educational agendas. One of the main principles is "Think Global, Act Local"; awareness of the global crisis is important, but action should be taken at the local level. This model provides the tools for each public education institution and the community to articulate their specific needs and match them to existing resources. According to MASHAV (2017), during the 2010–2016 period, close to 800 Kenyan educators from over 90 schools and institutions took part in training activities conducted both in Israel and in Kenya to put the model into practice. Additionally, experts undertook annual visits to Kenya in order to offer advice and counseling on deploying the expertise gained in Israel. During these visits, mobile courses were held in Nairobi, Kisumu and Mombasa for alumni and new participants.

The ESD program in Kenya offers many valuable learning environments for students, teachers, parents, community and other stakeholders in the following key areas:

1. Respect, value and preservation of the achievements of the past; appreciation of the wonders and the peoples of the Earth;
2. Living in a world where all people have sufficient food for a healthy and productive life;
3. Assessing resources, caring for and restoring the state of our planet;
4. Creating and enjoying a better, safer, more just world;
5. Being caring citizens who exercise their rights and responsibilities locally, nationally and globally;
6. Representing a new vision of education, a vision that helps people of all ages;
7. Understanding the world in which they live, addressing the complexity and interconnectedness of problems such as poverty, wasteful consumption, environmental degradation, urban decay, population growth, health, conflict and the violation of human rights that threaten our future;
8. Deployment of critical thinking and problem solving, leading to confidence in addressing the dilemmas and challenges of sustainable development;
9. Applying multi-method approaches in education: words, art, drama, debate, experience, different pedagogies that model the processes. Teaching that is geared simply to passing on knowledge should be recast into an approach in which teachers and learners work together to acquire knowledge and play a role in shaping the environment of their educational institutions.

Education for Sustainable Development has been gaining momentum since 2008 and there has been increased public awareness, and training to promote sustainable development in Kenya (KOEE 2019). This has been made possible by the many partnerships, collaborations and networks established locally and abroad to enhance the implementation and achievement of ESD objectives. At the local level, Regional Centers of Expertise (RCE) have been established in Universities to scale up awareness and deliver ESD objectives. Together with the Kenya Universities Green Network, higher educational institutions have taken ESD to a higher level by greening their campus environments. As part of their performance indicators, Ministries, Departments and Agencies are required to promote sustainability in their institutions and report annually on their performance. These efforts have made education for sustainable development synonymous with education for all in Kenya, disseminating it at all levels throughout the formal, informal and non-formal education systems, which all help learners to acquire the appropriate skills, attitude and knowledge.

In Kenya, ESD implies catalyzing and supporting the reorientation of the education and training systems (KOEE 2019), with a view to promoting teaching and learning that inculcates values, behaviours and lifestyles for sustainability stewardship. Sessional paper no. 11 of 2014 on education for sustainable development policy stipulates that ESD aims to facilitate education and learning that promotes equitable, efficient and sustainable utilization of the country's resources, and diverse, high-quality education that increases public awareness, leading to improved quality of life and productive livelihoods. ESD is perceived as way of life and a means towards sustainable utilization of resources for development.

Kenya's engagement with environmental and sustainability education in the light of GAP and ESD for 2030

The Ministry of Education (MoE) established a national ESD coordination desk in 2012 and collaborates with the Ministry of Environment and Natural Resources and the National Environment Management Authority (NEMA) to promote ESD activities. A national steering committee comprising representatives of government, civil society organizations and the private sector has been spearheading the implementation of ESD. ESD has also been embedded in the 2013–2018 National Education Sector Plan (NESP). MoE has established partnerships and networks with other organizations such as the Ministry of Health, Ministry of Environment and Natural Resources, UNESCO, UNEP, WFP, UNICEF and MASHAV, to ensure effective implementation (ESD Kenya Country Report 2005–2012).

Kenya's ESD Implementation strategy provides a facilitative environment and capacity for all sectors and stakeholders to make an effective contribution towards the achievement of sustainable development (Republic of Kenya 2008, NEMA 2012). The strategy aims to:

1. Enhance the role of education and learning to ensure equitable, efficient and sustainable utilization of the country's resources;
2. Promote quality education through formal and informal learning and public awareness to ensure better quality of life and productive livelihoods; and
3. Promote teaching and learning that inculcates appropriate values, behaviour and lifestyles to ensure good governance and sustainability.

The ESD strategy implementation framework also specifies the strategic objectives, activities, outputs, objectively verifiable indicators, means of verification and stakeholders involved in implementation. A national steering committee, with representation from the government sector, civil society and the private sector spearheads the implementation of ESD in Kenya. Sectoral ESD committees have also been established in order to address specific key sustainable development issues while regional and provincial ESD advisers and district ESD coordinators oversee ESD activities at the grassroots level. ESD activities in Kenya are being implemented and coordinated in line with the implementation strategy. This breaks down into seven strategies, namely advocacy and vision building, consultation and ownership, partnership and networks, capacity building and training, research and innovation, use of ICT and monitoring and evaluation.

Advocacy and vision building

The advocacy and vision building strategy calls for aggressive awareness campaigns to promote reflection on the root causes of unsustainable development in the context of social, environmental, cultural and economic ventures. The strategy also calls for awareness to be raised amongst the Kenyan population to encourage sustainable approaches to life and work. The vision-building aspects of the strategy also highlight the importance of awareness for the development of a sense of social responsibility, consciousness of individual actions and how they affect social interaction and development itself. Among other things, an 'ESD Media Training Kit' has been developed for this purpose, with both technical and financial support from UNESCO. And UNEP is supporting World Environment Day activities in Kenya, which creates awareness of sustainable development.

Consultation and ownership

The consultation and ownership strategy are crucial to the formulation and planning of local and national initiatives. The strategy requires consultation to include: transparent and timely provision of information on policy proposals and budgetary provisions by the various sectors; processes to solicit stakeholder input; legislative affirmation of—and commitment to—the ESD process; public awareness campaigns inviting feedback on the process and other ESD initiatives; and commissioning of research. At the RCE level, there is evidence that consultations with partners are ongoing; a number of these consultations have led to the formulation of ESD policies, for example, JKUAT, Pwani Universities, the Kenya Institute of Education (KIE) and RCE Nyanza, who have all developed institution-based ESD policies.

Partnership and networks

The partnership and networks strategy stemmed from the realization that the ESD process is too great and too complex for any single institution to promote on its own (ESD Kenya Country Report 2005–2012). While the Government's role focuses mainly on coordination and resource mobilization, the strategy advocates the establishment and strengthening of partnerships to ensure synergy and cooperation. It is believed that the strength and inclusiveness of the partnerships, networks and alliances formed by the stakeholders will determine the effectiveness of ESD implementation in Kenya. Two main activities are envisioned; the identification of partners and the development of working mechanisms.

The ESD process in Kenya is multi-sectoral with partnerships between the public sector, private sector, civil society organizations and the media. A number of RCEs have also conducted stakeholder analysis and developed databases, which include new partners in ESD implementation. Networks and linkages have been established with institutions such as the South African Development

Cooperation – Regional Environmental Education Programme (SADC-REEP) and RCEs in other countries such as RCE Graz, RCE London, RCE Kwa Zulu Natal, RCE Denmark, etc. Partnerships have also been formed, mainly between the NGOs, RCEs, development partners and the private sector, but also with London South Bank University, the United Nations University, the Nile Basin Initiative and Rwanda. A partnership between RCE Greater Pwani and REA Vipingo on research is a flagship example based on need.

The Kenya Education Sector Support Programme (KESSP) has contributed to a number of achievements in sustainable development in Kenya. It has: (i) provided a framework for the formulation of education policies and a sector-wide approach to costing; (ii) steered Kenya to speed up the achievement of EFA and the Millennium Development Goals (MDGs); (iii) enhanced the role of CSOs in the adoption and use of EFA goals; and, (iv) promoted the creation of national coalitions and networks (Pamoja Kenya, Elimu Yetu, the African Network Campaign on Education For All – ANCEFA, among others). In addition, the KESSP policy provides various opportunities to introduce ESD in formal education.

In 2009, the Ministry of Public Health and Sanitation and the Ministry of Education developed a national school health policy to enable the government to coordinate resource mobilization to improve children's health. Guidelines have been developed in order to implement this policy and to better integrate NGOs' *ad hoc* health education activities across schools. The Ministry of Environment and Mineral Resources and the Ministry of Forestry and Wildlife have developed the 2009 National Climate Change Response Strategy (NCCRS), and an investment framework programme for Kenya. Both frameworks emphasize the importance of climate change education.

The Kenya Government's Economic Recovery Strategy (2003) seeks to create 500,000 new jobs annually and, with its focus on growth, employment and poverty reduction, plays a key role in the promotion of ESD. The strategy promotes teaching and learning on the basis of skills training for young people, women and other vulnerable groups. Kenya's school curriculum has been revised to incorporate life skills. In light of the post-election violence of 2008, priority themes include conflict management, social cohesion and ethnic tolerance, survival skills, and gender equality, among others. The Kenya Institute of Education has developed, piloted and implemented the curriculum, and trained the teachers and education officers at national level.

The Kenya Institute of Education (KIE) collaborated with NEMA to introduce an ESD pilot in schools. This encouraged teachers deliver the curriculum through the use of projects. The pilot (two schools – Jamhuri Secondary and City Primary School) has established ecological gardens, compost sites and ESD resource centres. The schools were also provided with tools. In addition, KIE conducted a nationwide baseline survey on ESD and launched a 6-week teacher development course and a two-year teacher training course.

NEMA, a lead government agency, has collaborated with the private sector, formal and non-formal education institutions, NGOs, CBOS and religious groups on: awareness campaigns focusing on the root causes of unsustainable outcomes in social, environmental, cultural and economic development ventures. Locally, there has been extensive information sharing and feedback. At the international level, NEMA has established networks with SADC-REEP, WWF ESARPO and RCE Denmark. Officers in twenty Government Ministries and Departments have been trained on ESD, and research and innovation are being conducted in collaboration with institutions of higher learning and research institutions. In collaboration with KIE, NEMA has successfully conducted three pilot projects; (i) a green schools programme with 20 primary and secondary schools (ii) a botanical garden at Pwani University and (iii) environmental education packs to promote positive behaviour change among pupils. NEMA has also conducted effective monitoring and evaluation using continuous outcome mapping models; the ESD implementation of more than 100 primary, 100 secondary and 100 tertiary institutions has been monitored.

NEMA has spearheaded the formation of nine RCEs, and has supported CSOs such as the Wildlife Society of Kenya, the World Wide Fund for Nature (WWF), the African Fund for Endangered Wildlife (AFEW), Jacaranda Designs and UNESCO via the Chanuka Express ESD mobile outreach programme. NEMA has also partnered with the Kenya Organisation for Environmental Education (KOEE), the Wangare Maathai Institute and UNEP.

Capacity building and training

This strategy aims to ensure that ESD partners and stakeholders acquire and constantly improve their capacity and skills. The strategy identified the key areas for training as follows: communication and public awareness; the development of planning, management, evaluation and analysis skills in ESD initiatives; training and refresher courses for educators to promote values, attitudes and behaviours that will stimulate learning that embraces sustainable development; the development of all forms of appropriate content and materials (written, electronic, and audio-visual) on ESD; embracing of learner-centred teaching methods and facilitation techniques.

Despite this solid plan, the integration of ESD into the curriculum, as proposed by the ESD implementation strategy, is generally weak in all learning environments (ESD Country Strategy 2005–2012). Initiatives have been launched, in particular at tertiary level, under the MESA programme and NEMA. At lower educational levels, piecemeal curriculum reviews have been undertaken. However, ESD has tended to be viewed as another adjective to add to the curriculum. Further, although these reviews have been conducted in formal education, they cannot be carried out for the non-formal and informal sector as no written curricula exist here. Moreover, the Kenyan curriculum is essentially exam-oriented; pre-service and in-service teacher training is thus required in order to embed ESD in the curriculum. In addition, the curriculum is overloaded with content and learning outcomes, making the integration of ESD a highly complex task.

Research and innovation

The ESD implementation strategy anticipated that baseline studies, situational analysis, longitudinal studies and other sector specific/institutional demonstrations would be conducted to identify and further delineate the key ESD innovations to be addressed and put in place in thousands of local contexts across the country. The strategy also envisaged research to foster the integration of ESD into a multitude of learning situations and across a variety geographical divides, socio-cultural contexts and development programmes, in order to develop potential strategies for meeting the challenges involved. Three main research and innovation activities are:

1) Assessment of resource utilization with regard to ESD; NEMA provides resources (approximately KES 7 million per year) to selected RCE for a range of activities, but this is not based on any needs assessment. Innovations from CBOs and small community groups do not have adequate support for their projects. In addition, many of these innovations come from groups with only limited capacity to draft effective bid proposals to donors. There is need for a national needs assessment to provide the basis for resources utilization.

2) Documentation and dissemination of research findings and innovations; RCEs are making efforts to do this. The Mau Complex RCE is using the annual research week to showcase ESD research results, and a number of RCEs have attempted to produce brochures and/or posters listing their objectives and activities. Workshop presentations are the most common mode of dissemination.

3) Mainstreaming of ESD research and innovation; the Greater Pwani RCE is conducting a number of research projects and introducing a range of innovations, including the use of sisal waste as a substrate for mushroom production and biogas generation. Other examples include mariculture and the Pwani University botanical garden. The Western RCE is also conducting a series of research studies, and an International Training Programme (ITP) change project is looking at community engagement as a potential method of disseminating ESD.

According to the ESD Kenya Country Report (2001–2015), for the ESD strategy to be achieved, a move towards participatory action research—which involves the beneficiaries—needs to be prioritized. While NGOs and RCEs are encouraged to partner with the private sector and development partners, there is need for a national kitty that will use a call for proposals approach to allocate resources annually and by theme. There is a lack of systemized support for ESD innovation in Kenya. However, support from UNESCO, UNEP and one or two civil society organizations has provided some impetus to the ESD process. Wider support would be forthcoming if a common and binding vision for ESD were established. While RCEs are hosted within institutions of higher learning, documentation is ad hoc and dissemination of findings remains poor. ICT is the lifeline, networking and linking partners, storing data and enabling ESD stakeholders to share information. ICT offers new modes of communication and learning spaces, thus magnifying opportunities to explore global dialogue, experience sharing and mutual support as consensus is built around sustainable development.

Information and Communication Technologies (ICT)

ESD implementation began in traditional environmental/outdoor education, which aimed to get learners outside to allow them to experience the natural world and learn about it (Eze and Emmanuel 2013). The tools and processes offered by ICT enable information to be accessed, retrieved, stored, organized, manipulated, produced, presented and exchanged by electronic and other automated and mean that ICT has become an enabler for ESD implementation. ICT strategy incorporates hardware, software and telecommunications in the form of personal computers, digital cameras, phones, faxes, modems, CD and DVD players and recorders, radio and TV programmes, database programmes and multimedia programmes that facilitate communication (Eze and Emmanuel 2013). In the ESD context, ICT has the power to reach out to diverse audiences and trigger socio-economic development by sustainably addressing global environmental and social challenges. ICT helps to make the world a better place, providing information resources, tools and portals for educators, supplementing classroom-based activities and providing tools for distance/online learning.

Monitoring and evaluation

It is anticipated that the strategy for monitoring and evaluating the impact of the ESD process will be integrated into the process itself. The strategy covers three main activities:

a) *Development of M&E tools* – The Kenyan ESD implementation strategy provides status, communication, facilitative and results indicators to monitor the implementation of ESD at different levels. Outcome mapping is being used to assess the outcomes of activities and progress with implementation; provincial and district ESD implementation guidelines, developed for provincial and district environmental committees and other stakeholders, are also being deployed. Civil society organizations use institution-based M&E systems; they do not have separate tools to monitor ESD activities.

b) *Building of M&E capacity* – NEMA facilitated a capacity building workshop for primary stakeholders, focusing on outcome mapping. There is need for a similar training to be conducted at RCE level in order to reach more stakeholders.

c) *M&E itself* – NEMA coordinates the monitoring and evaluation of ESD activities in Kenya. Stakeholders use outcome mapping to provide the relevant data and submit it to NEMA for compilation. The product is the annual ESD Journal, posted on NEMA website. To date approximately 100 primary schools, 100 secondary schools and 100 tertiary institutions are having their ESD activities monitored. NEMA supported KIE's development of a nation-wide ESD baseline.

ESD stakeholders need to be trained to use the monitoring and evaluation tools developed for ESD. The strategy also needs to be reviewed so as to ensure that: (i) there are clear indicators for all stakeholders; (ii) it is aligned with related strategies in other sectors; and (iii) it is aligned with new developments in both the public and private sectors. Monitoring and evaluating change requires more than outcome mapping.

Regional Centres of Expertise (RCE)

RCEs are networks of existing formal, non-formal, and informal organizations facilitating ESD in local and regional communities. The RCEs are hosted by higher education institutions and, due to their unique synthesis of stakeholders from different sectors, are able to access pools of expertise from different fields in order to share information and experiences and promote dialogue through sustainable development partnerships. RCEs aim to promote the long-term goals of ESD, such as environmental stewardship, social justice, and improvement of the quality of life. In Kenya, NEMA has spearheaded the establishment of nine RCEs, namely, Nairobi, Mau Complex, Narok, Western Kenya, Nyanza, Rift Valley, Coast, Central Kenya and Upper Eastern, which are hosted regionally within universities and address local themes and issues.

The core functions of RCEs include; (i) re-orientation of education towards sustainable development, designing integrated sustainable development curricula that address issues relevant to the local context; (ii) increased access to the quality basic education that is essential in the regional context; (iii) delivery of training programmes for all levels of society as well as development of methodologies and learning materials; and (iv) taking the lead on advocacy and awareness raising, focusing on educators and the central role of ESD in achieving a sustainable future.

Environmental and sustainability education in Kenya: emerging issues and trends, and current and future needs

Broadly, sustainable development stakeholders in Kenya are increasingly embracing education, public awareness and training as ways of advancing sustainable development. The Government has incorporated education strategies, tools and targets into national sustainable development strategies, climate change plans and related economic frameworks such as the Green Economy. Further, partnerships, collaborations and networks, including RCEs, have been formed to enhance the implementation of ESD. One success of ESD has been its facilitation of interactive, learner-driven pedagogies that equip learners with the knowledge, values and skills required to promote sustainability. To this end, continued efforts are being made to train teachers and education officials and to encourage schools to practice ESD.

Despite these efforts, stakeholders have identified a number of challenges to the implementation of ESD in Kenya. These include the fact that the concept is not fully understood; inadequate teaching materials, especially at informal and non-formal levels; lack of educator competencies, non-inclusion of ESD in teacher pre-service training, inadequate capacity-building and training skills; inadequate partnerships amongst ESD stakeholders including with the media; limited research and innovation and a disconnect between ESD research and the needs of industry.

Further, it has been noted that there are limited resources for ESD implementation; use of ICT to enhance ESD is minimal; and ESD monitoring and evaluation tools are not well developed. At policy level, there is a lack of clearly articulated strategy and policy relating to ESD; and fragmented legal frameworks. Moreover, the current curriculum is too exam oriented and overloaded to allow integration of ESD.

Covid-19 has presented a real threat and has radically changed the world. Wolff (2020) notes that lockdown and lifestyle changes have made it clear to all that human life is vulnerable and nature is unpredictable; all of a sudden, social wellbeing is now challenging individual freedom, and even basic human needs such as nutrition and a network of social relations are no longer self-evident. The Covid-19 pandemic has created both challenges and opportunities in Kenya. The restrictions on movement and gatherings imposed within the country so as to curtail the spread of the diseases restricted formal and non-formal learning. ESD implementation slowed down and very little was achieved, especially between March 2020, when the first case of Covid-19 was detected in Kenya, and August 2021, when some of the restrictions were dropped. However, the pandemic was a blessing in some ways for the education sector in Kenya. Learning institutions, especially tertiary institutions and universities, have invested heavily in ICT infrastructure, distributing computers at subsidized rates to students, for example, and providing online teaching and learning platforms and studios where lecturers could record lecturers for learners. This brought about partnerships with ICT service providers such as Kenet, Safaricom and Airtel that enabled ICT support systems to be enhanced. Collaborations were established to provide the funds and facilities required to facilitate online teaching and learning. The Covid-19 pandemic has led many people to consider the need to switch changed to a more sustainable lifestyle, consuming fewer resources and considering both present and future generations. In the context of such change, education has an important role to play in framing a post Covid-19 ESD strategy.

References

Chaudhry, S. 2019. Politics of Land Excisions and Climate Change in The Mau Forest Complex: A case Study of the South-Western Mau Forest. Retrieved from https://www.researchgate.net/publication/332258607_Politics_of_Land_Excisions_and_Climate_Change_in_the_Mau_Forest_Complex_A_Case_Study_of_the_South-Western_Mau_Forest.

Council on Foreign Relations. 2008. Understanding Kenya's Politics. https://www.cfr.org/backgrounder/understanding-kenyas-politics.

Eze, I. and E.O. Adu. 2013. The utilization Of ICT in education for sustainable development. pp. 1514–1522. *In*: Bastiaens, T. and G. Marks (eds.). Proceedings of E-Learn 2013; World Conference on E-Learning in Corporate, Government, Healthcare, and Higher Education. Las Vegas, NV, USA: Association for the Advancement of Computing in Education (AACE). Retrieved March 3, 2022 from https://www.learntechlib.org/primary/p/115092/.

GoK. 2008. Kenya Vision 2030: A Globally Competitive and Prosperous Kenya. National Economic and Social Council (NESC), Nairobi.

GoK. 2015. Education for Sustainable Development Kenya Country Report 2005–2012. Retrieved from https://www.nema.go.ke/images/Docs/Guidelines/the%20status%20of%20esd%20in%20kenya%20draft%201%20better%20version.pdf.

GoK. 2018. The Third Medium Term Plan 2018–2022. Third Medium Term Plan 2018–2022 Transforming Lives: Advancing socio-economic development through the "Big Four".

Institute of Development Studies. 2021. Social Economic and Political Context of Kenya. Retrieved from http://interactions.eldis.org/unpaid-care-work/country-profiles/kenya/social-economic-and-political-context-kenya.

Kenya National Bureau of Statistics. 2014. Exploring Kenya's Inequality: Pulling Apart or Pooling Together. Retrieved from https://www.knbs.or.ke/?p=352.

KIPPRA. 2020. Kenya Economic Report 2020. https://kippra.or.ke/download/kenya-economic-report-2020-2/?ind=3126&filename=Kenya.

KOEE. 2019. Education For Sustainable Development in Kenya. https://koeeorg.wordpress.com/2019/10/07/education-for-sustainable-development-in-kenya/. Retrieved on 02.03/2022.

Michura, E.G. 2019. Sustainable Education in Kenya Model for Vision 2030 Retrieved from https://journaljesbs.com/index.php/JESBS/article/view/30107/56490.

NEMA. 2012. Education for Sustainable Development (ESD). Kenya Country Report 2005–2012.

Nyangena, W. 2009. Kenya Vision 2030 and the Environment: Issues and Challenges, Government Printer, Nairobi. Retrieved from file:///C:/Users/Sellah/AppData/Local/Temp/kenya20vision20203020and20the20environment.pdf.

Ototo, G. and R. Vlosky. 2019. Overview of Forest Sector in Kenya. Retrieved from https://www.researchgate.net/publication/328826868_Overview_of_the_forest_sector_in_Kenya.

Republic of Kenya. 2008. Education for Sustainable Development Implementation Strategy. Nairobi, Ministry of Environment and Mineral Resources.

Republic of Kenya. National Education Sector Plan, 2014. Retrieved from https://www.globalpartnership.org/sites/default/files/2014-03-Kenya-Education-Plan-2013-2018_0.pdf.

UN DESA. 2009. Creating an Inclusive Society: Practical Strategies to Promote Social Integration www.un.org/esa/socdev/egms/docs/2009/Ghana/inclusive-society.pdf.

UNDP. 2020. Human Development Report 2020. Retrieved from https://hdr.undp.org/en/content/latest-human-development-index-ranking?utm_source=EN&utm_medium=GSR&utm_content=US_UNDP_PaidSearch_Brand_English&utm_campaign=CENTRAL&c_src=CENTRAL&c_src2=GSR&gclid=Cj0KCQiA3fiPBhCCARIsAFQ8QzWTC7TnoKcsGdID4hFFB7967gOjszPbQPDr1BHhzq8FLbG9bWIC9S0aAtk3EALw_wcB.

UNEP. 2011. Decoupling Natural Resource Use and Environmental Impact From Economic Growth. Retrieved from https://wedocs.unep.org/bitstream/handle/20.500.11822/8775/Decoupling_Summary_EN.pdf?sequence=1&isAllowed=y.

UNESCO. 2019. Framework for The Implementation of Education For Sustainable Development (ESD) Beyond 2019.

Wandabi, D. 2019. Education for Sustainable Development in Kenya. A Paper by Kenya Organization for Environmental Education (KOEE). https://koee.org.

Wolff, L. 2020. Sustainability Education in Risks and Crises: Lessons from COVID 19. Sustainability 12(12): 5205. https://doi.org/10.3390/su12125205.

Chapter 2

A Review of the Staging and Enactment of Environment and Sustainability Education in South Africa

An Illustrative Case Study

Rob O'Donoghue and *Eureta Rosenberg**

Introduction

In South Africa, Conservation Education (1960s-onwards) transitioned into Environmental Education (EE) (1980s-onwards) that from the late 1990s onwards, further expanded into a wider and more inclusive field of Environmental and Sustainability Education (ESE). The latter included the advent of Education for Sustainable Development (ESD) as global imperative to achieve the Sustainable Development Goals (SDGs) in the context of the United Nations' widely-supported Agenda 2030, as well as green skills pathways for a more environmentally sustainable and inclusive economy.

The vantage point for this review of the trajectory of ESE in South Africa is the 'staging of risk,' as a metaphor for contemplating how environment and sustainability imperatives came onto a widening modernist institutional education stage. Education and communication responses to escalating risk initially came to be framed and

Rhodes University, South Africa.
* Corresponding author: e.rosenberg@ru.ac.za

enacted as interventions to create awareness, change attitudes, clarify values and thus to effect behaviour change to ameliorate risk. This early interventionist staging of risk is giving way to approaches that are less linear, less impositional and instrumentalist, and more inclusive, open-ended and centred on co-engaged learning and action.

The emergence of these 'participatory' approaches can at times reflect a similar instrumental logic, but are more commonly informed by constructivist, socio-constructionist and activity theories of learning and development (outlined for example by Vianna and Stetsenko 2006), which shape the staging of risk as an open question for deliberative learning actions (after Vygotsky, as outlined by Vianna and Stetsenko 2006). Within these framings, participation is a necessary socio-cultural process in the differentiating of knowledge and the acquisition of associated skills in the schooling system and in the transitioning of government, industry and communities towards a greener and more inclusive economy. And yet, an ontological grasp of increasingly complex webs of risk also requires real-world knowledge, much of which has been constituted by the sciences within the modernist project. This suggests a paradox that Sfard (1998) resolved for formal education settings by pointing out that learners need *both* knowledge acquired from others who know (teacher or text), and, alongside this, a process of participation in expanding the knowledge that is already known, and which contemporary environment and development issues show up as inadequate by itself (Santos 2007, Lotz-Sisitka 2017), in order to transform practices towards the achievement of sustainable development goals.

This review contextualises and analyses a case of ESE in South Africa where schooling and green economy transitioning are now key policy focus areas in the emerging and diversifying field of ESE. An open-ended vantage point is developed for reviewing the modernist staging of risk as a transformative agenda for this field.

The staging of risk-clarifying a perspective for reviewing ESE in contemporary modernity

Sørensen (2018) notes how the sociologist Beck (1992) identified the boundary at an emerging 20th century interface between science as a generator of new 'real-core' knowledge that uncovers risk and an associated assumption of a human social condition of non-knowledge in every-day life. This bounded interface gave rise to state education imperatives that stage risk to resolve the public's 'non-knowledge' or unawareness. Thus early education involved the functionalist staging of risk to mediate social life and effect behaviour change; education informed by science staged risk for learners and the public as issues to be made aware of and problems to be solved through behaviour change.

Through these processes ESE has proliferated. Climate change, for example, became visible as an emerging reality and a matter of concern through the generation of scientific theory constituted with empirical data.[1] Sørensen (2018) notes how

---

[1] Here, the review simply notes how scientific knowledge production came to reside at the mediating intersection between the knowledge it had constituted and the assumption of non-knowledge in society that has shaped the proliferation of ESE since the latter part of the 20th century.

'new risks are [initially] only visible to us through scientific theories, experiments and instruments' (quoting Beck 1992, 27, brackets added). Climate science as a knowledge-generating field produces knowledge exemplifying climate change as a real 'matter of concern' (after Archer) that is being brought within our grasp to understand. In this way climate change has become manifest in an array of complex social-ecological, economic and political perspectives and concerns both for and through its staging as risk in education and communication initiatives across the globe. The originating scientific knowledge is deployed and reconstituted as risk-animating concepts for deliberative learning transactions to foster learning-led change.

Beck noted how imperatives such as the shadow side of development initially constitute 'second-hand non-experience' (Beck 1992, 71–72). Earlier, Baudrillard (Poser 2017) explored how a 'simulacrum' in modernist social processes of communicative meaning-making, such as those constituting a hyper reality of abstracting detachment around which a referential value in relation to social-ecological and politico-economic realities forms, is restored in contexts of communal co-existence. Here modern education can be read as a realist narrative of referential re-inscription (reflexive modernization after Beck 1992) through evaluative engagement with problems that are becoming manifest as risk in the modern world. This allows us to contemplate education as emergent dialectic processes that enable participants to deliberate concerns as a learning-led process of change towards realising future sustainability within science-informed *and* participatory meaning-making processes of co-engaged learning to change.

Historical socio-political and economic contexts and associated educational approaches

To analyse ESE in South Africa, it is important to consider the *longue duree* of its social and economic life (for in-depth historical analyses, see Carruthers 1995, Khan 1994, O'Donoghue 1997). It is particularly important from a sustainable development perspective to consider a pre-colonial time when African share-croppers were farming grains by means of strong social networks (Keegan 1986), and tended livestock with situated knowledge of fire and disease control (O'Donoghue 1997). The colonization of the region brought the introduction of mechanization, which meant advancement for some, but impoverishment for the majority who, under colonial governments, lost access to land and water. The equal participation of Africans in the economic activity that followed the modernisation of agriculture and mining was precluded by European settlers with preferential access to technology (Keegan 1986). Pre-colonial times were not a utopia, but new kinds of deprivation, tenured labour and dependency on an inequitable economy ensued (Van Onselen 1996); leading to the birth of poverty and unemployment.

Colonial hunting practices for sport and trade contributed to the demise of vast herds of wild animals, which in turn led to the introduction of fenced conservation areas like the Kruger National Park (Carruthers 1995). Along with awareness of diminishing wildlife and nature areas came Conservation Education, strongly directed at African communities (Khan 1994). With the introduction of the Land Act in 1948, 'home lands' were established where growing numbers of Black Africans had to live on less productive land, bringing land degradation and, with it, the first documented instances of EE, in the form of awareness raising programmes (led, for example, by the Native Farmers Association) to reduce soil erosion in homeland areas (Khan 1994). The science of the time advocated the fencing off of grazing lands, and EE promoted associated behaviour change and sanctions against 'poaching' the remaining wildlife. This occurred while the system allowed settlers to own large tracts of land with 'game' considered theirs to hunt, and with relocated, impoverished households providing their own lived-experience reasons why the imposed solutions were not suitable (Khan ibid).

Systemic reasons for land degradation and the loss of biodiversity were largely overlooked. They went hand in hand with a segregated education system which saw the introduction of an inferior Bantu education designed to keep Black Africans in menial labouring positions, an ideal resource for agriculture and mining which aggregated capital in the hands of an elite. The local and situated knowledge of indigenous peoples, on how to survive, thrive and sustain themselves in Africa, seem to have been largely overlooked in EE programmes at the time, even though such knowledge was also necessary to sustain the European settlers (O'Donoghue 1997). A notable exception was conservationist Ian Player who worked closely with his Zulu counterpart Magqubu Ntombela, when they established the Wilderness Leadership School in 1955 (O'Donoghue ibid).

Emergence of formal environmental education and yet more challenging economic realities

Fast forward 40 years: In 1994, Nelson Mandela's African National Congress took over as the country's first democratically-elected government, which now theoretically gave Black South Africans equitable access to land, education and economic opportunity. Mandela introduced a Reconstruction and Development Plan to benefit those previously sidelined by apartheid's unequal development. This coincided with the restructuring of educational systems and curricula, which included the new democracy's first *White Paper on Education and Training* (Republic of South Africa 1994) making explicit provisions for EE that is linked to both the sustainable use of resources, and quality of life for all, that is, equitable and sustainable development:

> Environmental education, involving an inter-disciplinary, integrated and active approach to learning, must be a vital element of all levels and programmes of the education and training system, in order to create environmentally literate and active citizens and ensure that all South Africans, present and future, enjoy a decent quality of life through the sustainable use of resources (Republic of South Africa 1994, 22).

Members of the Environmental Education Association of Southern Africa, a multi-racial regional body established in the late 1980s in neighbouring Swaziland, drew on the outcomes of the 1992 United Nations Conference on Environment and Development which positioned environment and development as 'two sides of the same coin' (O'Donoghue 1993) and in particular, on the People's Charter developed by civil society organisations and indigenous peoples at the Rio Earth Summit.

But while there were narratives in the UN, in South Africa's new democratic government (e.g., the National Environmental Management Act, Republic of South Africa 1998), and the EE discourse in the country, about *integrating* environment and development, they were in practice treated as separate by most government policies and juxtaposed as such by industry and labour discourses. Instead of reconstructing the unequal economy, the effect of policy frameworks like Broad-based Black Economic Empowerment was most notably to shift ownership from White elite to Black elite in what remained an unequal, extractive economy, with hopes for Black owned businesses to emerge, but leaving the majority still outside the mainstream economy, and the numbers of unemployed rising.

Against these economic challenges which threatened the dream of a decent life for all, EE practitioners and teachers also kept alive niches of agency-building and sustaining livelihoods. The Eco-Schools programme, for example, run by the Wildlife and Environment Society and partners (www.wessa.org.za), supported food gardens, the recycling of waste, and restoration of local wetlands to protect biodiversity, among others, as action projects through which schools not only received recognition from the Foundation of Environmental Education as world-class EE, but were also recognised by authorities as significant community development initiatives (Rosenberg 2008). In the 1990s, such school-based activity was often treated as extra-curricular. This was however set to change.

Formalising environment and sustainability education in South Africa

During the 1990s, environmental educators worked with the new government on shaping the national curriculum (for an overview of this policy work, see Lotz-Sisitka et al. 2020). Following much debate it was concluded that environment and sustainability should feature across all subjects, and therefore in cross-cutting principles and learning outcomes. This was (and still is) evident. For example, the most recent version of the school curriculum, the National Curriculum and Assessment Policy Statement (CAPS Grades R-12, Republic of South Africa 2018) includes the following principle for all subjects:

> Human rights, inclusivity, environmental and social justice: infusing the principles and practices of social and environmental justice and human rights as defined in the Constitution of the Republic of South Africa. (p.5)

The curriculum also states, among the cross-cutting generic learning outcomes, that all learners should be able to:

- ... use science and technology effectively and critically showing responsibility towards the environment and the health of others; and to
- ... demonstrate an understanding of the world as a set of related systems by recognising that problem-solving contexts do not exist in isolation (p.4).

In line with this latter curriculum principle, a growing number of ESE research and training programmes started to shift the narrative away from calls for awareness-raising and behaviour change, which put the blame for environmental degradation, pollution and poverty at the door of the individual—positioned as 'an other' by the educator or scientists who staged risk so as to give effect to behaviour change through education. This instrumentalist approach has roots in the civilizing turn that assumed that problems such as litter or over-grazing were caused by ignorance or 'non-knowledge.' Programmes like the Gold Fields Environmental Education Course (Molose 2000), the Southern African Development Community Regional Environmental Education Programme, and the Masters in Environmental Education at Rhodes University, started to encourage educators to stage risk as systemic causes for development-and-environmental problems alongside community practices and individual action choices; and to situate the possibilities for change in a dialectic, iterative movement between recognizing structural socio-political and economic causes and exercising collective agency.

When a global narrative emerged that EE was too narrowly focused on protecting nature, without consideration for the development of people, South African environmental educators noted that they had actively worked to broaden this perspective. The new ESD discourse was nonetheless adopted by the government, with the proviso that the environment remained important in the consideration of what would make development more sustainable.[2]

Articulating and realizing the ESE-related curriculum principles and outcomes in schools proved to be challenging. For one, the curriculum underwent a swing from the outcomes-based and learner-centred approaches (learning through participation) introduced in 1994, to a stronger content focus in the 2018 CAPS (learning through acquisition). Also relevant for this review is that the staging of risk in a problem-centred discourse prevailed in the prescribed content of many subjects. The Life Sciences, for example (see: Schudel 2014), positioned people as separate from nature and the source of threats and risks to ecosystems and biodiversity, with the individual behaviour-change narrative prevailing. Subjects like Technology referred to the need to consider the environment when working on projects, but were silent on how technology or economic theory could provide solutions to environmental problems and risks. In Geography, teachers reported (Rosenberg 2008) that they found content such as alternative forms of development difficult to teach as South Africa presented few examples of these, and the curriculum required the use of local case studies. The curriculum for several subjects tasked teachers to explore 'contemporary social and environmental issues' with learners. The development trajectory of the country saw high levels of pollution and inefficiency continuing even as unemployment soared, energy provision failed and the economy went into decline, with mining and agriculture both shedding jobs. Learners found these realities with no clear solutions hard to face, depressing and disheartening, and many dropped the subject of Geography for exactly this reason (ibid).

ESE programmes like Eco-Schools encouraged teachers to 'bring environment and sustainability *out* of the curriculum' that is, to give expression to the environmental

---

[2] The connections between environment and sustainability were also recognised in the recent Berlin launch of ESD 2030, where the importance of EE was emphasised.

content that was already specified. In addition to providing supplementary scientific information to explain and expand curriculum content, the Eco-Schools programme started to provide auditing tools that educators could use with learners to explore their own, context-relevant 'matters of concern' and hence to bring participation and a sense of agency to the learning experience. This was also done in a university context (Togo and Lotz-Sisitka 2009, 2013) and 'giving away the tools of science' became a constructivist, democratically-inspired feature of citizen science/environmental monitoring programmes, in schools and communities. Teacher education with an ESE focus emerged as a strong feature in universities, with the education faculties of Rhodes, Stellenbosch and other universities supporting teachers, lecturers and education officials to study ESE practices (e.g., Du Toit 1999, Janse van Rensburg and Lotz-Sisitka 2000, Le Grange and Reddy 2000, Lotz 1996, Lotz-Sisitka 2002, Lotz-Sisitka and Raven 2001, Mbanjwa 2003, Nduna 2004, Naidoo 1993, Nsubuga 2010, O'Donoghue 1990, O'Donoghue 2000, Olvitt 2004, 2012, Ramsarup 2006, Schudel 2012, 2014, Wagiet 1996, Wagiet 1997). The insights from this body of work informed the development of *Fundisa for Change*, a multi-institutional programme for educators, the contemporary case example presented in this chapter.

Emergence of green economy discourse and skills planning

Among the many initiatives with an ESE focus that engage teachers, the *Fundisa for Change* programme (www.fundisaforchange.co.za) stands out as a 'foundation-building' feature of a number of national environmental skills plans, notably the Environmental Sector Skills Plan (Department of Environment Affairs 2010) and the Biodiversity Human Capital Development Strategy (South African National Biodiversity Institute 2010). These skills plans were developed at the start of the last decade as part of a broader government—civil society collaboration that was formalized in a National Environmental Skills Planning Forum, convened by the Department of Environment, Forestry and Fisheries, and a multi-partner national Green Skills System Building programme (www.greenskills.co.za).

The Green Skills programme, resourced in the period 2015–2018 through a fund focused on job creation, sought to develop the skills dimensions of the drive towards a green economy, a drive that started to circle back to integrating environment and development issues. In 2011 a Green Economy Accord (EDD 2011) committed industry, labour and government to a clean, green, low-carbon and employment intensive development path. Ten years later, in the context of Covid-19, this commitment was reiterated as South Africa's President launched a National Climate Change Adaptation Strategy and declared that while the economy required urgent recovery measures,

> we should not merely return to where we were before the pandemic struck. We are instead looking at actions that will build a new, inclusive economy that creates employment and fosters sustainable growth. An important aspect of this new economy is that it must be able to withstand the effects of climate change. A climate-resilient economy is necessary to protect jobs, ensure the sustainability of our industries, preserve our natural resources and ensure food security (Ramaphosa 2020).

How exactly would this be achieved? And what does this intent mean for ESE? In the Green Skills Programme several studies were undertaken to identify the skills needed to enable industries like mining, agriculture and manufacturing to shift towards a more sustainable and inclusive economy (for an overview, see Rosenberg et al. 2020). In the process of conducting these studies, a methodology was developed (*ibid*) to integrate environment and development concerns in a multi-level framework informed by Bhaskar's laminated ontology and critical realist epistemology in which absence and a dialectical pulse were key features of a transformational intent (*ibid*). These studies confirmed an observation by others (e.g., Cock 2014), that there are different approaches to a green economy evident in business, government and civil society and that not all of these would lead to transformative outcomes for the majority of South Africans or the environment. What exactly a green economy is, and how best to educate towards it, is still unresolved, but the case study presented next provides some pointers.

Implementation of ESE and the follow-up 'ESD for 2030' in the education system

What does the current situation of the implementation of ESE in South Africa look like? High level reviews synthesizing across a wider scope of initiatives (e.g., O'Donoghue 2016a and a comprehensive review of ESE policy research by Lotz-Sisitka et al. 2020) complement this chapter, but so as not to repeat these, and to make the conceptual overview more concrete, this section will be approached through a case study. *Fundisa for Change* (https://fundisaforchange.co.za/) was chosen for this case analysis for the following reasons:

- Its multi-partner and national scope and prominence—involving several government departments and almost all South Africa's 32 universities,
- A body of research on and evaluation of the programme (e.g., Songqwaru 2012, 2020), and
- The fact that it straddles schools-based work and emerging green skills developments.

During the UNESCO World Conference on ESD in May 2021 (UNESCO 2021) South Africa's Minister of Basic Education and Training highlighted *Fundisa for Change* as a flagship national ESD initiative. The government recognizes the programme as supportive of the national curriculum and national teacher education requirements. The analysis here is focused on the ways in which this programme demonstrates current conceptual and practical trends in ESE in South Africa.

Fundisa for change and green entrepreneurs' day

Fundisa for Change aims to build the capacity of teachers and teacher educators in three connected dimensions:

i) 'Knowing your subject', aimed at recognizing the environment and sustainability-related content in the school curriculum and extending related knowledge in order to teach it well

ii) 'Improving your teaching practices' - aimed at expanding pedagogy in ways that are more learner and learning-centred, and

iii) 'Improving your assessment practices' to better assess particularly higher order learning outcomes. This framing of capacity-building extends beyond simply making teachers aware of the need for ESE; it embeds ESE in the curriculum, rather than treating it as an 'add-on.'

Fundisa for Change educator courses have been customised for the Life Sciences, Marine Sciences, Geography and Climate Change, among others. The case reviewed is a recent (2021) version of the course, developed by three partners (Rhodes, Wits and North-West Universities) for the Commerce subjects, specifically Economics and Management Sciences (EMS), a compulsory primary school subject, and Business Studies, a high school elective. The course's focus—on sustainability and the SDGs, green economy and entrepreneurship—was timely, given South Africa's development context outlined earlier, and the fact that entrepreneurship is being widely promoted as a response to the failure of the economy to create employment for a growing youth population.

More than 100 educators signed up for the course, offered through Saturday morning synchronous online classes and asynchronous 'work-away' tasks both on and offline. The course is currently being evaluated. Early reports indicate that participants found the content on the SDGs, green economy and sustainability 'new knowledge' that they did not have before. The responses of teachers during interactive lectures on how green entrepreneurship features in the CAPS curriculum add additional insight. It became apparent that some teachers had in fact been working with 'green economy' concepts such as circular economies, waste reduction, sustainability and integration of social and environmental aspects of sustainability, without naming them as such. This is evident, for example in the context of the Grade 7 curriculum, where the EMS CAPS requires teachers to involve learners in project work, to participate in an 'Entrepreneurs Day', in which they need to use recyclable materials, among other sustainability considerations.

During online course sessions EMS teachers shared examples of how they routinely design projects for learners to create value out of discarded materials, using problem-solving and creative skills to produce new products out of old, that they then sell in order to raise a profit, thereby getting a taste for what it would be like to run a business, how to budget and calculate expenses and income, how to promote the business to customers, and how to take social and environmental considerations into account. While the learners reportedly greatly look forward to Entrepreneurs Day (also known as Market Day), teachers emphasised that most of the learning, along with the assessment, takes place in the preparations for and follow-up after the event.

In one particularly informative example, the teacher narrated how the learners produced promotional materials weeks in advance to generate support for Market Day in the local community; how profits were made by making and selling items from discarded materials, e.g., 'jean bags' sewn from old denims and pencil cases crocheted from plastic packets. Most striking was the fact that at this school, the profits are subsequently shared with the community during Heritage Day (a national holiday focused on cultural traditions) when the elderly and unemployed are invited to the school for a meal: 'We [teachers and learners] cook for them.'

These ESE aspects of the Market Day are a significant advance on awareness raising activities such as litter clean-ups on the school grounds and surrounds—instead of just picking up plastic packets, they are being turned into sellable products; they also expand on audits—which usefully identifies the sources of litter (or other local issues). Through Market Day, EMS learners...

- get a taste of working on solutions ('waste to wealth') and a chance to explore their own and collective agency
- gain subject-specific competencies such as calculating cost and profit, budgeting
- gain generic competencies such as planning, team work and communications; and
- exposure to ethics and relational values.

The teachers' accounts furthermore shed light on what might be significant contours of a practical, realistic and necessary green economy in South Africa. The school that engages learners in sharing profits back into the community by hosting a communal meal, is not obliged to do so by the curriculum. These teachers draw on their own knowledge of how community bonds are maintained and why they are important. Their Market Day is not the Silicon Valley version of an enterprise, but it has the hallmarks of sustainability: unsustainable consumption is reconfigured through the circular use of materials that would otherwise create waste; the importance of the surrounding community in making the business possible in the first place is recognized; ethical considerations and the valuing of positive relationships with that community—'giving back'—by using business profits to further sustain the learners' and schools' community.

The *Fundisa for Change* programme engages teachers to explore how such embedded learner projects, which resonate with teachers' understandings of the learners' communities and their matters of concern, can be used to achieve the intended curriculum outcomes ('results through relevance'), and how projects that make sense in the socio-cultural context of township and rural communities, can also develop academically-relevant competencies that resonate with several SDGs: sustainable production and consumption, poverty reduction, reduction of hunger, and decent work, among others.

ESD and Hand Prints for Care

An analysis of the resource materials used on the *Fundisa for Change* Green Economy and Entrepreneurship course further illustrates current features of ESE trends in South Africa. Teachers enjoyed using online tools for calculating one's carbon footprint, particularly when a South African version was found, with locally applicable calculations. The resource that demonstrates recent trends well, however, is a set of Hand Prints for Care activities, developed by O'Donoghue and colleagues as part of South Africa's participation in the multi-country ESD Expert Net. In this resource, a local 'matter of concern' is placed in the centre of a diagram, with links spreading out to any number of SDGs to which it applies (Figure 1).

The teaching and learning activity involves identifying a local, contextually relevant matter of concern with development and sustainability dimensions (Q1 in Figure 1) and then considering and investigating it in the light of the SDGs (and vice versa) (Q2), and then also to work out 'what can be done' (Q3–4). The hand-print is the positive partner to footprint; footprint calculations make us aware of our impacts, and hand-prints help us overcome despondency by looking at what we can do to reduce our negative impact/increase our positive impact (enabling participation, agency and competency development).

The range of Hand-Print resources guide schools as well as communities in tree planting and growing saplings, composting, worm-farming, vegetable gardening ('Have you grown your greens?') and recycling, among other sustainable livelihood activities. For the *Fundisa for Change* Green Economy and Entrepreneurship course, a lesson exemplar was developed in which teachers would explore with learners what to do about local matters of concern/contemporary issues like unhealthy diets, food insecurity and poor and destitute neighbours, in the context of Covid-19. Solutions proposed include finding out from elderly community members 'what is already known' to address such issues, that is, surfacing and learning about local and indigenous knowledge. Examples include the making of sourdough bread and *amasi*, an indigenous fermented milk product that improves the diet and can be produced sustainably at low cost. The curriculum encourages learners to address 'contemporary social and environmental issues' through a business idea, i.e., it forms the basis of business and enterprise development.

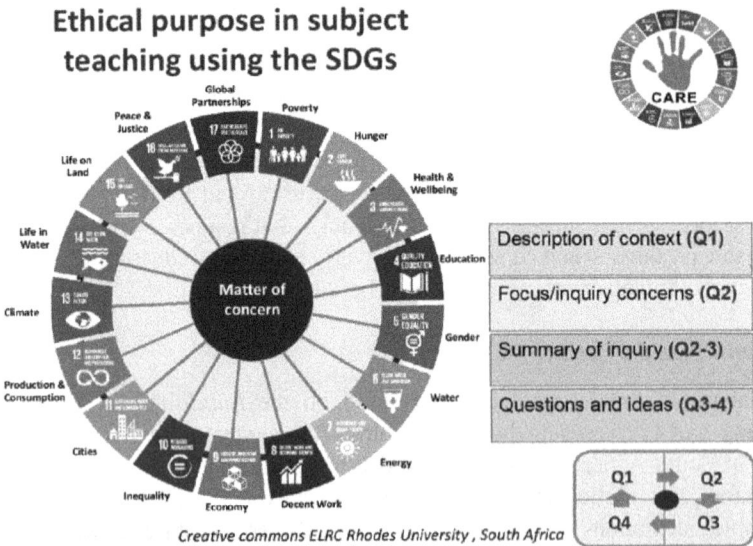

Figure 1. The SDG Wheel—Exploring local matters of concern in relation to global goals, as the basis for enquiry and active learning.

Sadly, in the high school curriculum, no 'green' business ideas were to be found in the participating schools. It would seem that the found ESE principles from primary school do not make it through to high school. The dearth of examples of green economy activities nationally, and associated lack of conceptual and policy coherence, could well be to blame.

Issues, trends and future ESE needs identified through the lens of staging of risk

The Entrepreneurs Day featured in the EMS curriculum has educators encouraging situated learning, creativity, problem-solving and collective agency among Grade 7 learners. Teachers were teaching the value of reducing waste, doing more with less and creating circularities in localised economies, as well as the value of care, social networks and community connections. Through the latter, they extended standard entrepreneurship training models that prioritise competitiveness, unique niches and a singular focus on business growth (Rosenberg 2021). In the Market Day activities teachers situated the agency of learners and communities as collectives, recognizing the impact of harsh economic realities but not being rendered hopeless and helpless by them. One of the key needs for the future is to reconfigure what green economy and green enterprises mean in contexts of long-standing poverty and degraded natural resources; how they actually develop, what learning is required and what people already know about how to sustain communities (Rosenberg 2021), rather than to teach local youth Western based theories of how businesses and entrepreneurs (should) develop.

The *Fundisa for Change* lesson exemplars as well as the Hand-Print Care materials demonstrate a trend (see Figure 2) in both the CAPS curriculum and among ESE practitioners to expand from awareness-raising and problem-focused approaches to sustainability issues, with exploration towards and engaging in solutions, turning local matters of concerns into opportunities for action, development, and

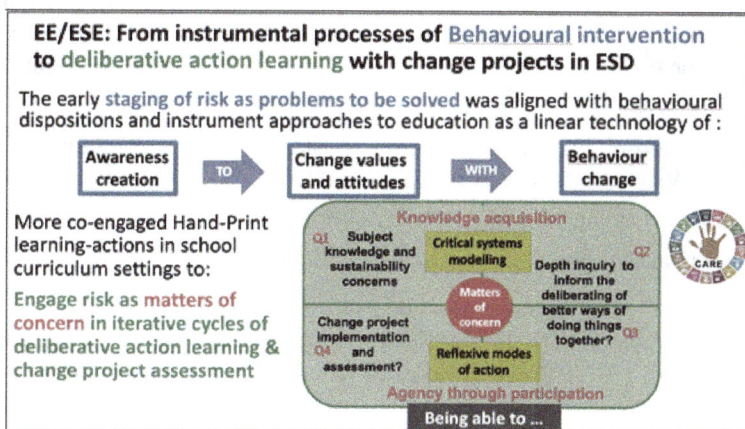

Figure 2. Trends in ESE reflected in the design of Hand-Print Care ESE SDG materials.

learning-led change. The earlier individual behaviour change focus is being expanded to processes of providing learners with opportunities to propose and participate in co-constructed actions, thereby emphasizing collective agency which is clearly needed if systemic environment and development issues with roots in a discriminatory past are to be successfully tackled, even at a local level.[3]

The staging of risk is thus reconfigured as systemic and collective responsibility and also extended with the notion of opportunity, as embedded in the notions of green economy, green enterprises, and making profits in collective entrepreneurial activities that address local needs and are also used to further extend community values (exemplified in cooking for the elderly and unemployed).

The case exemplifies broader trends away from assuming 'non-knowledge' and assumptions about unawareness, to recognition of existing knowledge (the knowledge of community members on preparing healthy foods, for example, and the teachers' knowledge of how to make and share profit in local communities). Recognition of the learners' own and shared experiences is an important aspect of embedded or situational learning (O'Donoghue 2017). This does not however exclude valuing new knowledge, as testified by the *Fundisa for Change* participants' delight at encountering carbon footprint calculators and new examples of creating 'wealth out of waste' (shared by fellow teachers); nor should it under-estimate the importance of schools creating opportunities for learners to gain new shared experiences.

It should not be interpreted, as the introduction of outcomes-based education in 1995 unfortunately did, as a trend away from bringing new knowledge into learning situations. It is rather a trend that acknowledges the importance of a variety of teaching strategies to support a variety of (active) learning opportunities: learning as knowledge acquisition *and* learning as participation (Sfard 1998), but also learning as expansion and transformation (Lotz-Sisitka 2016). As the critical pedagogue Paolo Freire noted, (2000, brackets added) '*Education as the practice of freedom [is] the means by which men and women deal critically and creatively with reality and discover how to participate in the transformation of their world*' (p.34). In such discovery, the acquisition of scientific sustainability knowledge is important, but not through a staging of risk that assumes the participant is void of relevant knowledge, awareness or experience on which to build. This emerging approach embodies a wider range of educational approaches, including:

- Constructivism and in particular Vygotskian socio-constructionist activity theory (Vianna and Stetsenko 2018)
- A 'new generation of critical theories' (Lotz-Sisitka 2016), and
- Dialectical critical realism (O'Donoghue 2016b, 2017).

While these theories and associated pedagogical innovations have constituted, in our view, an advancement on functionalist, instrumentalist approaches to ESE, the latter approaches of course also persist. Whereas the SDGs are now firmly ensconced

---

[3] Internationally, this development has also been explored by Rieckmann (2020) and Schank and Rieckmann (2019).

in national ESE narratives, the global goals have been difficult to explicitly include in curriculum practices. The SDG Wheel (Figure 1) was thus developed for pedagogical use as a curriculum tool for *Fundisa for Change* participants to expand their inquiry work with learners beyond somewhat narrowly inscribed conventions that have tended to ascribe problems to errant or ignorant individual behaviours. The SDG Wheel has been useful to expand and deepen educators' grasp of the scope of environment and sustainability concerns and the complex web of historical, socio-economic and political processes underpinning them.

Conclusion

From a cultural-historical perspective on knowledge generation, institutional power and education in modernity, this chapter provided a vantage point for tracking how ESE emerged with expanding scientific knowledge to inform the institutional staging of risk in widening education imperatives (Figure 2). The chapter contextualised a recent case of ESE, the *Fundisa for Change* teacher education programme on 'Green Economy and Entrepreneurship,' which depicted schooling and green economy transitioning as key policy areas in an expanding field of ESE in South Africa at this time. Imperatives for educational change initially emerged within the colonial to apartheid nation state and in the democratic era, evolved into widening programmes for the institutional mediation of social life. An early staging of scientific facts as the foundations for education as a process of awareness creation to effect change, is continuing into the present. Conservation education, EE and ESE were initially taken up as interventions to engage students and citizens in problems for the reflexive enactment of change to resolve the staged risk. These primarily constitute what Beck referred to as 'second-hand non-experience' and the social imaginaries they produce are not easily enacted as real-world change.

Informed by a levelling of earlier power gradients in the latter part of the 20th century, ESE was also accompanied by a participatory turn. The participatory turn appeared to have the potential to overcome the problem of the 'second-hand, non-experience' accommodation of abstract expert knowledge informing about risk. Educational activities developed as the recontextualization of abstract knowledge to signify risk as real concerns in a local context for activating reflexive learning actions and deliberative co-engagement. The problem was, however, not easily resolved within an entrenched circularity with an institutional reliance on empirical sciences into an interventionist logic for effecting change. Narrative engagement in risk has also not been widely translated into change; instead learners chose to 'drop' subjects that put too much emphasis on intractable development problems (for example).

What is starting to emerge as illustrated in the case example and review (Figure 2), is a reframing of the prevailing logic of intervention from education for change into co-engaged processes of transformative action learning, or learning-led change. The last decade has seen the beginnings of education expanding to accommodate a more equitable socio-economic agenda, with a recognition of the importance of the existing knowledge, experience and matters of concern of learners and their communities. There is some evidence of shifts towards better situated and more inclusive educational practices where ESE is emerging as socio-historical and evaluative critical process that involves a co-engaged staging of shared matters of concern, and educational activities that build collective agency—through new knowledge and participation—to address contemporary environment and development issues with deep historical roots. The contours of this oeuvre are being uncovered with dialectical critical realism after Bhaskar (O'Donoghue 2017) along with struggles for transitioning to an 'inclusive green economy' that is as yet poorly clarified and enacted. The review points to the need for ESE to also involve transformative processes, for example, re-thinking what green enterprises might entail, and how best to sustain South African communities and their environments. For such transformative and expansive social learning, it has been instructive to review how Grade 7 EMS teachers work with(in) the curriculum and the schools' communities to enact value-laden, community-based green enterprise education with learners, as an example of localised ESE addressing local matters of concern and in the process, global sustainable development goals.

References

Beck, U. 1992. Risk Society: Toward a New Modernity. Sage, London.

Carruthers, J. 1995. The Kruger National Park: A Social and Political History. University of Natal Press, Pietermaritzburg.

Cock, J. 2014. The 'Green Economy': A Just and Sustainable Development Path or a 'Wolf in Sheep's Clothing?' Global Labour Journal 5(1), DOI: https://doi.org/10.15173/glj.v5i1.1146.

Department of Environment Affairs. 2010. Environmental Sector Skills Plan and Strategy. DEA, Pretoria.

Du Toit, D. 1999. Through Our Eyes: Teachers Using Cameras to Engage in Education Curriculum Development Processes. MEd diss., Rhodes University, Makhanda.

Freire, P. 2000. Pedagogy of the Oppressed. Continuum, New York.

Janse van Rensburg, E. and H. Lotz-Sisitka. 2000. Learning for Sustainability: An Environmental Education Professional Development Case Study Informing Education Policy and Practice. Learning for Sustainability Project, Johannesburg.

Keegan, T.J. 1986. Rural Transformations in Industrializing South Africa: The Southern Highveld to 1914. MacMillan, Basingstoke.

Khan, F. 1994. Rewriting South Africa's Conservation History - The Role of the Native Farmers Association. Journal of Southern African Studies 20(4): 499–516.

Le Grange, L. and C. Reddy. 2000. Introducing teachers to outcomes-based education and environmental education: a western cape case study. South African Journal of Education 20(1): 21–25.

Lotz, H. 1996. The Development Of Environmental Education Resource Materials for Junior Primary Education through Teacher Participation: The Case of the We Care Primary Project. PhD diss. Stellenbosch University, Stellenbosch.

Lotz-Sisitka, H. and G. Raven. 2001. Active Learning in OBE: Research Report of the National Environmental Education Programme – GET Pilot Research Project. Department of Education, Pretoria.

Lotz-Sisitka, H. 2002. Curriculum patterning in environmental education: a review of developments in formal education in South Africa. *In*: Janse van Rensburg, E., J. Hatting, H. Lotz-Sisitka and R. O'Donoghue (eds.). EEASA Monograph: Environmental Education, Ethics and Action in Southern Africa: 97–120. HSRC Press, Cape Town.

Lotz-Sisitka, H. 2016. A review of three generations of critical theory: towards reconceptualsing critical HESD research. pp. 207–222. *In*: Baarth, M., G. Michelsen, M. Rieckmann and I. Thomas (eds.). Routledge Handbook of Higher Education Research for Sustainable Development. Routledge, London.

Lotz-Sisitka, H. 2017. Decolonising as future frame for environment and sustainability education. pp. 45–62. *In*: Corcoran, P., J. Weakland and A. Wals (eds.). Envisioning Futures for Environment and Sustainability Education. Wageningen Academic Publishers, Wageningen. 10.3920/978-90-8686-846-9_2.

Lotz-Sisitka, H., E. Rosenberg and P. Ramsarup. 2020. Environment and sustainability education research as policy engagement: (Re-) invigorating 'politics as *potentia*' in South Africa. Environmental Education Research 27(4): 525–553, DOI: 10.1080/13504622.2020.1759511.

Mbanjwa, S. 2003. The Use of Environmental Education Learning Support Materials in OBE: The Case of the Creative Solutions to Waste Project. MEd Diss., Rhodes University, Makhanda.

Molose, V. 2000. Materials in Flexible Learning Teacher Education Courses in Environmental Education: An Evaluative Case Study. MEd Diss., Rhodes University, Makhanda.

Naidoo, P. 1993. Collaborative Teacher Participation in Curriculum Development: A Case Study in Junior Secondary General Science. MEd Diss., Rhodes University, Makhanda.

Nduna, N. 2004. The Use of Environmental Learning Support Materials to Mediate Learning in Outcomes-Based Education: A Case Study in an Eastern Cape School. MEd Diss., Rhodes University, Makhanda.

Nsubuga, Y. 2010. The Integration of Natural Resource Management into the Curriculum of Rural Under-Resourced Schools. PhD Diss., Rhodes University, Makhanda.

O'Donoghue, R. 1990. Environmental Education, Evaluation and Curriculum Change: The Case of the Action Ecology Project 1985–1989. MEd Diss., University of Natal, Pietermaritzburg.

O'Donoghue, R. 1993. (ed.). The Environment, Development and Environmental Education: Environmental Education Policy Initiative (EEPI). Share-Net, Howick.

O'Donoghue, R. 1997. Detached Harmonies: A Study in/on Developing Social Processes of Environmental Education in Eastern Southern Africa. PhD Diss., Rhodes University, Makhanda.

O'Donoghue, R. 2000. Environment and Active Learning in OBE: NEEP Guidelines for Facilitating and Assessing Active Learning in OBE. Share-Net, Howick.

O'Donoghue, R. and E. Neluvhalani. 2002. Indigenous knowledge and the school curriculum: a review of developing methods and methodological perspectives. pp. 121–134. *In*: Janse van Rensburg, E., J. Hatting, H. Lotz-Sisitka and R. O'Donoghue (eds.). EEASA Monograph: Environmental Education, Ethics and Action in Southern Africa. EEASA/Human Sciences Research Council, Cape Town.

O'Donoghue, R. 2016a. Evaluation and Education for Sustainable Development (ESD): navigating a shifting landscape in Regional Centres of Expertise (RCEs). pp. 223–238. *In*: Baarth, M., G. Michelsen, M. Rieckmann and I. Thomas (eds.). Routledge Handbook of Higher Education Research for Sustainable Development. Routledge, London.

O'Donoghue, R. 2016b. Working with critical realist perspective and tools at the interface of indigenous and scientific knowledge in a science curriculum setting. pp. 159–177. *In*: Price, L. and H. Lotz-Sisitka (eds.). Critical Realism, Environmental Learning and Social-Ecological Change. Routledge, London.

O'Donoghue, R. 2017. Situated learning in relation to human conduct and social-ecological change. pp. 25–38. *In*: Lotz-Sisitka, H., O. Shumba, J. Lupele and D. Wilmot (eds.). Schooling for Sustainable Development in Africa. Schooling for Sustainable Development. Cham: Springer. 10.1007/978-3-319-45989-9_2.

Olvitt, L. 2004. The Adaptive Development and use of Learning Support Materials in Response to the First Principle of the Revised Curriculum Statement: The Case of Hadeda Island. MEd diss. Rhodes University.

Olvitt, L. 2012. Deciding and doing what's Right for People and Planet: An Investigation of the Ethics-Orientated Learning of Novice Environmental Educators. PhD diss. Rhodes University.

Poser, M. (ed.). 2017. Jean Baudrillard: Selected writings. https://antilogicalism.com/wp-content/uploads/2017/07/baudrillard.pdf.

Ramsarup, P. 2006. Cases of Recontextualising the Environmental Discourse in the National Curriculum Statement (R-9). MEd Diss. Rhodes University.

Reddy, C. 2004. Democracy and in-service processes for teachers: a debate about professional teacher development programmes. *In*: Waghid, Y. and L. Le Grange (eds.). Imaginaries on Democratic Education and Change. Pretoria: Southern African Association for Research and Development in Higher Education.

Republic of South Africa. 1994. White Paper on Education and Training. Notice 196 of 1995. Department of Education, Pretoria.

Republic of South Africa. 2018. Curriculum and Assessment Policy Statement (CAPS). Department of Basic Education, Pretoria.

Rieckmann, M. 2020. Emancipatory and transformative Global Citizenship Education in formal and informal settings: Empowering learners to change structures. Tertium Comparationis. Journal für International und Interkulturell Vergleichende Erziehungswissenschaft 26(2): 174–186. https://www.waxmann.com/index.php?eID=download&id_artikel=ART104545&uid=frei.

Rosenberg, E. 2008. Eco-schools and the quality of education in South Africa: Realising the potential. Southern African Journal of Environmental Education 25: 25–43.

Rosenberg, E., P. Ramsarup and H. Lotz-Sisitka. 2020. Green Skills Research in South Africa: Models, Cases and Methods. Routledge, New York.

Rosenberg, G. 2021. SETA Chair Research Programme: Entrepreneurship Case Studies and Programme Tools. Rhodes University Business School and Services SETA, Makhanda.

Santos, B.dS. 2007. Another Knowledge is Possible. Beyond Northern Epistemologies. Verso, London.

Schank, C. and M. Rieckmann. 2019. Socio-economically substantiated Education for Sustainable Development: Development of competencies and value orientations between individual responsibility and structural transformation. Journal of Education for Sustainable Development 13(1): 67–91. https://doi.org/10.1177/0973408219844849.

Schreiber, J. and H. Siege. 2017. Curriculum Framework: Education for Sustainable Development. Engagement Global, Bonn.

Schudel, I. 2012. Examining Emergent Active Learning Processes as Transformative Praxis: The Case of the Schools and Sustainability Professional Development Programme. PhD Diss. Rhodes University.

Schudel, I. 2014. Exploring a knowledge-focused trajectory for researching environmental learning in the South African curriculum. Southern African Journal of Environmental Education 30: 96–117.

Schudel, I. 2017. Deliberations on a changing curriculum landscape and emergent environmental and sustainability education practices in South Africa. pp. 39–54. *In*: Lotz-Sisitka, H., O. Shumba, J. Lupele and D. Wilmot (eds.). Schooling for Sustainable Development in Africa. Schooling for Sustainable Development. Cham: Springer. 10.1007/978-3-319-45989-9_3.

Sfard, A. 1998. On two metaphors for learning and the danger of choosing just one. Educational Researcher 27(2): 4. DOI: 10.2307/1176193.

Songqwaru, Z. 2012. Supporting Environment and Sustainability Knowledge in the Grade 10 Life Sciences Curriculum and Assessment Policy Context: A Case Study of the Fundisa for Change Teacher Education and Development Programme Pilot Project. South Africa. MEd Diss., Rhodes University.

Songqwaru, Z. 2020. A Theory-Based Approach to Evaluating a Continuing Professional Development Programme Aimed at Strengthening Environment and Sustainability Education. PhD Diss., Rhodes University.

Sørensen, M.P. 2018. Ulrich Beck: exploring and contesting risk. Journal of Risk Research 21(1): 6–16, DOI: 10.1080/13669877.2017.1359204.

South African National Biodiversity Institute. 2010. Biodiversity Human Capital Development Strategy. SANBI and the Lewis Foundation, Pretoria.

Togo, M. and H. Lotz-Sisitka. 2009. Unit-Based Sustainability Assessment Tool. A Resource Book to Complement the UNEP Mainstreaming Environment and Sustainability in African Universities Partnership. Howick: UNEP/Share-net.

Togo, M. and H. Lotz-Sisitka. 2013. Exploring a systems approach to mainstreaming sustainability in universities: a case study of place in South Africa. Environmental Education Research Journal 19(5): 673–693. doi:10.1080/13504622.2012.749974.

UNESCO. 2021. Ministerial Roundtable on the implementation of ESD for 2030. UNESCO World Conference on Education for Sustainable Development: Learn for our Planet. Act for Sustainability. 17–19 May 2021. Berlin. https://en.unesco.org/events/ESDfor2030.

Van Onselen C. 1996. The Seed is Mine: The Life of Kas Maine, a South African Sharecropper 1894–1985. Hill and Wang, New York.

Vianna E. and A. Stetsenko. 2006. Embracing history through transforming it: Contrasting Piagetian versus Vygotskian (activity) theories of learning and development to expand constructivism within a dialectical view of history. Theory Psychology 16: 81. DOI:10.1177/0959354306060108. http://www.sagepublications.com.

Wagiet, M. F. 1996. Teaching the Principles of Ecology in the Urban Environment: An Investigation into the Development of Resource Materials. MEd diss., Rhodes University.

Wagiet, R. 1997. Environmental Education: A Strategy for Primary Teacher Education. PhD Diss., Rhodes University.

Chapter 3

Trends and Challenges in Environmental and Sustainability Education in Uganda

James Musana * and *Ronald Bisaso*

Introduction

Globally, there is increasing focus on education as a vehicle for social, economic and environmental sustainability. Education systems are required to contribute to the transformation of society to improve people's quality of life through knowledge generation, innovation and the development of human capital.

This trend has been evident in Uganda through the ratification of international and continental protocols (the Sustainable Development Goals, African Union's Agenda 2063), and the embedding of sustainability in national development agendas such as the Uganda Vision 2040 and the National Development Plan III 2020–2025. The Ugandan Ministry of Education and Sports, and other ministries, agencies and other stakeholders have continued to support sustainability education initiatives in Uganda. However, to fully integrate sustainability in curricula, teaching, research and leadership among other areas, there is still need for a robust policy environment, and to raise awareness, develop capacity, provide financial resources to support strategies and put in place change champions such as leaders and environmental sustainability advocates (Ralph and Stubbs 2014). This chapter presents the existing trends and challenges in environmental and sustainability education (ESE) in Uganda.

East African School of Higher Education Studies and Development, College of Education and External Studies, Makerere University, Kampala, Uganda.
Email: ronald.bisaso@mak.ac.ug
* Corresponding author: musanajames@yahoo.com

The political, economic, cultural and social contexts and conditions/education system in Uganda

Uganda is a land-locked country located in East Africa, bordering South Sudan in the north, the Democratic Republic of Congo in the west, Kenya in the east, and Tanzania and Rwanda in the south. This geographical location makes Uganda dependent on other countries such as Kenya and Tanzania, which have direct access to the sea at Mombasa and Dar es Salaam respectively, in order to import valuable goods and services into the country. The high cost of transporting imported goods and services from the seaports to the country means that the costs of such goods and services are always higher than those provided or manufactured locally and fewer people can afford them. The total population stands at 43,746,516 (Musasizi 2018), with 51.2% of the population being female and 48.8% male (UNFPA 2017). The average annual population growth is 3.3% (Uganda Bureau of Statistics 2017).

Uganda has one of the youngest populations in the world, with 55% of the total population being aged between 0 and 15, 78% aged 30 and younger, and 2% aged 65 and older (Uganda Bureau of Statistics 2017a, b, UNFPA 2017, Irish Aid, Resilience and Economic Inclusion Team, Policy Unit 2018). This creates a significant discrepancy between the population of dependent and working individuals in the country, with the majority of individuals still being dependents. This presents a considerable challenge in terms of economic growth and development since children are dependent on their parents and guardians for almost everything and the rate of consumption is therefore higher than the level of production.

Uganda is predominantly an agricultural country, with 80% of the population dependent entirely on agriculture (Musasizi 2017, Irish Aid, Resilience and Economic Inclusion Team, Policy Unit 2018, Ggoobi et al. 2017). This means that the majority of the population is dependent on nature and is vulnerable to natural calamities such as drought and climate change; it is one of the reasons underlying the continuing high percentage of people 21.4%, living below the poverty line, i.e., less than US$1.90 per day (PWC Uganda 2018, World Bank 2016, USAID 2018).

Politically, Uganda has a multiparty system of governance. However, the country has never experienced a peaceful transfer of power from one president to another since gaining independence. This has remained an area of great concern. Currently, there are over 29 political parties in the country; however, only around 13 of them are active (Kabaasa et al. 2017).

According to Ssentongo (2014), the country has over 65 tribes and ethnic groups. Uganda is thus a multilingual country with no national language; the official language is English. This creates a communication gap between people from different regions, especially those who have not had any formal education. It also presents challenges in terms of national unity and the integration of people from diverse cultures.

Education in Uganda comprises three years of pre-primary education, seven years of primary education, six years of secondary education (divided into 4 years of lower secondary and 2 years of upper secondary school), and three to five years of post-secondary education (The Republic of Uganda 2017). Examinations are taken and certificates are awarded at each level (Teachers Initiative in Sub-Saharan

Africa-TISSA 2013, Uganda Bureau of Statistics 2016). Currently, there are around 7,210 pre-primary schools, 20,305 primary schools, 2,995 secondary schools, and 221 post-secondary education institutions (The Republic of Uganda 2017). In terms of ownership, educational institutions are categorized into two categories, namely government and privately-owned institutions. Higher education institutions in Uganda are further divided into three sub-sections: universities, Other Degree Awarding Institutions (ODAI) and Other Tertiary Institutions (OTI) (National Council for Higher Education – NCHE 2018). According to the Uganda National Council for Higher Education (a statutory body tasked with quality assurance and monitoring in the higher education sector) there are 9 public universities, 42 private universities, 10 ODAI, nine of which are private and only one public, and 160 OTI (NCHE 2018). Twenty-seven % (61) of the higher education institutions are publicly owned while 73% (163) are in the private sector (NCHE 2016). Since the majority of the country's institutions belong to the private sector, the provision of quality education could be compromised by the business interests of private investors, who might be more interested in money and the maximization of profit than the promotion of easy access to education and the improvement of quality.

The historical development of environmental and sustainability education in Uganda

The development of ESE in Uganda can be traced back to the pre-colonial, indigenous peoples of Uganda. This kind of education is popularly known as African Indigenous Education (AIE). Mugimu and Nakabugo (2009) characterize AIE as the education provided to young people by parents and elders in their communities before formal education was introduced by missionaries. On the other hand, Mushi (2009) has defined AIE as a process of passing the inherited knowledge, skills, cultural traditions, norms and values of the tribe or clan among tribal members and from one generation to the next. AIE offered young people the critical and practical skills they needed to survive and to live as responsible adults and productive members of society. Omolewa (2007) states that knowledge, skills, values and attitudes were passed on to young people through oral narratives and rites of passage such as circumcision, naming ceremonies and many other rites. Adeyemi and Adeyinka (2003) assert that oral narratives included proverbs, stories, folklore, myths, riddles, songs, dramas, histories, legends, and other forms of expression, which were frequently used to educate and entertain children as well as community members. Such narratives and rites were passed on from one generation to another, and initiated young people into the moral, philosophical, and cultural values of the community; the entire community was thus involved in educating the younger generation.

Education was organized on the basis of the gendered roles young people were expected to play in society. For example, boys observed and imitated what their fathers and uncles did and learned practical skills, which they developed as they matured into manhood and become heads of their own households. Those included activities such as agriculture, hunting, fishing, smithing, military training, and many other roles depending on how a particular ethnic group, clan or family derived its livelihood (Mugimu et al. 2009). Girls, on the other hand, were socialized by their mothers and aunts, and learned the roles of mother, wife, and other skills deemed appropriate to their gender (Adeyemi and Adeyinka 2003). Girls participated in activities such as weaving, cooking, gathering of food and firewood. One distinctive feature of the AIE system is that, by its very nature and method, failure was virtually nonexistent because everybody was trained on the job and was gainfully employed in the community (Adeyemi and Adeyinka 2003). According to Ssozi (2012), pre-colonial communities in Uganda lived within the limits of nature and respected the rules and spirituality associated with it. He cites the Ganda tribe in central Uganda, which attached much importance to certain plants and animal species, and for whom destruction of such plants, especially species that were sources of medicine or beneficial to the entire community, was an abomination. Additionally, the Baganda people attached cultural importance to particular animals and plants, and clan members were not supposed to eat or destroy any plant or animal considered to be a totem of the clan. This is also a common practice in other tribes and communities in Uganda. This kind of understanding and attachment to nature promoted and continues to promote a spirit of sustainability and biodiversity conservation. However, it should be noted that this kind of education focused mainly on clan and tribe issues and did not prepare its members to encounter the outside world; knowledge-sharing and dissemination to people outside a given tribe or community was limited.

Mugimu et al. (2009) assert that AIE was firmly established in Uganda until it was destabilized by the Western colonial powers in the 19th century. In the 1890s, missionaries introduced formal education in Uganda. Their main aim was to teach people how to read and write, going beyond the acquisition of vocational skills, the development of religious morals and the formation of character. When Uganda gained independence in 1962, the focus of education shifted to the generation of a critical mass of educated people who would advance the transformation of the socio-economic life of the people in a newly-independent country. Education was thus expected to train qualified personnel to replace many expatriates who were leaving the country at that time. Qualified people were also needed to boost the various economic activities in which the country had started to engage. Education was viewed as a tool that could give people confidence to believe in themselves, to believe that it was incumbent upon them to manage the affairs of their own country, and to believe that they were not inferior to the colonial masters. During the colonialist era, Ugandans had been relegated to low ranks in civil service, and they were therefore, keen for education to empower the local population to undertake self-rule and demonstrate that they were capable of solving their problems and sustaining their independence. The Castle Education Commission was appointed in 1963 to review the education system, which was still operating on the basis of the recommendations of the de Bunsen Education Committee of 1952 (Ssekamwa 1997). The Commission recommended that enrolment in formal education be increased in order to increase the number of graduates, and as a result many schools were constructed to accommodate the increasing number of new students.

Since the era of colonialism, there has been a deliberate attempt by the government and other education stakeholders to address issues of sustainable development, in particular, by equipping learners with the practical skills that will enable them to live a sustainable life. Various commissions have been set up in the colonial and post-colonial contexts to streamline Uganda's educational system and work towards sustainability. These commissions included the Phelps-Stokes Commission in 1925, the Thomas Education Committee in 1940, the 1952 Bernard de Bunsen Education Committee, the Castle Education Commission in 1963 and the Uganda National Education Policy Review Commission in 1987 (Kamya 2019). Emphasis was also put on the teaching of practical skills which could help learners become self-reliant and live a sustainable life. The Uganda National Education Policy Review Commission, chaired by Prof. W. Senteza Kajubi, strongly advocated the balancing of academic and practical skills, for example, terming it Basic Education for National Development (BEND). The government is currently trying to switch from a content-based to a competence-based curriculum because education continues to be mainly academically-oriented. It is hoped that the competence-based curriculum will help learners to become critical thinkers and have the skills to solve problems that are essential to human wellbeing and ecological sustainability (Ssozi 2012).

Furthermore, the government and various educational institutions are also finding ways in which ESE can be implemented at all levels of education. Uganda has a sustainable development plan, for example, spearheaded by the Ministry of Water and Environment (Ministry of Water and Environment 2015). The Ministry oversees strategies, policies and regulations relating to the environment, including the National Environment Management Authority (NEMA), which is the national authority overseeing environmental issues and management. In response to the launching of the UN Decade of Education for Sustainable Development (DESD) (2005–2015), the Ugandan government's Ministry of Education and Sports developed a strategy for the implementation of ESD in education. Many educational institutions are now promoting sustainable development, and have formed environmental protection and sustainable development associations such as, for example, Higher Education for Sustainable Development (HEfSD) (Franco et al. 2018). The Uganda Martyrs University's improving livelihoods programme has increased incomes, improved food security and water conservation, and led both to more sustainable livelihoods and to better relationships between the university and its neighboring communities (Tilbury 2009). Makerere University conducts research on eco-systems. Currently, the broad aims of education in Uganda are aligned with the United Nations' Agenda 2030.

The understanding of environmental and sustainability education in Uganda

Uganda cannot afford to remain aloof when it comes to matters of ESD, especially at a time when there is much criticism of the education system for producing graduates without the practical skills, ability to be innovative, values, attitudes and work ethics that are required to promote sustainable development (The Republic of Uganda 2011). In 2010, the Uganda National Commission for UNESCO (UNATCOM) developed an ESD implementation strategy for the education system. This strategy was developed with the participation of a cross-section of stakeholders from the educational, economic, environmental and social sectors in meetings, workshops and seminars (UNATCOM 2010, 2016). It provides guidance on how ESD should be implemented and it takes into account national aspirations and goals as set out in the Constitution, the country's vision 2040 and the Third National Development Plan. At the end of 2016, a new national ESD policy was introduced, and this provides the framework for refocusing ESD in both the formal and informal education sectors (Uppsala University 2017). The new ESD policy takes into account not only national aspirations with regard to sustainable development but also the regional and international development agenda such as the East African Community Vision 2050, the Africa Union Agenda 2063 and the global 2030 agenda (The Republic of Uganda 2017).

As stipulated both in the old and new ESD strategies, the major aim of ESD is to promote values and practical processes that provide the knowledge, skills, attitudes and competencies to nurture sustainable development at all levels of education and society (UNATCOM 2010, 2016, UNESCO 2017a, b, The Republic of Uganda 2017). This is being achieved through the refocusing of the existing curriculum to incorporate the various aspects of sustainable development. It is also being enhanced by the promotion of learner-centered rather than teacher-centered methods to enable learners to get to grips with sustainable development issues in a way that is appropriate for their age and level of education. Furthermore, ESD is also understood as including improved access to quality education at all levels. The Ugandan Ministry of Education and Sports leads on all matters concerning ESD and works in partnership with other regulatory bodies, agencies, educational institutions, politicians, civil society and all other education stakeholders. Educational institutions are recognized as key players and change agents when it comes to instilling and disseminating positive attitudes, and promoting public awareness and understanding of ESD.

The implementation of environmental and sustainability education and the ESD global action programme (GAP) in Uganda

Uganda has made some progress towards the implementation of the Sustainable Development Goals (SDGs) (The Republic of Uganda 2016, United Nations 2016). The country is one of the first to have aligned its national planning processes with the SDGs' Agenda 2030 (United Nations Development Group 2016, The Republic of Uganda 2020). It is estimated that 69% of the SDGs have already been integrated into the Second National Development Plan and other key government policy documents (Abebe 2017). A case in point is the Constitution of Uganda, which explicitly spells out the commitment of the people of Uganda to building a better future by establishing a socio-economic and political order based on the principles of unity, peace, equality, democracy, freedom, social justice and progress. Secondly, Uganda's Vision 2040 aims at transforming Uganda from a peasant country to a modern and prosperous one within a period of 30 years. Thirdly, the overall goal of the third National

Development Plan (2020–2025) is to transform Uganda from a lower-income to a middle-income country by 2025, by strengthening its competitiveness to generate sustainable wealth, employment and inclusive growth. The country's commitment to sustainable development is further demonstrated through its ratification and implementation of international, continental and regional development agendas such as the 2030 Agenda, the Africa Union Agenda 2063 and the East African Community Vision 2050 (The Republic of Uganda 2018, Owori 2017).

A variety of laws and policies have also been put in place, in accordance with the three pillars of sustainable development, to keep the country on the road to sustainable development. These include environmental policies such as the National Environment Management Policy (1994); the National Policy for the Conservation and Management of Wetlands (1995); National Oil and Gas Policy (2008); the Renewable Energy Policy (2007), and the Urban Policy (2011), among others, a few. Economic policies include the Uganda Revenue Authority Act (1991); the Income Tax Act (Cap 340); the Value Added Tax Act (Cap 349), and the East African Customs Management Act, to mention but a few. Finally, there are social policies such as the USE/Universal Post Primary Education Training policy (2007); Second National Health Policy (2010); the HIV/AIDS Policy (2011); National Gender Policy (2007); National Employment Policy (2011); the National Equal Opportunities Policy (2007), and the National Child Labor Policy (2007); the National Council for Disability Act (2003); the Persons with Disabilities Act (2006); the Genital Mutilation Act (2010); the Universal Primary Education Act (1997); the Prevention of Trafficking of Persons Act (2009); and the Workers' Rights to Labor Unions Act No. 7 (2006) (The Republic of Uganda 2016). However, some of these laws have not been fully implemented due to factors such as political influence, financing, and inadequate human resources.

The Ministry of Education and Sports of Uganda has created a number of departments and commissions in order to address ESD issues. The Directorate of Educational Standards, for example, is responsible for ensuring the quality of education. The National Council for Higher Education regulates university and tertiary education and the Universal Primary and Secondary Education (UPE and USE) focuses on issues of access and equity. Attempts are still ongoing to mainstream ESD at all levels of the education system (UNESCO 2016a). Revision of education curricula at primary and secondary education levels is underway to address issues pertinent to ESD, such as ensuring that curricula have local, national and international relevance and equipping learners with sustainability knowledge and skills. The government is working with private partnerships to try and increase access and equity, and improve the quality, relevance, effectiveness and efficiency of education at primary, secondary, technical and tertiary levels (UNATCOM 2010). Access to education has increased, mainly at the primary level, but quality remains a challenge, especially in schools in hard-to-reach areas (Initiative for Social and Economic Rights 2019). Affirmative action to increase women's access to higher education, for instance 1.5 free entry points, has enhanced gender equity and hence increased the opportunities for women to participate in sustainable development. The government has also introduced a higher education loan scheme for students wanting to undertake higher education but unable to afford it.

Education institutions have also made attempts to integrate ESD into teaching and learning. ESD clubs have been created, for example, bringing together teachers and students to discuss and take action on ESD issues, and some teachers have been trained to bring ESD into the curriculum (Aguti 2012). Other efforts include workshops involving university management, staff, students and the wider community, and ESD sensitization week, which aim to increase participants' awareness and interest (Lotz-Sisitka et al. 2015). ESD postgraduate diploma, master's degree and short courses have been developed in some schools and education faculties including at Makerere University (Aguti 2012). These courses cover the basic concepts, ESD pedagogy and other issues related to sustainable development. These and many other examples indicate Uganda's commitment to ESE.

Uganda is one of the Global Action Program (GAP) key partner states on the African continent, and its role is to reinforce ESD in national and international education and sustainable development policies. The GAP has played an important role in the country: Uganda works with other countries and international organizations, for instance, to promote dialogue and coordinate responses to climate change environment (UNESCO 2017, 2020). As Leicht et al. (2018) pointed out, Uganda has created partnerships with various governments and international agencies in order to implement ESD, and engage in advocacy, research, capacity-building and stakeholder training on its principles. Uganda has also established a number of ground-breaking ESD initiatives in partnership with various countries, in order to address some of the greatest health-related challenges the world is facing, such as Covid-19 virus (Payyappallimana and Fadeeva 2018, Schreiber and Siege 2016, The Federal Republic of Germany 2021). ESD issues are included in professional development programmes for teachers and teacher educators in order to nurture high quality professionals and build capacity (SWEDESD 2012, McKeown and Hopkins 2014, Chavatzia 2015, UNESCO 2016b). However, although a great deal of effort has been invested in ESD capacity in the public sector, very little has been done in the private sector, despite its role in the transition to sustainability (Ssozi 2012, UNESCO 2018). Furthermore, financial constraints have meant that many institutions, especially those located in remote areas, have not put training in place (Aguti 2012).

Uganda also has the aim to promote leadership, policy advocacy, environment and sustainable entrepreneurship, including youth and conservation leadership programmes and hip-hop youth in Northern Uganda. The youth and conservation leadership programmes work with university environmental clubs and associations on sustainable projects to provide students with mentoring and prepare them to address sustainability issues on campus and with local communities (UNESCO 2016c). The hip-hop youth programme in Northern Uganda aims to reduce and bridge the socio-cultural and economic gap between young people in Uganda. It helps young people to acquire artistic and entrepreneurial skills (UNESCO 2015). According to the Ugandan Ministry of Education and Sports (2017), the country's education strategies and plans are part of its commitment to GAP. The ministry believes that the country will achieve the United Nations Sustainable Development Goals, especially Goal 4 (UN-SDG 4) in 2030 through the transformation of its education systems, unleashing innovation, prioritizing inclusion and expanding financing. National and ministerial plans are committed to integrating ESD systematically into education policies. The follow-up programme, ESD for 2030, is important in Uganda because ESD has been retained as a target in the country's education goals for 2030.

Challenges affecting environmental and sustainability education in relation to the sustainable development goals in Uganda

Despite the number of Uganda's achievements with regard to sustainable development, the country still faces many daunting challenges in this area. These include poverty, demographic structure with a high dependency ratio, inequality, service delivery bottlenecks, high rates of unemployment especially for youth, reliance on natural resources and agriculture, low agricultural productivity, low levels of access to clean and modern energy, natural resource degradation, and vulnerability among various segments of the population, such as women, children, persons with disabilities, indigenous communities, hard to reach populations, and other vulnerable groups (The Republic of Uganda 2011, 2016a, 2017a).

There are also challenges relating to lack of strong policies focusing on whole institution approaches to the integration of ESE across all education institutions and all educational levels in the country. In addition, little has been done to raise awareness, develop capacity, provide financial resources to support strategies and put in place change champions such as leaders and environmental sustainability advocates (Ralph and Stubbs 2014). Higher education institutions in Uganda are facing a range of challenges that hinder the integration of ESE into teaching and learning activities. Academic staff, for instance, tend to stick rigidly to their area of expertise and are not prepared to cross boundaries with other disciplines, to borrow information, values, knowledge, attitudes and skills that are relevant to their teaching and learning environment (Nabayego et al. 2015, Kakembo and Barymak 2017, Ssozi 2012). This tendency to get trapped in the rigid observance of disciplinary boundaries limits thinking, prevents the emergence of new ideas and impedes the identification of solutions to current sustainable development challenges (UNESCO 2010). This has in turn led to problems such as poor integration of ESE in all courses, inadequate knowledge and expertise on ESE issues, negative attitudes toward ESE on the part of students and staff, lack of staff skills and competencies to implement ESE, lack of awareness of ESE and key sustainability issues, low staff morale with regard to the integration of ESE in curricula, disciplines being treated as though they are not interconnected and interdependent, and poor teacher training on ESE related issues (Ssentongo and Byaruhanga 2015, Kizza and Tumwebaze 2018, Mamdani 2009). Such challenges led Ssentongo and Byaruhanga (2015) to postulate that the mainstreaming of ESD in education systems should rise above the traditional boundaries, which tend to limit communication between different disciplines, faculties, schools, departments and their academics. All these challenges have a negative impact on the delivery of the SDGs in Uganda because education is both a goal in itself and a means of attaining all the other SDGs (UNESCO 2017). Therefore, if sustainable development issues are not integrated into the education system effectively, it will remain a challenge for Uganda to attain the SDGs.

Environmental and sustainability education in Uganda: Emerging issues, trends and future needs

Student participation in ESE activities is taking root in a number of universities including Makerere University, Uganda Martyrs' University and Busitema University. At Uganda Martyrs' University, for example, postgraduate students have started a Green Campus sustainability initiative named after Professor Wangari Maathai, who was a promoter of ESE. Students engage primarily in tree planting on campus and in the surrounding areas, and have set up a nursery for fruit trees and other kinds of indigenous trees. Through the Ministry of Education and Sports, the government is trying to revise the curriculum, shifting from a content-based to a competence-based focus to meet the needs of learners and society in general. This is in response to public criticism that formal education system is too theoretical, emphasizing cognitive rather than affective and psycho-motor skills. Graduates of the country's formal education system sometimes lack the practical skills, values and ethics that are required to survive and manage resources in a challenging world. The plan is to accelerate change further by increasing the number of vocational institutions catering to those who do not continue with secondary and tertiary education (The Republic of Uganda 2016).

Various universities, including Makerere University, Kyambogo University, Mbarara University, Busitema University and Nkumba University are attempting to mainstream ESE into their curricula. Nkumba University, for example, is trying take a whole-institution approach, integrating environmental and sustainability issues into all university disciplines, faculties, schools, institutes, programmes and courses. Environmental and sustainability issues are also integral part of university policies, plans and management practices, and of student activities.

The Government of Uganda continues to address the challenges associated with ensuring quality and inclusive education at all levels of education. This is in line with its commitment to achieving SDG goal 4 on quality education by 2030, ensuring that all school age children receive free, equitable and quality primary and secondary education leading to relevant and effective learning outcomes (The Republic of Uganda 2017). This will also enhance the ability of education to produce graduates with required skills to meet the market requirements, which currently remains a challenge.

With regard to future needs, there is a need to integrate values of AIE into the school curriculum so that children and students who miss learning at home can learn at school and vice versa. There is also need to refocus the curriculum, teachers' practices, teaching methods and learning aids to ensure that teaching and learning activities are relevant to students' local context, culturally appropriate, age and gender sensitive, inclusive of all learners, and take global concerns into account. There is also a need to emphasize lifelong learning and continuous professional development for all educators to ensure that they are continuously empowered. The multi- and trans-disciplinary nature of ESE requires people who are skilled and comfortable teaching across disciplines. Educators thus need to develop competences that can enable them go beyond the boundaries of their own specializations and build an appreciation of the complex nature of sustainable development.

There is also a need to bring on board all education stakeholders if ESE is to be integrated into the country's education system. Leaving out some vital stakeholders may result in opposition and practices contrary to mainstreaming as sustainability values may not be accepted by those who are not involved with the promotion and implementation of ESE. This can be done, for example, by engaging educational institutions, the National Council for Higher Education (NCHE), the National Curriculum Development Centre (NCDC), teachers and the Ministry of Education and Sports of Uganda in curriculum reviews at a range of levels that include ESE issues. Finally, there is a need for educational institutions to benchmark their activities against other institutions and organizations, drawing on best practice to inspire staff and students. Using specific examples of success may make teaching easier than drawing on theoretical explanations that change is possible through training sessions and workshops.

Conclusion

In order to present the existing trends and challenges in environmental and sustainability education in Uganda, this chapter has highlighted the political, economic, cultural and social contexts of the country; the historical evolution of environmental and sustainability education; the understanding of ESE and how it is being implemented. Through the GAP, Uganda has been able to cooperate with various educational stakeholders, both inside and outside the country and the country's partnerships have had some success in implementing ESE and moving towards the attainment of the SDGs. However, since Uganda is a developing country, a number of challenges continue to impede steady progress and the implementation of ESE and the SDGs in all sectors, and education in particular. These include the high rate of population growth, which has created a high dependency ratio because over 50% of the population is below the age of 18, and still depends on their parents and guardians for their livelihood. A young population means that the country has great potential in terms of human resources; however, this comes at a cost that is currently difficult for the country to manage given that 21.4% of people live below the poverty line, i.e., on less than US$1.90 per day (PWC Uganda 2018: 5, World Bank 2016: x & xii, USAID 2018: 1). Implementation of ESE requires a great deal of resources: educators have to be trained, awareness needs to be raised, there needs to be sufficient infrastructure to cater for learners'—and the education sector's—varied needs. However, due to the small budget allocated to the education sector, it may not be easy to achieve full integration of ESE and attain the SDGs by 2030. The ESE model is very important for the attainment of the SDGs and for Uganda; however, ESE content and teaching methods need to be tailored to learners' varied needs and the country's needs, which vary by geographical region. Attention should be paid to an internationally-recognized model of ESE, but it also needs to be relevant to local people and not merely copy and paste what is being done elsewhere, believing that this will also work in Uganda. Finally, the achievement of ESE and the SDGs will require active participation by all education stakeholders.

References

Abebe, J.O. 2017. Accelerating the Implementation of Agenda 2030 on Sustainable Development Approaches by African Countries in the Localization and Implementation of Sustainable Development Goals (SDGS). UN Women East and Southern Africa, Nairobi, Kenya.

Adeyemi, M.B. and A.A. Adeyinka. 2003. The Principles and content of African traditional education. Educational Philosophy and Theory 35(4): 425–440.

African Peer Review Mechanism (APRM). 2017. Uganda Country Self-assessment Report. New Partnership for Africa's Development.

Aguti, J.N. 2003. A study of In-Service Distance Education for Secondary School Teachers in Uganda: Developing a Framework for Quality Teacher Education Programmes. University of Pretoria, South Africa.

Ahimbisibwe, F. 2018. Uganda and the Refugee Problem: Challenges and Opportunities. Institute of Development Policy (IOB). University of Antwerp.

Chavatzia, T. 2015. 56th IEA General Assembly 2015: Challenges and Opportunities in Measuring Education for Sustainable Development and Global Citizenship Education. UNESCO, Paris France.

Diisi, J. 2017. Land Cover Trends in Uganda. Uganda National Forestry Authority.

Franco, I. O. Saito, P. Vaughter, J. Whereat, N. Kanie and K. Takemoto. 2018. Higher education for sustainable development: Auctioning the global goals in policy, curriculum and practice. Sustainability Science 1–22.

Ggoobi, R., B.M. Wabukala and J. Ntayi. 2017. Economic Development and Industrial Policy in Uganda. Friedrich-Ebert-Stiftung, Kampala.

Initiative for Social and Economic Rights. 2019. Status of Implementation of SDG 4 on Education: Is Uganda on Track? The Initiative for Social and Economic Rights.

Irish Aid, Resilience and Economic Inclusion Team, Policy Unit. 2018. Uganda Country Climate Risk Assessment Report. Embassy of Ireland in Uganda.

Jowi, J.O. and M. Obamba. 2015. Research and Innovation Management: Comparative Analysis of Ghana, Kenya, Uganda. Programme on Innovation, Higher Education and Research for Development (IHERD).

Kabaasa, B.B., Y. Kiranda and E. Kitamirike. 2017. Uganda's political outlook post the 2016 elections: a review of the process and implications for the future of multi-party democracy. Journal of African Democracy and Development 1(2): 187–201.

Kakembo, F and R.M. Barymak. 2017. Broadening perceptions and parameters for quality assurance in university operations in Uganda. JHEA/RESA 15(1): 69–88.

Kamya, B. 2019. Analysis of the different education policy reforms in Uganda (1922–2000). RAIS Conference Proceedings. 311–317.

Kizza, J. and H. Tumwebaze. 2018. Intricacies of internationalization of higher education for sustainable development in developing countries: A case study of Kyambogo University. International Journal of Arts and Entrepreneurship 7(11): 24–36.

Leicht, A., J. Heiss and W.J. Byun. 2018. Issues and Trends in Education for Sustainable Development. UNESCO, Paris France.

Lotz-Sisitka, H., A. Hlengwa, M. Ward, A. Salami, A. Ogbuigwe, M. Pradhan, M. Neeser and S. Lauriks (eds.). 2015. Mainstreaming Environment and Sustainability in African Universities: Stories of Change. Grahamstown: Rhodes University Environmental Learning Research Centre.

Mamdani, M. 2009. Scholars in the Market Place: The Dilemmas of Neo-liberal Reforms at Makerere University 1989–2005. HRSC Press, Cape Town South Africa.

McKeown, R. and C. Hopkins. 2014. Teacher Education and Education for Sustainable Development: Ending the DESD and Beginning the GAP. Toronto: York University.

Ministry of Health. 2018. Statistical Review of Progress to Inform the Mid-Term Review of the Uganda Health Sector Development Plan 2015/2016–2019/2020. Kampala, Uganda.

Ministry of Water and Environment. 2015. Uganda National Climate Change Policy. Ministry of Water and Environment.

Mugimu, C.B. and M.G. Nakabugo. 2009. Back to the Future? The indigenous education curriculum in Uganda. International Studies in Education 10: 15–22.

Musasizi, J. 2017. Deforestation in Uganda: Population increase, forests loss and climate change. Trends in Green Chem. 3(3): 22–40.

Musasizi, J. 2018. Deforestation in Uganda: Population increase, forests loss and climate change. Environ Risk Assess Remediat. 2(2): 46–50.

Mushi, P.A.K. 2009. History of Education in Tanzania. Dar-es-Salaam: Dar-es-Salaam University Press.

Nabayego, C., M. Kiggundu, N. Itaaga and A.M. Muwagga. 2015. Informal product-based training as a strategy for nurturing patriotism for Uganda's development through university education. World Journal of Educational Research and Reviews 2(2): 014–021.

Nampala, M.P. 2018. A critical evaluation of the national health system of Uganda. Biomed. J. Sci&Tech Res. 5(5): 1–6.

National Council for Higher Education. 2018. The State of Higher Education and Training in Uganda 2015/16. Kampala, Uganda: National Council for Higher Education, Ministry of Education & Sports.

Omolewa, M. 2007. Traditional African modes of education: Their relevance in the modern world. International Review of Education 53: 593–612.

Owori, M. 2017. Pro-poor orientation of the 2017/18 Uganda Budget: What will the 'industrialisation' focus mean for the poorest and most vulnerable people? Development Initiatives.

Payyappallimana, U. and Z. Fadeeva. 2018. Innovation in Local and Global Learning Systems for Sustainability Ensure Healthy Lives and Promote Well-being for All Experiences of Community Health, Hygiene, Sanitation and Nutrition Learning Contributions of the Regional Centres of Expertise on Education for Sustainable Development, UNU-IAS, Tokyo, Japan.

PWC Uganda. 2018. Uganda Economic Outlook. Pricewaterhouse Coopers Limited.

Ralph, M. and W. Stubbs. 2014. Integrating environmental sustainability into universities. Higher Education 67: 71–90.

Schreiber, J. and H. Siege. 2016. Curriculum Framework Education for Sustainable Development. Engagement Global gGmbH, Bonn, Germany.

Siraje, K. and H.S. Mbowa. 2018. Effect of education on sustainable development in East African universities: a case of two universities in Uganda and Rwanda. Journal of Research Innovation and Implications in Education 2(4): 23–30.

Ssekamwa, J.C. 1997. History and Development of Education in Uganda. Fountain Publishers.

Ssentongo, J.S. 2014. Living with ethnic difference in Uganda: Reflections on realities and knowledge gaps with specific reference to Kibaale. In Cross-Cultural Foundation of Uganda. Managing Diversity: Uganda's Experience. Kampala: Cross-Cultural Foundation of Uganda.

Ssentongo, J.S. and A. Byaruhanga. 2015. Mainstreaming education for sustainable development in Uganda martyrs university: a critical analysis of the strategy. In: Lotz-Sisitka, H., A. Hlengwa, M. Ward, A. Salami, A. Ogbuigwe, M. Pradhan, M. Neeser and S. Lauriks (eds.). Mainstreaming Environment and Sustainability in African Universities: Stories of Change. Grahamstown: Rhodes University Environmental Learning Research Centre.

Ssozi, L. 2012. Embedding education for sustainability in the school curriculum: the contribution of faith based organisations to curriculum development. Journal of Sustainability Education 3: 2151–7452.

SWEDESD. 2012. Unfolding the power of ESD: Lessons learned and ways forward. Report of the Conference: The Power of ESD-Exploring Evidence & Promise Visby 24–26.

Teachers Initiative in Sub-Saharan Africa-TISSA. 2013. Teacher Issues in Uganda: A Diagnosis for a Shared Vision on Issues and the Designing of a Feasible, Indigenous and Effective Teachers' Policy. Ministry of Education and Sports.

The Federal Republic of Germany. 2021. German Sustainable Development Strategy. German Federal Government.

The Ministry of Education and Sports of Uganda. 2017. Education and Sports Sector Strategic Plan 2017/18–2019/20. Kampala, Uganda.

The Republic of Uganda. 2011. Uganda's Position Paper on RIO+20. National Planning Authority, Kampala, Uganda.

The Republic of Uganda. 2016. Review Report on Uganda's Readiness for Implementation of the 2030 Agenda. Kampala, Uganda.

The Republic of Uganda. 2017a. State of Uganda Population Report. National Population Council (NPC), Kampala Uganda.

The Republic of Uganda. 2017b. Education Abstract. Education Policy and Planning Department, Ministry of Education & Sports, Kampala, Uganda.

The Republic of Uganda. 2018. Roadmap for Creating an Enabling Environment for Delivering on SDGs in Uganda. Office of the Prime Minister, Kampala, Uganda.

The Republic of Uganda. 2020. Third National Development Plan (NDPIII) 2020/21–2024/25. National Planning Authority.

The Uganda National Commission for UNESCO. 2010. Uganda: Education for Sustainable Development Implementation Strategy. Kampala, Uganda.

The Uganda National Commission for UNESCO. 2016. Global Citizenship Education Capacity Building Workshop for the Integration of Global Citizenship Education (GCED) into the Curriculum in Uganda Kampala: Workshop Report. Kampala, Uganda.

Tilbury, D. 2009. Higher Education for Sustainability: A Global Overview of Commitment and Progress, 18–28. The Global University Network for Innovation.

Uganda Bureau of Statistics. 2016. The National Population and Housing Census 2014–Main Report. Kampala, Uganda.

Uganda Bureau of Statistics. 2017a. The National Population and Housing Census 2014–Education in the Thematic Report Series, Kampala, Uganda.

Uganda Bureau of Statistics. 2017b. Statistical Abstract. Kampala, Uganda.

Uganda Bureau of Statistics. 2017c. Uganda National Household Survey 2016/17. Kampala, Uganda.

UNESCO. 2015. Post-2015 Dialogues on Culture and Development. UNESCO, Paris France

UNESCO. 2016a. Action for Empowerment: Guidelines for Accelerating Solutions Through Education, Training and Public Awareness. UNESCO, Paris France.

UNESCO. 2016b. Profile Booklet: Key Partners of the Global Action Programme on Education for Sustainable Development. UNESCO, Paris France.

UNESCO. 2016c. Roundtable: Sustainability Starts with Educators. The United Nations Educational, Scientific and Cultural Organization. Paris, France.

UNESCO. 2017. Education for Sustainable Development: Partners in Action. UNESCO, Paris, France.

UNESCO. 2017a. Education for Sustainable Development Goals Learning Objectives. UNESCO, Paris, France.

UNESCO. 2017b. Education for Sustainable Development (ESD) and Global Citizenship Education (GCED): Transforming and Sustaining our World through Learning. Regional Technical Workshop for Eastern Africa, July 4–7, 2017 | Nairobi, Kenya.

UNESCO. 2017c. A Review of Education for Sustainable Development and Global Citizenship Education in Teacher Education. UCL Institute of Education, Paris, France.

UNESCO. 2018. Profile booklet: Key Partners of the Global Action Programme on Education for Sustainable Development. UNESCO, Paris, France.

UNESCO. 2020. Education for Sustainable Development: Partners in Action, Global Action Programme (GAP) Key Partners' Report (2015–2019). The United Nations Educational, Scientific and Cultural Organization. Paris, France.

UNFPA. 2017. Worlds Apart in Uganda: Inequalities in Women's Health, Education and Economic Empowerment. United Nations Population Fund, Kampala Uganda.

United Nations. 2016a. Compendium of National Institutional Arrangements for Implementing the 2030 Agenda for Sustainable Development. UN-Department for Economic and Social Affairs.

United Nations Development Group. 2016. The Sustainable Development Goals are Coming to Life-Stories of Country Implementation and UN Support. The UNDG Sustainable.

Uppsala University. 2017. Visby Recommendations for Enhancing ESD in Teacher Education. Swedish International Centre of Education for Sustainable Development (SWEDESD).

USAID. 2018. Uganda: Nutrition Profile. The Integrated Regional RHITES Programs.
World Bank. 2016. The Uganda Poverty Assessment Report. The World Bank Group, Washington DC.
World Bank. 2017. Industry, Innovation, and Infrastructure: Build Resilient Infrastructure, Promote Inclusive and Sustainable Industrialization, and Foster Innovation. Atlas of Sustainable Development Goals.
World Bank. 2018. Tracking SDG7: The Energy Progress Report. World Bank Publications.

Asia

Chapter 4

The Status of Environmental and Sustainability Education in India

Sudeshna Lahiri

Introduction

The Millennium Development Goals (MDGs) set during the United Nations Millennium Summit in September 2000 were replaced and upgraded by the Sustainable Development Goals (SDGs) at the United Nations Conference in 2012. These global goals aim to address poverty, inequality and climate change. Along with other countries, India signed the Agenda 2030 at the United Nations Sustainable Development Summit on 25 September 2015. The document released by the Research and Information System for Developing Countries (RIS) clarifies that India's adoption of the SDGs is comprehensive and emphasizes the five Ps: people, planet, prosperity, peace and partnership (RIS 2016). Goals number 6, 7, 11, 12, 13, 14, and 15 deal specifically with the environment and sustainability, and the remaining goals also address sustainability in general.

The challenges facing the Indian peninsula with the delivery of the SDGs relate primarily to the country's varied climate, geography and socio-economic strata. Despite having the status of a 'soon-to-be' developed country, India is struggling with the challenges of unemployment, economic growth, food, water and energy security, disaster mitigation and poverty alleviation. Khan (2019) reports that the Union of India Government tried to align the SDGs with its National Development

Department of Education, University of Calcutta, India.
Email: sledu@caluniv.ac.in

Agenda by bringing out key policy documents. However, the mapping of schemes and programmes of the union ministries and departments onto the SDGs gave rise to concerns that there was no framework at the level of districts/local governments. The policy think-tank, the National Institution for Transforming India (NITI Aayog), and a few State Governments, have taken the initiative with regard to planning, implementing and monitoring the SDGs. However, this has not yet proved feasible in practical terms due to inadequate funds at various levels of governance.

The pro-active approach taken by India to the fulfilment of the SDGs is embedded in its constitution, which emphasises the importance of the environment and sustainability. Article 48A states, 'The State shall endeavour to protect and improve the environment and to safeguard the forests and wild life of the country.' Further, Article 51A (g) highlights that is vital, '[t]o protect and improve the natural environment including forests, lakes, rivers and wild life, and to have compassion for living creatures.' Compassion and co-existence with nature is well recognized by the Constitution of the Republic of India. These articles obviously call for the raising of awareness among the population across different education levels. However, the barrier to widespread environmental education (EE) at the school level is its inclusion as 'Environmental Studies' which provides children with scientific facts and information on global environmental problems (Siddiqui and Khan 2015). Pupils need to know about their responsibilities to prevent the 'damage' that they are reading about as scientific facts. Hence, the school curriculum and teacher education programmes are two of the major mechanisms for delivering environmental and sustainability education (ESE). The main objective of this chapter is to discuss contemporary delivery of ESE in India pursuant to the SDGs. The subsequent sections trace back Indian practices through the history of policies aiming to make EE an essential part of the Indian curriculum. It also sets out the programmes at national and local level aiming to raise awareness in line with ESE. In every course of studies, the syllabus is being revised to include EE in the curriculum. However, this is in the context of Environmental Science programmes within higher education and Environmental Studies programmes in primary education. These programmes have not yet been re-named 'Environmental and Sustainability Education' (ESE) in line with the Agenda 2030 for Sustainable Development.

Changing concepts of sustainability

Sustainable development is defined differently by different researchers and in different reports. For example, the Brundtland Commission Report defines it as development that caters to the needs of the present without compromising the needs of future generations (United Nations 1987). Similarly, the Johannesburg World Summit on Sustainable Development (United Nations 2002) identifies three pillars of sustainable development: economic, social and environmental development. Ahmed (2010) takes a systemic approach to sustainable development and a systemic view of the world, society and people's lives for effective and appropriate solutions to address and mitigate the problems.

For the last two decades, ESE has straddled the gap between EE and Education for Sustainable Development (ESD). EE has overextended itself with multiple meanings and concepts (Jenkins 1994, as cited in Iyengar and Bajaj 2011) covering curricular practices in a wide range of locales and classrooms. Iyenger and Bajaj have defined EE as including a cognitive component (knowledge, awareness), an affective component (attitude, behaviour), an action-based component (skills), and an impact component (sustainability, equity) with the aim of creating culturally and socially aware citizens who are respectful of human rights and participate in the promotion of a well-balanced environment. EE has variously been defined as: nature study and conservation; education for planetary citizenship; and more recently ESD, which also emphasizes the ethics, attitudes, and behaviours necessary to sustain the planet for future generations. ESD is broader in scope than EE, with its discourse and concepts covering three key spheres:

1) the environment (including water and waste)
2) society (including employment, human rights, gender equity, peace, and human security), and
3) the economy (including poverty reduction and corporate responsibility and accountability) (Iyenger and Bajaj 2011).

The concept of ESE in higher education attracts the following questions (Terlević et al. 2015):

- How are sustainable development and sustainable spatial development represented?
- How do students and teachers understand sustainable development?
- How can the sustainable spatial development paradigm be integrated into the curriculum and delivered?

The current trend of ESE may hence only be understood by reference to the historical background, educational provision and current discourses.

The context of ESE in India

The need to know more about the environment is now centre-stage, since human beings are not living in harmony with nature. Not only is reckless human behaviour endangering our habitat, there is also an imbalance in the equity of availability of natural resources in overcrowded cities.

Socio-cultural contexts

The need for EE or ESE has existed ever since the industrial revolution in India. However, environmentally-friendly behaviour and practices have been part of the Indian way of life since the Aryan era (Almeida and Cutter-Mackenzie 2011). That article cites the description in the hymn *Purusha sukta* in the *Rig-veda* of the process of creating human beings as an essential element of the universe. Ancient India practiced ESE, establishing scholarship and the gurukul (educational institution), in the forest and away from the city and human habitats (Gupta 2007). Teaching might have been delivered through lectures and rote learning; however, the educational practice was naturalism and close observation of nature. Gupta emphasizes that schools or *pathshalas* were in indigenous settings and close to nature. India has had a deep-rooted culture, including environmental and sustainability practices, since time immemorial. Beliefs and lifestyles in India also reflect sustainability. Its mythology associates nature with spirituality and its philosophy endorses the ecological cycle as part of the three natural attributes of *prakriti* or nature: creator (Brahma)—preserver (Vishnu)—destroyer (Mahesh). Shiva (2005) highlights the importance of the natural world in Hindu culture, which is aware of and connected with nature, worshipping flora, fauna, rivers, oceans, and mountains, and including religious/social observations with an ecological significance. Indian civilization has encouraged worshiping nature and preserving it for the next generation. As cited in Iyenger and Bajaj (2011), Sharma (2006) and Ravindranath (2007), the ancient Hindu texts place an emphasis on environmental protection, restricting the over use of natural resources, advocating a harmonious symbiosis with the natural environment, and encouraging sustainable development of ecosystems.

Economy

At its worst, the industrial revolution has over-extended some geographical regions and given rise to many other problems as a result of chaotic lifestyles, such as ground water shortages, delayed monsoons, breakdowns in sanitation systems, and over-flow and mismanagement of household and industrial waste. India is characterized by the over-population of certain cities due to employment generation. These cities are changing their structure, development and technology to accommodate their populations; the cities are expanding and forests have been shrinking, leading to climatic changes. Interestingly, the cities are not only attracting migrants looking for job opportunities, but also a younger population attending the growing number of educational institutions.

As indicated by the Ministry of Statistics and Programme Implementation (MoSPI), India's progress on the MDGs at national and state/union level is mixed (MoSPI 2014, cited in Kamepalli and Pattanayak 2015). Migration from rural to urban areas since 1921 had been constant and immense, and has placed burdens on the available natural resources. Industrial growth and development have led to overcrowded cities, which have started expanding, eroding the forests. The United Nations population division put the projected population of India at close to 1.380 billion for 2019 (The World Bank n.d.), making it home to one-sixth of the world's citizens.

How can India can live up to its commitments to the SDGs? Pandey (2017) argues that India is among the world's least wasteful economies. Its per capita emissions of carbon dioxide (2.0 tons per person) were about one-eighth of those of the US (16.6) in 2019 (Levin and Lebling 2019). Moreover, the Indian government's Voluntary National Review (VNR) presented to the High-Level Political Forum on Sustainable Development, New York, on its implementation of SDGs (United Nations High Level Political Forum [UN-HLPF] 2017), shows that India has encouraged sustainable and climate-adaptive agriculture (SDG 2) by, among other things, promoting organic farming and issuing soil health cards to farmers. India has also implemented SDG 14 by developing a coastal ocean monitoring and prediction system and an oil spill management system to monitor marine pollution. With regard to SDG 17, India collaborates on global technical support in a number of areas, including the development of data collection and monitoring and evaluation methodologies.

Global policies, local implementation

Showing concern for the environment on a global political platform, in 1972, Indira Gandhi, the then Prime Minister of India, declared at the United Nations Conference on the Human Environment (UNCHE) at Stockholm that "India did not intend to deprive the environment for long and we do not wish to deprive the environment any further and yet we cannot for a moment forget our large number of impoverished citizens" (Srivastava 2011). At the request of member countries, India prepared three reports on the state of the environment, focusing on: environmental degradation and control in India; problems relating to human settlement in India; and rational management of natural resources (Tandon 2008). These reports were important to understanding how the population explosion adversely affects nature and ecosystems. With the Supreme Court's 2003 directive to make EE compulsory at every level of education, the National Council of Educational Research and Training (NCERT) prepared and published the EE syllabus for Grades 1–12 in 2004 (Iyenger and Bajaj 2011).

Following the Supreme Court's mandates and recommendations for policy formulation in 1991 and 2003, changes were made to the National Curriculum Framework (National Commission for Education Research and Technology [NCERT] 2005). To consolidate the role of EE in sustainable development after the Stockholm conference in 1972, the Ministry of Environment and Forests established the Center for Environmental Education (CEE) as a centre for excellence in August 1984 (CEE n.d.). It not only ran the ESE programme, but also promoted sustainable development in education (RESET n.d.). The institutes set up by the Ministry of Human Resources and Development (MHRD) and non-governmental organizations—The Energy and Resources Institute (TERI); Bharati Vidya Peeth (BVP); the Centre for Science and Environment (CSE); the World Wide Fund for Nature (WWF); the National Council for Science Museums (NSCM) and the National Council of Education, Research and Training (NCERT)—all run ESD programmes for student teachers, principals, school administrators and policy-makers.

Historical development of ESE in India

ESE was a mode of teaching and learning process that was very familiar to Indian thinkers. Nobel Laureate Rabindra Nath Tagore believed that nature was the best teacher and had a naturalist philosophy with regard to educational practice: nature would transmit the necessary knowledge and pupils could not be taught by any external force (Aggarwal and Chaudhary 2015). To experiment and put his educational philosophy into practice, Tagore established his school at Santiniketan in December 1909 and the Visva-Bharati University in 1918 in order to enable teaching and learning to take place in the midst of nature (Chattopadhyay 2018). The curricula encouraged sustainability and the syllabi for a range of subjects aimed to help students live in symbiosis with nature. Similarly, Mahatma Gandhi incorporated the environment into education in a movement known as Nai Taleem, or Basic Education, founded in 1937 (Almeida and Cutter-Mackenzie 2011). ESE was a way of life practiced in civilizations in the Indus Valley through the ages, even before the formal education system began. Where Gandhian philosophy emphasises that 'the aim of teaching science in primary school should be in the physical and biological environment' (CEE 2010, as cited in Iyenger and Bajaj 2011), Nobel laureate Tagore confirmed a more deterministic environmental philosophy (Paramanik 2018). India has produced many globally-acclaimed thinkers and philosophers, who have propounded, preached and practiced environmentally friendly education.

In April 1989, the National Council of Educational Research and Training (NCERT) produced a draft document entitled Development of Guidelines and Syllabus based on the Teacher Education Curriculum Framework for Elementary and Secondary Teacher Education (NCERT 1991). It was released in two volumes (I & II), dealing with elementary and secondary schools respectively. This may have been the first time Environmental Studies was formalized in syllabi. EE had been recognized by the National Policy in Education (NPE 1986), but it was not made compulsory at any level of education (National Policy in Education 1986, n.d.). Environmental activist and litigator Mahesh Chandra Mehta petitioned the Supreme Court of India to enforce a decision in the year 1991 requiring the teaching of environmental studies to be compulsory at all levels of education in India (Environmental Education n.d.). The landmark decision of the Supreme Court order made EE a mandatory part of the fundamental duties of citizens to 'protect and improve the natural environment.'

Following a Supreme Court 2003 directive making EE teaching compulsory at every level of education, the NCERT in 2004 prepared and published an EE syllabus for Grades 1–12 (Iyenger and Bajaj 2011). With the Supreme Court having set out a mandate and policy recommendations in 1991 and 2003, the National Curriculum Framework was changed (National Commission for Education Research and Technology 2005). To ensure the role of EE in sustainable development, the Indian Ministry of Environment and Forests established a centre of excellence, the Center for Environmental Education, in August 1984 (CEE n.d.). Its textbooks, classroom discussions and research include examples of man-made disasters to create awareness. The case studies form the basis of the programme that is offered at various levels of higher education. Pande (2001) as cited in Iyenger and Bajaj (2011), for example, reports the study of the 1984 gas tragedy in Bhopal and how learning is drawn from the scale of intervention in an Indian context.

The Himalayan state of Uttaranchal developed a course introducing school students to local environmental and livelihood issues. Mukund (1988), as cited in Iyenger and Bajaj (2011), reports the adoption by the Hoshangabad Science Project of low-cost and innovative changes to the curriculum involving the use of the children's immediate physical environment. Later, Anitha (2004) as cited in Iyenger and Bajaj (2011), explains a project concerning a local site, Srikrishnapuram, which assisted social science and science teachers dealing with the environment-related topics in the curriculum by producing a manual that addressed immediate, local environmental issues. Policy-makers and various commissions have thus made sincere efforts to make environment and sustainability an integral part of the curriculum.

Indian ESE policies and policymakers

Policy documents and regulations have recommended reforms at every level of education in order to raise students' awareness of the environment, environmental issues and sustainability.

The Curriculum for the Ten Year School: A Framework. The first National Curriculum Framework (NCF) in 1975, titled The Curriculum for the Ten Year School: A Framework, has boosted the teaching of environmental studies, adding it to the primary curriculum alongside science and mathematics (NCERT 1976). A document released by the NCERT (1976) collected recommendations from expert groups. The main expert group was brought together by the Ministry of Education and Social Welfare to propose a curriculum for 10+2 level students at school. This is a revolutionary framework that aims to provide motivation for the teaching of environmental studies.

National Policy on Education (1986): The National Policy on Education (NPE 1986) recognizes the need for environmental awareness. Protection of the environment is one of the core areas identified on which the NCERT will shortly develop model syllabi and standard instructional packages. This reflects the National Policy on Education, NPE 1986, which stressed the need to create awareness of environmental issues within education by assimilating them into syllabi for different subjects, and also within all sections of society (GoI-MHRD 1998). Knowledge of environmental and sustainable education is relevant in various sections of the society, and also includes the practice of sustainable agriculture.

National Curriculum for Elementary and Secondary Education: The revised National Curriculum for Elementary and Secondary Education (NCESE) reflects on-going concerns about the protection of the environment and the conservation of natural resources (NCERT 1988). These concerns are outlined in the policy document on the destruction of the environment and overuse of (non-renewable) environmental resources, which highlights how humanity is upsetting the ecological balance through over-ambitious development programmes. The school curriculum should therefore encourage pupils to protect and conserve resources for future generations. The school curriculum should:

- Create awareness to counter the pollution arising from the increasing wealth of the population.
- Highlight actions that can be taken to protect and care for the environment; reduce pollution, and conservation for future generations.

At the primary level (Grades I and II), science is the key focus of environmental studies. The social sciences syllabus for Grades I and II should introduce pupils to the environment as a whole without distinguishing between the natural environment and social aspects.

National Curriculum Framework (NCF): The National Curriculum Framework (NCF 2005) is the overall framework for school curricula. It has also reviewed the National Curriculum Framework for School Education (NCFSE 2000), recommending that science teaching be modified. The objective is to ensure students analyse everyday experience, concerns and issues relating to the environment in every subject, and undertake outdoor project work. Among many other core areas, nature conservation should be a common feature of every school subject. In this regard, the curriculum aims to sensitize children to the environment and help them understand the need to protect it. The framework highlights how the rapid development of new technologies and faster-moving lifestyles has led to environmental degradation, which once again draws attention to the need to nurture and preserve the environment.

National Curriculum Framework for Teacher Education (NCFTE): This important document was published in 2009/2010. It laid down a detailed framework, setting out the different dimensions of teacher education and its symbiotic relationship with school education (NCTE 2009). The NCFTE re-designed citizenship education to incorporate environmental issues and environmental protection, and reflect the fact that the school curriculum and teaching and learning practices should take account of sustainable development. In the present ecological crisis, it recommended that teachers and students should to be educated to change their patterns of consumption patterns and their attitudes to natural resources in the context of highly commercialized and competitive lifestyles. Whenever Environmental Studies (EVS) teaching is part of ESE, courses should include philosophy and the theory of knowledge, since alongside the sciences, social sciences and EE, these are the foundations of EVS.

National Study on Ten Year School Curriculum Implementation: This report, prepared by Yadav (2011), makes the following observations:

- In primary classes, science is known as General Science in five States and Union Territories. In 23 States and Union Territories, it is known as Environmental Studies; while Gujarat, Dadra & Nagar Haveli, Daman and Diu call the subject 'Environment.'
- Social science is named 'Environmental Studies' in 29 States and Union Territories. However, the nomenclature 'Social Studies' is used in Andaman & Nicobar Islands, Sikkim, Uttar Pradesh and Jammu & Kashmir.
- Environmental Sciences are taught in an integrated approach (integrated into offered disciplines) by 23 States and Union Territories. In Haryana and Tamil Nadu, a disciplinary approach is used to teach the subject.

National Education Policy (NEP): On 29 August 2020, the cabinet approved the National Educational Policy (NEP) 2020 and its vision for education. At the school level, the NEP offers a wide range of curricular elements for school students to choose between, integrating skills and competencies with subject knowledge (MHRD 2020); these include environmental awareness. Environment concerns include water and resource conservation, not polluting or littering and other issues. Under 'Knowledge of India,' knowledge about the future includes the environment. The higher education curricula need to include EE to offer holistic and multidisciplinary education; and should have the flexibility to innovate. Moreover, the NEP strongly recommends cutting-edge research in certain areas including environment and sustainable living (MHRD 2020). The Ministry of Education (erstwhile, MHRD) aims to act as a facilitator to promote quality academic research in all fields of work through a new National Research Foundation. The robust research ecosystem that this will create should make a serious contribution the approach to climate change in an increasingly uncertain world.

Understanding of environmental and sustainability education in India

Indian higher education has slowly developed a new area of study, widening the scope from EE and Sustainability Science (SS) to, more specifically, Environmental and Sustainability Education (ESE). The new strand of study is interdisciplinary and trans-disciplinary in nature (Khuman et al. 2014), engaging with the geographical differences that give rise to communities with diverse cultures and traditions. It aims to prepare future leaders to appreciate the value of changes in global, social and human systems, and to tread the path of sustainability when implementing policies. In the meanwhile, Gupta (2007) coins one more term, namely Education for a Sustainable Future (ESF). This approach fosters synergetic inter-relationships between ecology, economics and social equity. It is multi-disciplinary in nature, learner-centred and participatory. The major components of environmental and sustainability education (ESE) in classrooms are:

- *Knowledge*: Learners should have basic knowledge of the natural sciences, social sciences, and humanities. Basic knowledge should also include the practices of the land in the form of festivals, rituals and chores.
- *Values*: The core of ESE is respect for present and future generations, cultural differences and diversity, and the natural environment.
- *Issues*: Learners should identify and understand the complexities of issues relating to the environment.
- *Perspective*: Learners need to consider the issue in terms of the past, present and future, identifying its roots and what the prognosis may be in the years ahead.
- *Skills*: In order to live sustainable lives and have sustainable livelihoods, learners need to have practical skills; these should be practiced even after the learners leave school.

ESE is based not only on theoretical concepts but on problem-solving rooted in real-life situations (Khuman et al. 2014). Local and regional bodies have introduced curriculum innovations, frameworks and programmes to deliver ESE. Evaluations of programmes by different reports and researchers are cited in Iyenger and Bajaj (2011).

In its policy document, Curriculum Framework for Quality Teacher Education, CFQTE 1998, the National Council of Teacher Education (NCTE) rightly highlighted the importance of EE for the conservation and improvement of the natural environment (NCTE 1998). The policy illustrated the general objectives for teacher education programmes that should be aligned with the context of education and the issues that concern it. The objectives included the sensitization of teachers and teacher-educators to evolving issues relating to ecology and the environment. Besides the objectives laid in CFQTE 1998, it has been realised further that ESE equips individual with the cognitive and affective skills and promotes the behaviours necessary to ensuring a sustainable future. In support of offering the ESE graduate programme, Tomar (2014) articulates that the said programme brings sustainable development issues into teaching and learning and promotes critical thinking, problem-solving and collaborative decision-making. Though many universities offer conventional teaching, Khuman et al. (2014) realised that undergraduate and the graduate programmes in sustainability education may also be offered via distance learning. The realisation was doubtful regarding the lack of success of school education in promoting EE.

The National Education Policy (NEP 2020) has given school students immense flexibility in their choice of individual curricula, certain subjects, skills, and competencies. Skills that can be chosen include: environmental awareness (including water and resource conservation, and avoidance of pollution and littering); and current affairs and knowledge of critical issues for the local community, the states, the country, and the world (MHRD 2020). This document sets out the vision of the MHRD (now the Ministry of Education) that 'Knowledge of India' should include a clear sense of India's future aspirations with regard to the environment and ensure that students' ethical decision-making ability includes respect for the environment. The NEP 2020 also proposes that technological professionals should be trained on health and environmental issues; and on sustainable living (MHRD 2020, 54). The policy further envisages the fruitful use of Artificial Intelligence (AI) and disruptive technologies in higher education to transform lifestyles and classroom teaching, covering areas such as clean and renewable energy, water conservation, sustainable farming, environmental preservation, and green initiatives (p.69).

Who is eligible for these graduate and master degree programmes on ESE? Khuman et al. (2014) identify the major candidates for ESE study as follows:

- *Group I*: The general population: Highly-educated citizens working in specialized professions.
- *Group II*: Policy-makers, planners and deliverers, including political leaders.
- *Group III*: Academics and scientists who can bridge the gap between groups I and II.

The ESE graduate programme needs to be offered to different layers of the population from general to specialised. Needless to say, the duration and volume of content delivered will need to be determined by the trainees' knowledge, experience and expertise. The syllabi of the graduate and master's degree programmes may include biological and landform diversity in different agro-ecologies; the key role of the climatic conditions and changes in the India as subcontinent; diversity in culture and traditions; economic liberalization; the extinction of traditional cultures and customs, and many other related contemporary issues. Programme content may include physical, biological, socio-cultural, economic, geological and geographical diversity, landform and structural diversity, and ecosystem diversity. It may also cover policies and programmes at various levels of governance.

Previous discourse also focused on scientific phenomena; however, these programmes may also include environmental psychology and eco-feminism. When ESD is studied as Sustainability Science, it has a defined meaning and scope. As cited in Khuman et al. (2014), Kates et al. (2001) define the discipline as seeking to promote understanding of the interaction between nature and society. Khuman et al. also cited Pillai's view (2008) that the subject should aim to train a cadre of youth who would be aware of the deep-seated socio-cultural roots underlying a unique philosophical outlook on environment and development. When ESD is taught in schools, it may cover the environment, environmental problems and our responsibility to protect it. Guha (2013), again cited in Khuman et al. (2014), believes that political leaders and senior bureaucrats also need to be environmentally aware.

Implementation of ESE and the Global Action Program (GAP) in India

Education is the most effective vehicle and instrument at the disposal of all countries for implementing and delivering Agenda 2030, making real changes in lifestyles and raising awareness of the limited natural resources available to humanity. In this regard, the UNESCO Global Action Programme (GAP) on ESD identifies five areas: advancing policy; transforming learning environments; building capacity in education; empowering youth; and accelerating sustainable solutions at the local level (gaia education n.d.). India is a member of the GAP Partner Networks. In order to advance policy (Partner Network 1), the Centre for Environment Education (CEE) intends to improve public awareness and understanding of the environment through a range of innovative training programmes promoting the conservation and sustainable use of nature and natural resources, through the provision of educational materials, and through capacity-building in the field of ESD. The CEE is a centre for advanced studies and continually pursues partnerships with a variety of international agencies in order to identify and report back on best practices with regard to climate change education and mitigation (UNESCO 2018). It is driving two major national ESD initiatives: the Paryavaran Mitra (PM) (friend of environment) and the MoEF's National Green Corp (NGC) programme (Bangay 2016). ESE teaching takes a pragmatic approach that is student-centric and requires students to learn through projects (Table 1).

In 2008–2009, a campaign titled 'Kaun Banega Bharat Ka Paryavaran Ambassador/Pick Right' was conducted targeting 200,000 schools across India, and aiming to raise awareness of environmental issues. The nation-wide *Parivartan Mitra* initiative aims to create a network of young leaders from schools across the country in order to address the challenges of environmental sustainability, and

Table 1. Comparison of ESD initiatives working towards the SDGs.

Paryavaran Mitra	Elements of teaching	National Green Corps
Curriculum-centric approach	Learning approach	Eco-club approach
Responsibility of subject teachers	Role of teachers	Teachers responsible for Eco-club
Participation by every student in every class and grade	Student participation	Select students participate from different classes and grades
Classroom activities and projects linked to the curriculum	Suggested teaching-learning approach	Club approach – extra/co-curricular activities and projects
Teaching and learning materials linked to the curriculum	Teaching Learning Material (TLM)	Supplementary materials
Middle school level (Grades 6–8)	Grades to be included	Secondary and post-secondary class students (Grade 9 and above)
Thematic focus: water and sanitation; energy; waste management; biodiversity and greening; culture and heritage	Thematic focus	Thematic focus: biodiversity conservation; water conservation; energy conservation; waste management; land use planning and resource management

instil environmental citizenship through changes in behaviour and the promotion of action at individual, school, family and community levels (Centre for Environment Education n.d.). Under the aegis of the Ministry of Environment and Forests (MoEF) and the Government of India, the National Green Corps (NGC) was launched in 2001–2002 and established NGC School Eco Clubs in around 120,000 Indian schools. The NGC Eco-Club in each school takes on 30–50 green cadets in the form of a Green Corps (Ministry of Youth Affairs and Sports-Government of India n.d.), whose activities are linked to culture and the cultural aspects of nature conservation and the environment. The challenge facing the CEE is how to translate ESD initiatives in schools into examinations and examination results in an educational system that is largely examination-centric.

Various government organisations, non-governmental organizations (NGOs) and agencies run programs to deliver ESE and initiatives promoting sustainable lifestyles (Table 2).

Different agencies, governmental and non-government organizations in India are actively running courses, programmes and awareness campaigns aimed at achieving the SDGs. However, they need to collaborate on aggressive social media campaigns so as to attract participants. The National Curriculum Framework (NCF-2005) recommends the inclusion of environmental issues and environment-related topics, and the Central Board of Secondary Education (CBSE) has included a wide range of topics related to environmental issues and sustainable development in its syllabi. However, these topics address concerns at the national level, rather than discussing local issues (Bangay 2016).

Table 2. A summary of organisations and programmes delivering ESE.

Wildlife Institute of India (WII)	Established in 1982; offers training programmes, academic courses and advice on wildlife research and management (Ministry of Environment, Forest and Climate Change n.d.).	• World Heritage Biodiversity Programme, Keoladeo National Park, Rajasthan • India World Heritage Biodiversity Programme, Manas National Park, Assam
Centre for Environment Education (CEE)	Established in 1984 as a Ministry of Environment and Forests centre of excellence to promote environmental awareness (CEE n.d.).	• Paryavaran Mitra • National Green Corps
The Energy and Resources Institute (TERI)	Established in 1974 by Darbari S Seth; aims to make a difference in perceptions of energy and help minimize pollution and harmful environmental consequences (TERI n.d.).	• Promotion of water, sanitation and hygiene education and water quality monitoring, Mangalore, Karnataka • Project SEARCH (Sensitization, Education and Awareness on Recycling for a Cleaner Habitat) • Green Olympiad
World Wide Fund (WWF)	Conceived in April, 1961, and established in September, 1961 in Morges, Switzerland (WWF n.d.).	The whole school ESD approach, Sunderbans, West Bengal Participatory video workshop, Sunderbans, West Bengal Sustainable Fishery Business, Sunderbans, West Bengal
The Bombay Natural History Society (BNHS)	Established in 1883 by nature-loving residents in Bombay, to exchange knowledge and display natural history specimens (BNHS n.d.).	Amur Falcon Project, Nagaland
Department of Environment, Government of NCT of Delhi	Government of National Capital Territory of Delhi	Anti-Fire Cracker Campaign
Government of India		National Clean Air Programme in 2019
Department of Agriculture, Cooperation & Farmers Welfare, Government of India	The Department of Agriculture (as it was known during the British era) was renamed the Ministry of Agriculture in 1947. In 1951, the Ministry of Agriculture was combined with the Ministry of Food to form the Ministry of Agriculture and Food.	Green Revolution umbrella scheme National Mission for Sustainable Agriculture

Emerging ESE issues and trends in India

Unfortunately, environmental and sustainability research in India is mostly based on secondary data, i.e., available literature, observation, and interviews, and there is a lack of evaluative studies (Bangay 2016). Moreover, the field needs to be updated in the context of the impact of Covid-19. While the management and disposal of Covid waste are important issues, a Harvard study suggests that air pollution may be affecting COVID-19 mortality (Harvard TH Chan n.d.).

In recent years, energy consumption has increased exponentially due to over-population and a rapidly-growing economy with fuel combustion leading to rising CO^2 emissions. International agencies are keeping track of fuel combustion in each country in metric tons (T) and an emerging trend in ESE at local level is the moderation of fuel usage in order to conserve the planet for future generations and ensure good air quality. This megatrend on the environmental impact of human activity arises from the serious sustainability challenges presented by demographic, socio-economic and technological changes. A number of countries have classified the impact (ImPACT) of the megatrend as follows: overall environmental impact (Im) is a product of the total population (P), production per person or affluence (A), consumption patterns (C) and the efficiency of producers as determined by technology (T) (Waggoner and Ausubel 2002 as cited in UN 2013). More recently, EE and sustainability development teaching have also begun to make a reference to environmental sociology and ecopsychology.

Conclusion

India was a signatory of the SDGs in 2015 and has thus committed to achieving the 17 SDGs and 169 associated targets. The various tiers of government (federal, state and local) are engaging with the goals and deploying programmes to fulfilment them. The NITI Aayog has been tasked with monitoring the implementation of SDGs. EE, as it is currently constituted in the country, hence needs to be upgraded to ESE, keeping the MDGs and SDGs at the centre of all programmes. Indian university campuses lack ESE frameworks and planning, and rigid bureaucracy impedes the delivery of ESE in academic institutions. Many of them may well have concerns that need to be addressed before they can offer the programmes. For example, at what level of school education (primary, secondary or senior secondary) should ESE be offered? Will it be optional? Or should it be offered as a built-in (optional or compulsory), course within a programme of study? Will it be offered as a specialization in higher education? A key concern is the way the EE curriculum treats environmental issues, awareness and pedagogy without considering the MDGs and SDGs as a whole. Since the SDGs and responsibility for the environment and sustainability are interconnected, ESE should be taught at every level of education in a multi-cultural society. The curriculum must also include EE; sustainable development; traditional practices; international and national policies; eco-psychology and eco-justice and many more emerging areas of study. In sum, ESE is currently only a nascent feature on Indian campuses and more research is needed to develop it as a discipline before EE can be transformed into ESE.

References

Aggarwal, N. and M. Chaudhary. 2015. A study of educational thoughts of Rabindra Nath Tagore in present era. International Journal of Education and Science Research 2(2): 64–67. http://www.ijesrr. org/publication/19/IJESRR%20V-2-2-13.pdf.

Ahmed, M. 2010. Economic dimensions of sustainable development, the fight against poverty and educational responses. International Review of Education 56: 235–253. DOI 10.1007/S11159-010-9166-8.

Almeida, S. and A. Cutter-Mackenzie. 2011. The historical, present and futureness of environmental education in India. Australian Journal of Environmental Education 27(1): 122–133.

Bangay, C. 2016. Protecting the future: The role of school education in sustainable development—an Indian case study. International Journal of Development Education and Global Learning 8(1): 5–19.

BNHS. n.d. The founders. https://www.bnhs.org/who-we-are.

CEE. n.d. About Us. https://www.ceeindia.org/about-us.

Centre for Environment Education (CEE). n.d. http://www.climatenetwork.org/profile/member/centre-environment-education-cee.

Centre for Environment Education. n.d. Paryavaran Mitra. https://paryavaranmitra.in/.

Chattopadhyay, T. 2018. Rainbow of education: Thoughts of Rabindranath Tagore. International Journal of Creative Research Thoughts (IJCRT) 6(2): 502–507. http://www.ijcrt.org/papers/IJCRT1893229.pdf.

Environmental Education. n.d. http://www.ecology.edu/environmentaleducation.html.

Gaia Education. n.d. UNESCO Global Action Programme. Retrieved from https://www.gaiaeducation.org/about/unesco-global-action-programme/.

GoI-MHRD. 1998. National Policy on Education 1986. https://www.education.gov.in/sites/upload_files/mhrd/files/upload_document/NPE86-mod92.pdf.

Gupta, A. 2007. A Study of Existing Perceptions/Understanding of Education for Sustainable Development of Indian Teachers. Prithvi Innovations: Lucknow.

Harvard TH Chan. n.d. Corona Virus and Air Pollution. Retrieved from https://www.hsph.harvard.edu/c-change/subtopics/coronavirus-and-pollution/.

Iyengar, R. and M. Bajaj. 2011. After the smoke clears: toward education for sustainable development in Bhopal, India. Comparative Education Review 55(3): 424–456 (Aug. 1, 2011).

Kamepalli, L.B. and S.K. Pattanayak. 2015. From Millennium to Sustainable Development Goals and need institutional restructuring. Current Science 108(6): 1043–1044.

Khan, J.A. 2009. Opinions: Challenges in Implementation of Sustainable Development Goals in India. https://www.cbgaindia.org/blog/challenges-implementation-sustainable-development-goals-india/.

Khuman et al. 2014. Sustainability science in India. Current Science 106(1): 24–26.

Levin, K. and K. Lebling. 03 December 2019. CO_2 Emissions Climb to an All-Time High (Again) in 2019: 6 Takeaways from the Latest Climate Data. Retrieved from https://www.wri.org/blog/2019/12/co2-emissions-climb-all-time-high-again-2019-6-takeaways-latest-climate-data#:~:text=Average%20per%20capita%20emissions%20were,the%20United%20States%20(16.6).

MHRD 2020. National Education Policy 2020. MHRD, Government of India. Retrieved from https://www.mhrd.gov.in/sites/upload_files/mhrd/files/NEP_Final_English_0.pdf.

Ministry of Youth Affairs and Sports, Government of India. n.d. National Green corps. https://fitindia.gov.in/events/national-green-corps/.

National Policy on Education 1986. n.d. Retrieved from https://www.mhrd.gov.in/sites/upload_files/mhrd/files/upload_document/npe.pdf.

NCERT. 1975. The Curriculum for the Ten-Year School – A Framework. New Delhi, NCERT.

NCERT. 1976. The Curriculum for the Ten Year School: A framework. https://ncert.nic.in/pdf/focus-group/NCF_10_Year_School_eng.pdf.

NCERT. 1988. National Curriculum for Elementary and Secondary Education: A Framework. NCERT: New Delhi. Retrieved from https://ncert.nic.in/pdf/focus-group/NCESE_1988.pdf.

NCERT. 1991. Elementary Teacher Education Curriculum: Guidelines and Syllabi. New Delhi: NCERT. Retrieved from http://14.139.60.153/bitstream/123456789/2671/1/ELEMENTARY%20TEACHER%20EDUCATION%20CURRICULUM-D8353.pdf.

NCERT. 2005. National Curriculum Framework 2005. Retrieved from https://ncert.nic.in/pdf/nc-framework/nf2005-english.pdf.

NCTE. 1998. Curriculum Framework for Quality Teacher Education. Member Secretary, National Council for Teacher Education. Retrieved from http://14.139.60.153/bitstream/123456789/2174/1/Curriculum%20framework%20for%20quality%20teacher%20education%20D10151.pdf.

NCTE. 2009. National Curriculum Framework for Teacher Education: Towards Preparing Professional and Humane Teacher. Member-Secretary, National Council for Teacher Education. Retrieved from https://ncte.gov.in/Website/PDF/NCFTE_2009.pdf.

Pandey, A. 2017. What India can teach the world about sustainability. World Economic Forum. Retrieved from https://www.weforum.org/agenda/2017/10/what-india-can-teach-the-world-about-sustainability/.

Paramanik, A. 2018. Environment, education and Rabindranath Tagore. Research Review International Journal of Multidisciplinary 3(8): 531–534.

Praveen, H. and N. Nasreen. n.d. Status of environmental education at secondary school level in India. https://paryavaranmitra.in/Praveen-Nasreen%20-%20Status%20of%20EE_4.1.pdf.

RESET. n.d. Education for Sustainable Development. Retrieved from https://en.reset.org/knowledge/advancing-sustainable-development-through-education-india#:~:text=India's%20developmental%20strategic%20framework%20is%20based%20on%20a%20five%20year%20planning%20system.&text=In%20order%20to%20promote%20the,as%20part%20of%20the%20curriculum.

RIS. 2016. India and Sustainable Development Goals: The Way Forward. New Delhi: Research and Information System for Developing Countries. Retrieved from http://ris.org.in/pdf/India_and_Sustainable_Development_Goals_2.pdf.

Shiva, V. 2005. India Divided: Diversity and Democracy under Attack. New York: Seven Stories Press.

Siddiqui, T.Z. and A. Khan. 2015. Environment education: An Indian perspective. Research Journal of Chemical Sciences. 5(1): 1–6. Retrieved from http://www.isca.in/rjcs/Archives/v5/i1/1.ISCA-RJCS-2014-179.pdf.

Srivastava, J. 2011. 'Norm' of Sustainable Development Predicament and the Problématique. India Quarterly 67(2): 93–110. Retrieved from https://doi.org/10.1177/097492841006700201.

Tandon, U. 2008. Population growth and sustainable development. Journal of the Indian Law Institute 50(2): 209–219.

TERI. n.d. Mr. Darbari Seth: The Visionary Founder of TERI. Retrieved from https://www.teriin.org/darbari-seth-visionary-founder-teri.

Terlević, M., A. Istenič Starčič and M. Šubic Kovač. 2015. Sustainable spatial development in higher education. Urbani Izziv 26(1): 105–120.

The World Bank. n.d. The Population, Total-India. Retrieved from https://data.worldbank.org/indicator/SP.POP.TOTL?locations=IN.

Tomar, A. 2014. Good Practice Stories on Education for Sustainable Development in India. United Nations Educational, Scientific and Cultural Organization: Japan. Retrieved from https://unesdoc.unesco.org/ark:/48223/pf0000232544_eng.

UNESCO. 2018. UNESCO Global Action Programme on Education for Sustainable development. Retrieved from https://www.dataplan.info/img_upload/7bdb1584e3b8a53d337518d988763f8d/246270e.pdf.

United Nations. 1987. Report of the World Commission on Environment and Development: Our Common Future (Brundtland commission report). Oxford University Press, New York.

United Nations. 13 Jun 2013. Global trends and challenges to sustainable development post-2015. In UN, World economic and social survey 2013: Sustainable development challenges, New York: United Nations. https://doi.org/10.18356/ba9e9a36-en.

United Nations High Level Political Forum. 2017. India Voluntary National Review Report on the implementation of Sustainable Development Goals. Retrieved from http://niti.gov.in/writereaddata/files/India%20VNR_Final.pdf.

WII. n.d. About Wildlife Institute of India. Retrieved from https://wii.gov.in/.

WWF. n.d. History. Retrieved from https://www.worldwildlife.org/about/history.

Chapter 5

The Contours of Education for Sustainable Development in Japan

Yoko Mochizuki

Introduction

> *Japan, a country poor in natural resources, has grown to be what it is today on the strength of its human resources. It has attached paramount importance to education as the basis of development.*
>
> *My government, together with Japanese non-governmental organisations, has proposed that the United Nations declare a Decade of Education for Sustainable Development . . .*
>
> *In the process of achieving economic growth, Japan experienced a period of serious pollution which caused ill health and even the loss of lives . . .*
>
> *The greatest contribution we can make to the realisation of sustainable development is to share the lessons we have learned so that our friends will not repeat the grim experience.*

Excerpts from a speech made by the then Primer Minister of Japan, Junichiro Koizumi, at the World Summit on Sustainable Development, 2 September 2002, Johannesburg, South Africa (Ministry of Foreign Affairs Japan 2002)

In 2002, the United Nations (UN) General Assembly adopted a resolution declaring a UN Decade of Education for Sustainable Development (DESD) from 2005 to 2014; the basis was a Japanese government proposal made in partnership with Japanese non-governmental organisations (NGOs). Given that the DESD was a Japanese proposal, there have been notable efforts by diverse Japanese stakeholders to mainstream Education for Sustainable Development (ESD) in various fields. The first

Université Paris Cité.
Email: yoko.mochizuki@u-paris.fr

section of this chapter provides snapshots of recent and ongoing efforts to integrate ESD and environmental and sustainability education into national-level policies, laws, regulations and other frameworks dealing with the economy, the environment, and education. The chapter then traces the evolution of environmental education (EE) and development education in Japan in response to domestic and international events, issues and trends, in order to lay the foundations for understanding the social and political contexts within which Japanese ESD policy and practice have developed and the particular conceptions of ESD have emerged.

Current political, economic, cultural and social contexts

Japan, a democratic country in East Asia with a population of 125 million, has the world's third-largest economy after the United States and China. Japan's economy was the world's second-largest from 1968 until 2010, when it was overtaken by China. As this chapter will make clear, Japan has played an important role in driving sustainable development policy in the international community since its remarkable rise to the status of a major economic power. For example, the Japanese government has contributed to the global implementation of ESD through its significant financial contribution to the lead agency of ESD—the UN Educational, Scientific and Cultural Organisation (UNESCO)—and to the UN University (UNU) headquartered in Tokyo, which, in 2003, established an ESD programme with funding support from the Ministry of the Environment, Japan. Aside from the Sustainable Development Goals (SDGs), Japan's current National Action Plan on ESD (Inter-ministerial Meeting on ESD 2021, 4) names three international frameworks related to sustainable development, namely the Paris Agreement on climate change, the Sendai Framework for Disaster Risk Reduction 2015–2030, and the UN Decade of Ocean Science for Sustainable Development 2021–2030. This signals key sustainability issues for Japan as one of the largest greenhouse gas emitters in the world and an island country susceptible to increasingly frequent and intense natural disasters. Other major challenges include a dwindling birth rate, an ageing population, and an exhausted countryside that are threatening the sustainability of local communities.

ESD in the national SDG framework, national environmental plan and CSR

Following the adoption of the 2030 Agenda for Sustainable Development, the Government of Japan developed a national framework to implement the SDGs. In 2016, it established the SDGs Headquarters, headed by the Prime Minister, to advance the national implementation of SDGs and to promote international cooperation to achieve them more broadly. Every year, the government formulates an SDG Action Plan, in which ESD is positioned as a human resource development initiative to support the achievement of the SDGs. As far as environmental policy is concerned, the Fifth Basic Plan for the Environment, approved by the Cabinet in

April 2018, establishes six cross-cutting Priority Strategies[1] drawing on the SDGs. It calls for 'Integrated Improvements on Environment, Economy and Society (II2ES),' viewing environmental policies as opportunities to create 'simultaneous solutions' for interlinked challenges as well as 'new avenues for growth' that ensure quality of life into the future. In this context, the Environmental Plan proposes the new concept of a regional recycling and symbiosis zone (also referred to as 'local SDGs') to maximise the vitality of each region and encourage each region to form a self-reliant and decentralised society. The Plan notes the central importance of developing human resources who understand the relationship between the environment, the economy and society and who are able to connect and network specialists from diverse fields at a regional level, to identify and utilize each region's unique natural and cultural resources.

The corporate sector in Japan is also taking a proactive approach to the adoption of the SDGs. In November 2017, Keidanren (Japan Federation of Economic Organizations) revised its Charter of Corporate Behaviour to put the achievement of the SDGs at the forefront, stating that companies have a role to play in building a sustainable society. Keidanren has long advocated the development of 'Society 5.0' which aims to optimise people's lives and society as a whole by maximising the use of innovative technologies such as the Internet of Things (IoT), Artificial Intelligence (AI) and robotics. In its preface to the Charter it states that Society 5.0 is in line with the principles of the SDGs. The Fifth Basic Plan for the Environment also identifies responses to technological innovation such as AI as one of the economic challenges that are inseparably linked with environmental and social challenges. Indeed, explicit linking of technological advancement to the achievement of the SDGs is not unique for Japan. Rather, it is reflective of the rapid rise of AI as a defining instrument of power in the international order, both in terms of hard power (military applications) and soft power (exerting economic, political and cultural influence) (Miailhe 2018). It is a noteworthy trend that resuscitates, or rather reinforces, technocratic approaches to sustainability transitions, which view technological advancement as the key to their success. Technocratic approaches have long been criticised in environmental and sustainability education literature (Gough and Scott 2008), and this new emphasis on technology, and in particular digital technology, deserves more attention from environmental and sustainability education researchers. In point of fact, the aforementioned National ESD Action Plan identifies the promotion of AI and Digital Transformation (DX) as the country's SDG policy, along with achieving gender equality and the green society (including through 2050 Carbon Neutral) (Inter-ministerial Meeting on ESD 2021, 3). Japan's SDG Action Plans and its Basic Plan for the Environment both essentially see the role of education as being

---

[1] The six interdisciplinary, cross-cutting priority strategies are: (1) development of a green economic system for sustainable production and consumption; (2) improvement of the value of national land; (3) sustainable community development using local resources; (4) enabling people to live healthy and prosperous lives; (5) development and dissemination of technologies that support sustainability; and (6) demonstration of Japan's leadership through international contributions and the building of strategic partnerships.

to develop human capital, thereby largely limiting it to enhancing the country's economic competitiveness.

ESD in formal and non-formal education

Over the DESD, Japan's Ministry of Education, Culture, Sports, Science and Technology (MEXT) has promoted formal ESD through measures to integrate ESD into education policy and the curriculum in Japan as well as through financial contributions to UNESCO, while the Ministry of the Environment has promoted community-based ESD through non-formal ESD measures and financial contribution to the UNU. This basic demarcation of responsibility in the Japanese government shapes the overall ESD landscape in the country. So far, the UNU has acknowledged eight local multi-stakeholder networks as Regional Centres of Expertise (RCE) on ESD in Japan (Chubu, Greater Sendai, Hokkaido Central, Hyogo-Kobe, Kitakyushu, Okayama, Omuta, and Yokohama), which are part of the expanding global network of over 170 RCEs.[2]

Mainstreaming ESD in formal education

In December 2016, the Central Council for Education released a report on the improvement of Courses of Study (national guidelines for teaching that are revised every ten years)[3] which states: 'Education for Sustainable Development (ESD) is a fundamental principle in the next revision of the Courses of Study' (MEXT 2016). This led to the March 2017 revisions of the Kindergarten Education Guidelines and the Courses of Study for Elementary and Junior High School Education, which include references to 'fostering the creators[4] of a sustainable society' and stipulate the incorporation of ESD-related content into each subject area. The Courses of Study for upper secondary schools were also revised in March 2018 to include similar measures.

In Japanese schools, many ESD activities are undertaken in the 'Period for Integrated Studies' (*sougouteki na gakusyu no jikan*). 'Integrated Studies' was introduced in Japanese school curricula from primary to high school in 2000, with the aim of encouraging integrated approaches to cross-cutting real-life issues. The Period of Integrated Studies has provided a significant institutional space for the introduction of ESD in the school curriculum. Integrated Studies is allocated 70 teaching hours per year at each grade in primary school (Grade 1–6) and lower secondary school (Grade 7–9), except for Grade 7, where it has 50 hours, and upper secondary school (Grades 10–12), where it has 87–174 hours (Ohagi 2019).

Another notable characteristic of ESD in Japan is the utilization of the UNESCO Associated Schools Project Network (ASPnet). Since 2008, MEXT has supported

---

[2] https://www.rcenetwork.org/portal/.

[3] The Courses of Study are "Gakushu-Shidou-Youryou" in Japanese. They are akin to national curriculum frameworks or guidelines in other countries. They detail learning objectives, topics to be covered, pedagogical approaches and other principles for each subject area.

[4] The Japanese word "tsukurite (創り手)" here literally means "creators". It can also be translated as "agents" in the sense that they actively shape a sustainable future.

measures to promote school-based ESD through ASP.net. The number of ASPnet schools in Japan has grown dramatically—from 20 in 2006 to 913 in 2014, far exceeding the target number of 500 in the final year of the DESD (Inter-ministerial Meeting on the UNDESD 2014, 7, Suzuki 2019). MEXT further supported the creation of ASPUnivNet, a network of 18 universities to support primary and secondary schools applying to become ASPnet schools and implementing ESD activities after having been recognised as such. Japan's highly centralised education system no doubt enabled an exponential increase in the number of ASPnet schools. However, these designated UNESCO schools—now exceeding 1,000—account for merely 2% of the elementary and secondary schools in Japan, and further dissemination of ESD in formal education is an ongoing challenge.

The revised Courses of Study required all elementary and secondary schools in Japan to address ESD and the SDGs from April 2020. In order to raise awareness of ESD further and support schools with ESD implementation, the Japanese National Commission for UNESCO developed a manual for promoting ESD in 2016, based on good practices in ASPnet schools; this was revised in 2018. The 50-page manual details five steps for introducing ESD in schools:

1) Articulation of the school principal's policy on school management and ESD goals
2) Preparation and implementation of a study plan (and lesson plans)
3) Development of a whole-school approach to ESD
4) Development of partnerships with community organizations, universities, businesses and other local stakeholders; and
5) Dissemination and reflections on ESD practice (Japanese National Commission for UNESCO 2018).

Promoting environmental education

In 2003, shortly after the 2002 UN General Assembly decision to launch the DESD, Japan enacted the Law for Enhancing Motivation on Environmental Conservation and Promoting Environmental Education. This Law highlighted the importance of the promotion of environmental conservation and environmental education for the building of a sustainable society. The revised 2011 law, which came into force in October 2012, was named 'An Act on the Promotion of Environmental Conservation Initiatives through Environmental Education and Other Measures.' It added the promotion of cross-sector collaborative efforts to its objectives, enabling local governments to formulate action plans to promote EE and use cross-sector collaboration as a framework. The Law also added the enhancement of EE in school education. In order to strengthen the foundations of EE, it further included accreditation/certification for collaborative action facilitators and the development of EE materials, in a system under which the national government receives and registers applications from businesses and other entities for training EE leaders.

Historical background: environmental education and development education

Most academic studies and policy documents on ESD refer to the global milestones of sustainable development—such as the 1987 report of the World Commission on Environment and Sustainable Development (commonly known as the Brundtland Report) or Agenda 21, established at the 1992 UN Conference on Environment and Development (the 'Earth Summit') as a global blueprint for sustainability. Chapter 36 of Agenda 21 is widely considered a foundational ESD text, as it recommended that 'environment and development education should deal with the dynamics of both the physical/biological and socio-economic environment and human (which may include spiritual) development, should be integrated in all disciplines, and should employ formal and non-formal methods' (UN Conference on Environment and Development 1992). Rather than treating ESD as an externally induced reform in response to Agenda 21, this section traces the history and attempts to bridge Japanese environmental and development education movements and current ESD discourses and practices. Before doing so, it considers the role the Japanese government played in putting sustainable development on the global agenda.

The report of the World Commission on Environment and Sustainable Development (1987), *Our Common Future*, provided the well-known definition of sustainable development as 'development that meets the needs of the present without compromising the ability of the future generations to meet their own needs' (43). Whereas it is generally known in Japanese ESD circles that the DESD was a Japanese proposal, it is less well known that the Brundtland Commission was also a Japanese proposal and that the commission was supported financially by Japan, which covered half the total cost of US$6m (the rest being provided by Canada, Denmark, Finland, the Netherlands, Norway, Sweden and Switzerland) (Hashimoto 2007). Saburo Okita, former Minister of Foreign Affairs and the Japanese member of the Brundtland Commission, regarded support for the commission as Japan's responsibility because it was a country that had achieved economic prosperity. Ironically, Japan was at the height of the bubble economy when the Brundtland report came out in 1987, and the report was seen by Japanese politicians as an effort to address global environmental problems, such as destruction of rain forests and desertification far from home, rather than to encourage reflections on unsustainable development within Japan (Hara 2007). It is obvious from this account that there is a gap between the Japanese state's commitment to progressive initiatives and worthy causes—demonstrated through its generous financial support extended through the UN mechanisms—and what it promotes domestically.

Going further back in history, during Japan's rapid industrialisation from the late nineteenth century to the mid-twentieth century, the desire to 'catch up' with the West prompted a disregard for environmental costs and a lack of regulatory safeguards. A series of industrial disasters came to be recognised as social problems in the late 1960s, including cadmium poisoning that caused a painful bone condition called *itai-itai* disease, life-threatening smog from petroleum and crude oil refineries in Yokkaichi, and two separate releases of methyl mercury into waterways in

Kumamoto and Niigata prefectures, which were linked to *minamata* disease. These four events came to be known as the major environmental scandals of the twentieth century in Japan (Hasegawa 2004). Current Japanese environmental policy and regulations are a direct consequence of the environmental disasters of the 1950s and 1960s. After years of corporate denial and government inaction, an accumulation of scientific evidence connecting chemical emissions to disease outbreaks—plus a growing willingness of victims to file lawsuits against the companies involved—led in the 1970s to the first national regulations prioritising human health over economic development. Indeed, the Fifth Basic Plan for the Environment, approved in 2018, set out a vision of Japan as a global exemplar, based on its history of overcoming pollution, its advanced environmental technologies, and its ethos of *mottainai* and traditions of living in harmony with nature.

Amid and following the high-growth period of the Japanese economy (1954–1973), Japan saw a rise in citizens' movements objecting to environmental pollution and development policies and projects. These movements mobilised local residents, including school teachers, who were both actual and potential victims of pollution and who advocated for their own and others' rights to life and a healthy environment. They employed a variety of methods, including field surveys (to assess the harmful effects of environmental pollution) and co-learning among teachers, experts, students and local residents. This led to an endogenous form of EE—known as *kogai-kyoiku* or pollution education—which contributed to social movements to protect citizens from various forms of environmental harms and subsequent costs to human wellbeing. In the late 1960s, highly localised pollution education came to be incorporated voluntarily into the school curriculum by progressive teachers in Yokkaichi, Kawasaki, Minamata and elsewhere (Ando 2015, 55). This endogenous form of environmental education, however, was largely replaced by a less radical approach in the 1980s, as discussed below.

The 1970s were an important period in which environmental issues moved higher up the national and international agenda. In 1971, the Environmental Agency was created in Japan. This preceded the 1972 UN Conference on the Human Environment in Stockholm, which called for EE to be used as a means to address environmental problems in its Recommendation 96. In 1975, this recommendation was addressed at the International Environmental Workshop in Belgrade, Yugoslavia, which adopted a global EE framework known as the Belgrade Charter. In the mid-1970s, there were already tensions between those who advocated *kogai-kyoiku*, based on the experience of fighting diseases caused by pollution, and those who called for a more apolitical form of environmental learning geared towards fostering pro-environmental behaviour and nature-based learning (Ando 2015). Internationally, the official launch of EE is attributed to the world's first inter-governmental conference on it in Tbilisi, Georgia (then part of the USSR) in 1977. By the mid-1980s, the term *kankyo-kyoiku* had been popularised among the Japanese public, reflecting the emergence of a range of civil society organisations promoting nature-based learning (Ibid.). In 1990, the Japanese Society for Environmental Education (JSFEE) was established. Osamu Abe at Rikkyo University in Tokyo, who served as the JSFEE President between 2009 and 2015, has played an instrumental role in the promotion of ESD in Japan in his capacity as a representative of the Japan Council on the UN DESD (ESD-J) (discussed in the next section).

In contrast to EE, which has both endogenous and exogenous origins, development education is an imported concept in Japan. Internationally, it built on early efforts to raise awareness of the so-called 'Third World' by NGOs based in Europe and North America in the 1960s. In the 1980s, development education in Japan came to address not only North–South inequality but also domestic issues relating to multiculturalism and international understanding in the wake of an influx of foreign workers. A turning point in Japanese development education was 1989—the year the Berlin Wall came down—when Japan's Official Development Assistance (ODA) disbursement reached US$9 bn, making Japan the world's largest bilateral donor (Kamibeppu 2002, 57). In the same year, the Courses of Study were revised to include EE and education for international understanding in the national curriculum. Although Japan maintained its status as the largest bilateral donor between 1989 and 2000 (with the exception of 1990), in 1991 the country entered the 'lost decade' when the Japanese asset price bubble burst. This lost decade coincided with a decade of important international conferences following the end of the Cold War, which broadened the concept of development. In light of these domestic and international developments, in 1997 the Development Education Council of Japan (DECJ, established in 1982) redefined development education as education that aims to foster individual learners' capacities to understand various development issues, deliberate on desirable forms of development and participate in shaping a more just global society (DEAR 2004, 4, author translation). In 2003, the DECJ gained legal status and it was re-named the Development Education Association and Resource Centre (DEAR). Haruhiko Tanaka (2005), who served as the DEAR representative between 2002 and 2008, has observed in retrospect that Japanese development education has essentially been ESD since development education was redefined in 1997.

Understanding of ESD in Japan

As early as June 2003, ESD-J was established as a consortium of civil society organizations and individuals involved in various social issues, from the environment to development to human rights to peace and gender, in order to build a shared understanding of ESD and deliver it in partnership with the government, local authorities, companies and educational institutions. Edgar González-Gaudino, a prominent Mexican environmental educator and a member of an international DESD advisory group set up by UNESCO, observed in 2005 that 'One de facto problem that the implementation of the [DESD] faces is that apparently only we environmental educators have become involved in debating its pros and cons' (González-Gaudino 2005, 244). This was patently not the case in Japan, where discussions on ESD not only involved environmental educators, but also development education academics and practitioners, and experts on education for international understanding, human rights education, peace education, adult education and so on. ESD's relevance to value-based and action-oriented education movements beyond EE in Japan is evident from the establishment of ESD groups in various academic associations, including

Gender Education　　　Peace Education

Development Education　　　　　Human Rights Education

Essence of

ESD

Multicultural Education　　　　　Environmental Education

Welfare Education　　XX Education

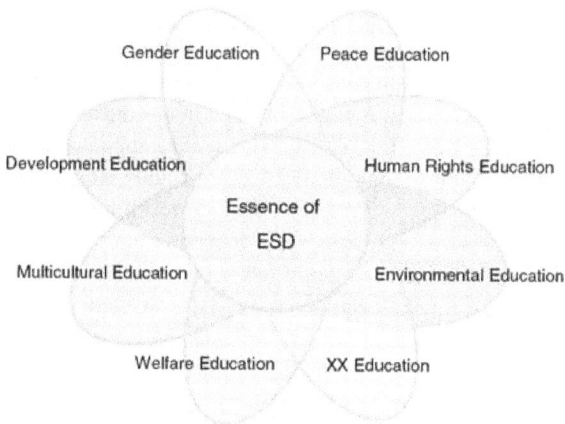

Figure 1. 'Flower of ESD' model by ESD-J. *Source*: Nagata 2017, 35, Figure 3.

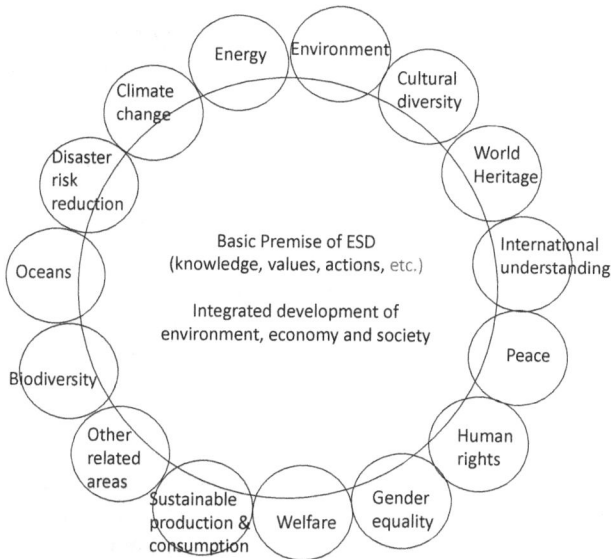

Energy　Environment

Climate change

Cultural diversity

Disaster risk reduction

World Heritage

Basic Premise of ESD
(knowledge, values, actions, etc.)

International understanding

Oceans

Integrated development of
environment, economy and society

Peace

Biodiversity

Other related areas

Human rights

Sustainable production & consumption　Welfare　Gender equality

Figure 2. Illustration of the ESD concept by the Japanese Government. *Source*: Japanese National Commission for UNESCO 2018, 6 (Translated by the author).

the JSFEE, the Japan Association for International Education and the Japan Society for the Study of Adult and Community Education.

　　Figure 1 shows the mainstream understanding of ESD as integrating various 'adjectival educations,' originally developed and widely disseminated by ESD-J, while Figure 2 is the visual representation of the concept of ESD developed by the Japanese government. Similarly, guidance for school teachers on how to implement ESD in alignment with the national curriculum was developed by the National Institute for Educational Policy Research (NIER 2012), identifying six core concepts for ESD—(1) diversity, (2) interdependence, (3) finitude (living on the finite planet), (4) fairness/equity, (5) cooperation/partnerships, and (6) responsibility—and key competencies to be emphasized in ESD teaching, such as critical thinking and collaboration.[5] All these Japanese models assume that (1) shared competencies can be developed through various types of adjectival education; and (2) whatever the entry point to ESD, there is a shared vision of a sustainable society which is more economically viable, ecologically sound, and inclusive and equitable. These approaches are in line with the UNESCO definition of ESD as education that 'empowers learners to take informed decisions and responsible actions for environmental integrity, economic viability and a just society, for present and future generations, while respecting cultural diversity' (UNESCO 2014, 12). The UNESCO definition also acknowledges that ESD is 'intended to encompass all activities that are in line with the [ESD] principles irrespective of whether they themselves use the term ESD or—depending on their history, cultural context or specific priority areas—environmental education, sustainability education, global education, development education, or other' (UNESCO 2013, Annex I, 2).

Discussing NIER's (2012) guidance, which aims to assist schools teachers to design and implement ESD in different subject areas, Kadoya and Goto (2013) argued that 'ESD involves a wide range of contents and all subjects, and all themes and topics can be related with ESD' (49). Although Nagata (2017) critiques these Japanese approaches for their emphasis on mainstreaming ESD within existing education systems, for underplaying the transformative nature of ESD and for being what he calls 'shallow' as opposed to 'deep' ESD, NIER's approach resonates with the 'embedding' approach proposed by UNESCO MGIEP (2017), which is based on the premise that ESD cannot be mainstreamed in schools unless it is addressed in core subjects such as mathematics, languages and science.

In addition to the models of ESD put forward by civil society (ESD-J) and the Japanese government, what characterizes the Japanese understanding of ESD is its emphasis on community-based approaches. Partly as a result of the strong ownership of the DESD exercised by the Japanese state, local governments, NGOs, academics and individuals, Japanese ESD is characterised by top-down directives from the centralised state and a variety of bottom-up responses. Top-down and bottom-up discourses on ESD in Japan often converge, however, taking the view that ESD constitutes a *machizukuri* or community development process (Mochizuki 2010, 2017a). Many factors account for this, including the creation of institutional

---

[5] The results of the survey undertaken by the government for the Leader Teachers' Training for EE and ESD indicated that only about 20% of teachers reported that they were familiar with ESD in 2008. In response, NIER was requested in 2009 to undertake research on "disseminating ESD" to all Japanese schools. The outcome of this research was an 8-page leaflet detailing the basic principles for integrating and implementing ESD in schools. It defined the abilities and attitudes to be emphasized in ESD-oriented teaching as: (1) the ability to think critically, (2) the ability to predict the future and plan for it, (3) the ability to be multifaceted and responsible, (4) the ability to communicate, (5) the attitude to cooperate with others, (6) the attitude to respect/appreciate connections, and (7) the willingness to participate. The framework was developed partly based on the book "Teacher's Guide for Environmental Education" produced by NIER in 2007 (Kadoya and Goto 2013).

mechanisms for broad-based ESD promotion such as an inter-ministerial platform for ESD implementation, comprising 11 ministries and agencies, ESD-J, and UNU's RCEs.

However, this convergence in the discourse reflects a deep-seated sense of crisis about the future of Japan. ESD is deployed both in the context of Japan's nation-building project (whose aims include improving Japan's self-image as a respectable country that upholds the ideals of ESD) to maintain its position as a leading world economy, and in the country's arduous efforts to reclaim what has been lost in the process of rapid development, establishing a new model of development that will enhance well-being for all. Many community-based environmental and sustainability education practices (see, for example, Ji and Fukamachi 2017, Iwabuchi and Takemoto 2017) provide a fundamental critique of the modernisation and developmentalism that have peripheralised rural parts of Japan and played a role in turning the country into a 'super ageing' (Muramatsu and Akiyama 2011) society. One thread that weaves many ESD activities together is concern about the loss of vitality in local communities—both rural and urban—and exploration of how ESD can supplement community revitalisation efforts by restoring an appreciation of the rich natural and cultural resources of local communities.

Japanese ESD discourses and practices encompass a wide continuum of approaches to social change, with top-down conservative approaches at one end of the spectrum and bottom-up activism, vibrant voluntary initiatives and active civic engagement at the other. Observing the trend towards institutionalisation and depoliticisation of EE, Ando (2015) calls for a rethinking of *kankyo-kyoiku* by revitalising *kogai-kyoiku* as 'education for environmental justice' (68–69). This call for the politicisation and radicalisation of EE resonates with Huckle and Wals' (2015) call for 'global education for sustainability citizenship,' which emphasises the exploration of 'structural causes' of social and environmental injustice and 'reformist and radical solutions' (495). A less radical, yet reformist, approach is also advocated by educators implementing ESD in schools. Based on his experience as a former principal of Nagatadai Primary School in the City of Yokohama, an ASPnet school that won awards for its exemplary ESD practice, Masaharu Sumita (2019) characterizes ESD as 'life lessons' (*inochi no jugyou*) and suggests that ESD needs to be underpinned by notions such as 'self-respect' and 'caring for others,' and that in the long run this will lead to the creation of a life-affirming society where children are happy about having been born and elderly people are happy about having lived a long and fulfilling life. What has distinguished the Nagatadai Primary School's approach from other schools' good practice is not typical school-based ESD activities such as partnerships with the local community on agricultural experiences for students or guest lectures on micro-plastics by external experts, but their unique efforts to improve the professional autonomy and working conditions of teachers.

In recent years, issues such as 'overwork death,' 'work-life balance' and mental health have emerged as problems affecting different sectors, including education. According to a national survey conducted in 2006 covering 2,160 public primary and lower secondary schools and around 50,000 teachers, the average length of break taken by teachers per day was around 10 minutes (Suzuki 2007). In his book *Creating a Colourful School*, Sumita (2019) emphasizes the need for school leaders to develop a leadership style that does not suppress but appreciates the opinions and initiatives of teachers and highlights the importance of teachers' well-being and comfort for the implementation of ESD. The school developed a number of initiatives to enhance teacher autonomy and well-being, including workshops on better time management, increased teacher involvement in decision-making about the yearly curriculum, and surveys on improvement of working conditions and concrete ideas for shortening work hours (such as shorter, more regular meetings and merging multiple school newspapers into one). In a survey conducted one year after these reforms were put in place, close to 80% of the Nagatadai teachers reported that their working conditions had improved (Ohagi 2019).

Implementation of ESD in the post-DESD period

Given that Japan was the country that proposed the DESD in Johannesburg in 2002, its government instituted various measures to promote ESD policy and practice, as outlined in the previous sections.[6] Building on the DESD, the government in 2016 released a national action plan on the implementation of ESD under the Global Action Programme (GAP) (Inter-ministerial Meeting on the UNDESD 2016). The national curriculum guidance, one of the key initiatives under this plan, was revised by MEXT in 2016 and 2017, requiring all K-12 schools to address ESD and the SDGs (see Current Contexts above). Another important initiative under the plan was the establishment of a nation-wide ESD promotion network which aims to allow people anywhere in Japan to learn about and implement ESD. The network links the national ESD Resource Centre in Tokyo (established in April 2016),[7] eight regional ESD Resource Centres (established between July and September 2017), and ESD partners at local level including universities, local education boards, local UNESCO associations, NGOs, local nature centres and local climate protection centres. ESD partners, registered by the network on a voluntarily basis since November 2017, are expected to support the ESD activities of local stakeholders such as school teachers, community learning centre (CLC) staff members, and NGOs. As of September 2019, approximately 100 ESD partners had been registered across the country.

The first phase (2015–19) of the National Action Plan was completed at the end of the 2019 Japanese fiscal year (March 2020). Following the UNESCO decision to launch a new global framework, ESD for 2030[8] for the period 2020–2030, Japan launched a new National Action Plan on ESD in May 2021, aligning with the UNESCO framework and building on previous plans (Inter-ministerial Meeting on ESD 2021).

Although Japan is generally considered a 'champion country' for ESD along with Germany, there is much scope to improve ESD activities. A paper by NIER,

---

[6] There are many accounts of Japanese policy development with regard to national ESD promotion during the DESD (see, for example: Abe 2009, 2012, Sato and Nakayama 2013, Interministerial Meeting on the UNDESD 2009, 2014, Mochizuki 2017a).

[7] https://esdcenter.jp/english/esd-resource-center-of-japan/.

[8] https://unesdoc.unesco.org/ark:/48223/pf0000370215.locale=en.

for example, looked at surveys by the Asia-Pacific Cultural Centre for UNESCO (ACCU) and the analysis of ESD good practices by ASPnet schools in Japan and identified three challenges for the implementation of ESD in Japanese schools:

1) Enhancing teacher understanding of the concept of ESD;
2) Building sustainable relationships with organisations and individuals outside of school that can support ESD implementation; and
3) Linking school-based curriculum management with ESD and integrating ESD values and principles across subjects (Ohagi 2019).

About half the respondents to the ACCU survey indicated that (a) there was not enough time to practise ESD and (b) they could not secure sufficient time for ESD, making these the third and fourth most often-cited reasons for the limited implementation of ESD in Japanese schools after teachers' lack of understanding and the elusiveness of ESD as a concept ('The ESD concept is too broad and difficult to understand') (Ibid.). This suggests, as Sumita (2019) proposed, that teachers and staff need improved working conditions if they are to engage meaningfully with ESD.

With regard to the ESD Promotion Network, while the number of ESD practitioners has been slowly increasing, a significant proportion of the general public is aware neither of ESD nor the SDGs and is taking no action to help shape a more sustainable society. The ESD Resource Centres are starting to explore strategies for utilizing existing resources and schemes such as teacher training activities (which are organized regularly by boards of education) to meet the need to train teachers to implement ESD in all schools as mandated by the new Courses of Study. There is much scope for promoting ESD for citizens in general through lifelong learning and through on-the-job training for professionals. As the UNU has been advocating through its promotion of RCEs, the role of higher education institutions and businesses needs to be further enhanced. Although the concept of the SDGs, unlike ESD, has been mainstreamed in the corporate sector in Japan, the private sector needs to go beyond CSR to redefine the model of development.

Environmental and sustainability education in Japan: emerging issues and trends, and current and future needs

Despite its role in supporting the Brundtland Commission and ESD globally, there is no denying that the Japanese government is more or less pursuing conventional patterns of development, even after experiencing the 3.11 'triple disaster' of the Great East Japan Earthquake, Tsunami, and Fukushima nuclear accident that struck the country on 11 March 2011. The 2020 Tokyo Olympic Games, which were postponed until 2021 due to the pandemic, were designated the 'Recovery Olympics', reflecting Japan's wish to mark its recovery from the 2011 disasters, but this also led to criticism that Japan is on a 'business as usual' path, prioritising the economy. There was a rise in anti-Olympic activism, with Fukushima's recovery being too distant to be celebrated and the risks associated with the site of the nuclear accident having been ignored (Boykoff and Gaffney 2020). There are also examples of ESD practice that are framed in opposition to central Japanese government policy, exploring ways to improve policy on specific issues such as radiation education (Goto 2017), indigenous rights recovery (Noguchi 2017) and disaster recovery projects in the post-3.11 context (Mochizuki 2017b).

The 3.11 triple disaster in 2011, and more recently the Covid-19 pandemic, have revealed the limitations of the current model of progress in painful ways. However, with its emphasis on cultivating ESD competencies, which are also important for ensuring economic growth, the promotion of ESD in schools continues to be set largely in the context of ensuring the strength of the nation within the global economy. Building back better and transitioning to sustainability are not primarily about scientific and technological solutions. Rather they require informed and empowered citizens, consumers, workers, students, educators and parents who have a deep understanding of their place within the ecosystem—people who know that they survive and thrive through their connectedness to each other and the ecosystems that support them. What needs to be challenged are not just superficial solutions to deep-seated and interconnected challenges but also uncritical and non-systemic thinking, and unreflective conformity with 'business as usual,' both on the part of decision-makers and local communities. It is the latter that will suffer the consequences of unsustainable development. This is where ESD has a crucial role to play.

Conclusion

Given that the DESD was a joint proposal by the Japanese government and NGOs, ESD has been extensively discussed by diverse stakeholder groups and a variety of initiatives have been put in place to promote ESD in Japan. Japanese ESD discourses and practices encompass a wide range of approaches to social change, from incremental, top-down, mechanistic approaches to activism demanding ecological and social justice. This goes significantly beyond the legacy of *kogai-kyoiku* or pollution education, which underpinned the original Japanese proposal for the DESD. While Japanese scholars tend to critique Japanese ESD policy and practice in the context of the ESD principles put forward by UNESCO, some European and North American scholars have characterised ESD—as defined and promoted by UNESCO—as 'technocratic' (Gough and Scott 2008) and policy-driven, dismissing it as complicit with the forces of neo-liberalism and globalisation (Jickling and Wals 2008, Torres 2009, Selby and Kagawa 2011, Huckle and Wals 2015). Although Japan does exhibit technocratic tendencies in its approach to ESD, with its emphasis on developing human resources to create a sustainable future (always including economic prosperity), there is no denying that ESD has inspired many educators, academics, civic groups and local communities to challenge the status quo and pursue an alternative path. What may be particularly distinctive about the promotion of ESD by Japanese stakeholders is that their actions, whether top-down or bottom-up, are often driven by a sense of responsibility—sometimes coupled with a sense of national pride—emanating from the fact that the DESD was Japan's idea. The notion that Japanese ESD policy and practice should be exemplary underlies top-down initiatives focused on boosting ESD's visibility as well as biting criticisms of such top-down approaches. While this sense of responsibility and national pride in upholding the ideals of ESD has definitely had the positive effect of making Japanese stakeholders hold themselves accountable for ESD implementation, it has also served to blur important distinctions between the self-strengthening, nation-building project and the need to harness the potential of ESD to achieve societal transformation. There is a continued need for transformative and self-reflexive forms of ESD practice in Japan, as there is elsewhere.

References

Abe, O. 2009. Jizokukano na kaihatsu no tame no kyoiku ESD no genjo to kadai (Current status and perspectives of ESD). Kankyo Kyoiku (Environmental Education) 19(2): 21–30.

Abe, O. 2012. Nippon – ESD no genjo to kadai (Japan – the current state and challenges of ESD). pp. 65–87. *In*: Abe, O. and H. Tanaka (eds.). Asia-taiheiyo chiiki no ESD: 'Jizokukano na kaihatsu no tame no kyoiku' no shintenkai (ESD in the Asia-Pacific Region: New Developments of Education for Sustainable Development). Akashi Shoten, Tokyo, Japan.

Ando, T. 2015. Kogai kyoiku kara kankyo kyoiku he saikou (Rethinking a shift from pollution education to environmental education). pp. 51–74. *In*: Sato, K. (ed.). Chiiki-gakushu no sozo: chiiki saisei e no manabi wo hiraku (Dynamics of Community-based Learning for Social Revitalisation). University of Tokyo Press, Tokyo, Japan.

Boykoff, J. and C. Gaffney. 2020. The Tokyo 2020 Games and the end of Olympic history. Capitalism Nature Socialism 31(2): 1–19.

Development Education Association and Resource Centre (DEAR). 2004. Kaihatsu kyoikutte naani? (What is development education?). DEAR, Tokyo, Japan.

González-Gaudiano, E. 2005. Education for sustainable development: configuration and meaning. Policy Futures in Education 3(3): 243–250.

Goto, S. 2017. An investigation into fairness and bias in educational materials produced by the Japanese government to teach school children about nuclear power and radiation. pp. 99–118. *In*: Singer, J., T. Gannon, F. Noguchi and Y. Mochizuki (eds.). Educating for Sustainability in Japan: Fostering Resilient Communities after the Triple Disaster. Routledge, London, UK.

Gough, S. and W. Scott 2008. The politics of learning and sustainable development. *In*: Farrell, R.V. (ed.). Education for Sustainability: Encyclopedia of Life Support Systems (EOLSS). Retrieved from http://www.eolss.net.

Hara, T. 2007. Bubble-keizai no sanaka de tadashiku rikai sarenakatta Brundtland Hokoku (Message of the Brundtland Report did not get across in Japan in the midst of the bubble economy). Gakusai (Interdisciplinarity) 20 <www.isr.or.jp/gakusai/20/>. Accessed 15 February 2016.

Hasegawa, K. 2004. Constructing Civil Society in Japan: Voices of Environmental Movements. Trans Pacific Press, Melbourne, Australia.

Hashimoto, Z. 2007. Brundtland-Iinkai to Nippon iin Okita Saburo sensei: Iinkai setsuritsu no keii nado (Four years with the Brundtland Commission and its Japanese member Mr Saburo Okita: background of establishing the commission). Gakusai (Interdisciplinarity) 20 <www.isr.or.jp/gakusai/20/>. Accessed 15 February 2016.

Huckle, J. and A.E.J. Wals. 2015. The UN Decade of Education for Sustainable Development: business as usual in the end. Environmental Education Research 21(3): 491–505.

Inter-ministerial Meeting on ESD. 2016. Wagakuni ni okeru 'jizokukano na kaihatsuno tame no kyoiku (ESD) ni kansuru Global Action Programme' jissi keikaku (National Action Plan for Implementing the Global Action Programme on ESD [ESD National Action Plan]). Jizokukano na kaihatsu no tame no kyoiku ni kansuru kankei shocho renraku kaigi (Inter-ministerial Meeting on ESD) (https://www.env.go.jp/press/files/jp/29478.pdf). Accessed 1 March 2021.

Inter-ministerial Meeting on ESD. 2021. Wagakuni ni okeru 'jizokukano na kaihatsuno tame no kyoiku (ESD) ni kansuru jissi keikaku (National Action Plan for Implementing ESD [Second ESD National Action Plan]). Jizokukano na kaihatsu no tame no kyoiku ni kansuru kankei shocho renraku kaigi' (Inter-ministerial Meeting on ESD). <https://www.mext.go.jp/content/20210528-mxt_koktou01-000015385_2.pdf> Accessed 12 February 2022.

Inter-ministerial Meeting on the UNDESD. 2006 (Revised 2011). Waga kuni ni okeru kokuren jizokukano na kaihatsu no tame no kyoiku no junen jissi keikaku (Japan's Action Plan for the UNDESD), Kokuren jizokukano na kaihatsu no tame no kyoiku no junen kankei shocho renraku kaigi (Inter-ministerial Meeting on the UNDESD). <www.cas.go.jp/jp/seisaku/kokuren/keikaku.pdf> Accessed 24 November 2015.

Inter-ministerial Meeting on the UNDESD. 2009. UNDESD Japan Report: Establishing Enriched Learning through Participation and Partnership among Diverse Actors. <www.mofa.go.jp/policy/environment/desd/report0903.pdf>. Accessed 1 March 2021.

Inter-ministerial Meeting on the UNDESD. 2014. UNDESD Japan Report. <www.cas.go.jp/jp/seisaku/kokuren/pdf/report_h261009_e.pdf> Accessed 1 March 2021.

Iwabuchi, T. and N. Takemoto. 2017. The Tohoku Green Renaissance Project – networking green rebuilding activities after a mega-disaster. pp. 216–228. *In*: Singer, J., T. Gannon, F. Noguchi and Y. Mochizuki (eds.). Educating for Sustainability in Japan: Fostering Resilient Communities after the Triple Disaster. Routledge, London, UK.

Japanese National Commission for UNESCO. 2018. Manual for Promoting ESD (Jizoku Kano na Kaihatsu no tame no Kyoiku Suishin no Tebiki) [online]. <https://www.mext.go.jp/unesco/004/__icsFiles/afieldfile/2018/07/05/1405507_01_2.pdf> Accessed 1 March 2021.

Ji, B. and K. Fukamachi. 2017. Can civil society revitalise dying rural villages? The case of Kamiseya in Kyoto prefecture. pp. 156–215. *In*: Singer, J., T. Gannon, F. Noguchi and Y. Mochizuki (eds.). Educating for Sustainability in Japan: Fostering Resilient Communities after the Triple Disaster. Routledge, London, UK.

Jickling, B. and A.E.J. Wals. 2008. Globalisation and environmental education. Journal of Curriculum Studies 40(1): 1–21.

Kadoya, S. and M. Goto. 2013. The past, present and future of ESD in Japan: how to develop and disseminate ESD at School with the network of the local community. NIER Research Bulletin 142: 47–58. <https://www.nier.go.jp/kankou_kiyou/kiyou142-105.pdf>.

Kamibeppu, T. 2002. History of Japanese Policies in Education Aid to Developing Countries, 1950s–1990s. Routledge, New York, USA.

MEXT. 2016. Yochien, Shogakko, Chugakko, Kotogatto oyobi Tokubetsushien Gakko no Gakushu Shido Yoryo nado no kaizen oyobi hitsuyouna hosaku nado ni tsuite (toshin) (Chukoshin 197-go) (On Improvement of Courses of Study for Kindergartens, Elementary Schools, Junior High Schools, High Schools and Special Needs Schools and the Measures Needed for Improvement, Report 197 by the Central Council of Education). <https://www.mext.go.jp/b_menu/shingi/chukyo/chukyo0/toushin/1380731.htm> Accessed 15 February 2021.

Miailhe, N. 2018. Géopolitique de l'Intelligence artificielle : le retour des empires ? Politique étrangère 2018/3 (Autumn Issue): 105–117.

Ministry of Foreign Affairs Japan (MOFA). 2002. Speech by Prime Minister Junichiro Koizumi at the World Summit on Sustainable Development on 2 September 2002, Johannesburg, South Africa. <www.mofa.go.jp/policy/environment/wssd/2002/kinitiative2.html> Accessed 15 February 2020.

Mochizuki, Y. 2010. Global circulation and local manifestations of education for sustainable development with a focus on Japan. International Journal of Environment and Sustainable Development 9 (1/2/3) (special issue on sustainable development and environmental education): 37–57.

Mochizuki, Y. 2017a. Introduction: top-down and bottom-up ESD—divergence and convergence of Japanese ESD discourses and practices. pp. 1–24. *In*: Singer, J., T. Gannon, F. Noguchi and Y. Mochizuki (eds.). Educating for Sustainability in Japan: Fostering Resilient Communities after the Triple Disaster. Routledge, London, UK.

Mochizuki, Y. 2017b. Postscript: Reflections on visions of rebuilding Tohoku and the future of ESD as a response to risk in Japan (with Makoto Hatakeyama). pp. 264–274. *In*: Singer, J., T. Gannon, F. Noguchi and Y. Mochizuki (eds.). Educating for Sustainability in Japan: Fostering Resilient Communities after the Triple Disaster. Routledge, London, UK.

Muramatsu, N. and H. Akiyama. 2011. Japan: super-aging society preparing for the future. The Gerontologist 51(4): 425–432.

Nagata, Y. 2017. A critical review of Education for Sustainable Development (ESD) in Japan: Beyond the practice of pouring new wine into old bottles. Educational Studies in Japan: International Yearbook 11: 29–41.

Noguchi, F. 2017. A radical approach from the periphery: informal ESD through rights recovery for indigenous Ainu. pp. 201–215. *In*: Singer, J., T. Gannon, F. Noguchi and Y. Mochizuki (eds.). Educating for Sustainability in Japan: Fostering Resilient Communities after the Triple Disaster. Routledge, London, UK.

Ohagi, A. 2019. Current Climate of ESD in Japan. National Institute for Educational Policy Research (NIER), Tokyo, Japan. <https://www.nier.go.jp/English/educationjapan/pdf/20190408-02.pdf> Accessed 1 March 2021.

Sato, M. and Nakayama, S. 2013. Development of DESD-IIS and Japanese contribution to DESD. pp. 19–34. *In*: Okayama ESD Promotion Commission and UNESCO Chair at Okayama University (eds.). Education for Sustainable Development (ESD) and kominkan/Community Learning Centre (CLC). Okayama University Press, Okayama, Japan.

Selby, D. and F. Kagawa. 2011. Development education and education for sustainable development: are they striking a Faustian bargain? Policy & Practice: A Development Education Review 12: 15–31.

Sumita, M. 2019. Karafuru na Gakko Zukuri (Creating a Colourful School). Gakubunsha, Tokyo, Japan.

Suzuki, K. 2019. Japan's ESD Activities with Emphasis on the Post-DESD Activities. Paper presented at 10th WEEC 2019, 3–7 November 2019, Bangkok, Thailand.

Suzuki, N. 2007. Kyoin kinmu jittai no houkoku: kyouin kyuuyo kakikaku ni mukete akiraka ni natta koto (Report on the survey of teacher work conditions: findings for teacher pay reform). Benesse Educational Research & Development Center (BERD) No. 9. <https://berd.benesse.jp/berd/center/open/berd/backnumber/2007_09/ren_suzukinao_01.html> Accessed 1 March 2021.

Tanaka, H. 2005. Kaihatsu kyoiku to jizokukano na kaihatsu no tame no kyoiku (ESD) – Sanka-gata shakai ni muketa shakai kyōiku no yakuwari (Development education and education for sustainable development (ESD): the role of social education for a participatory society). Studies in Adult and Community Education 49 (special issue on social education and lifelong learning under globalisation): 199–211.

Torres, C.A. 2009. Education and Neoliberal Globalisation. Routledge, New York, USA.

UN Conference on Environment and Development. 1992. Agenda 21, Chapter 36. Promoting Education, Public Awareness and Training. A/CONF.151/26.

UNESCO. 2013. Proposal for a Global Action Programme on Education for Sustainable Development as Follow-up to the UN Decade of Education for Sustainable Development (DESD) after 2014 (37c/57) <https://unesdoc.unesco.org/ark:/48223/pf0000224368> Accessed 1 March 2021.

UNESCO. 2014. Roadmap for Implementing the Global Action Programme on Education for Sustainable Development. <https://unesdoc.unesco.org/ark:/48223/pf0000230514> Accessed 1 March 2021. UNESCO, Paris, France.

UNESCO MGIEP. 2017. Textbooks for Sustainable Development: A Guide to Embedding. UNESCO Mahatma Gandhi Institute of Education for Peace and Sustainable Development (MGIEP), New Delhi, India.

World Commission on Environment and Development. 1987. Our Common Future. Oxford University Press, Oxford, UK.

Europe

Chapter 6

Environmental and Sustainability Education in Austria

Franz Rauch,[1,]* *Christiana Glettler,*[2] *Regina Steiner*[3] and *Mira Dulle*[1]

Introduction

This chapter traces the development of education and sustainable development in Austria. We start by giving an overview of Austrian culture, and the country's political situation and education system, before moving on to the development of Environmental and Sustainability Education (ESE).

From a historical perspective, environmental education became part of the school system in the 1970s. In the following years the discourse focused mainly on global learning, political education, ecological education and peace education. In the 1990s, education for sustainable development became part of the debate following the first World Summit in Rio, and was embedded in the curriculum in the form of environmental education and global citizenship education. One typical aspect of this in Austria is the connection between ESE and action research, primarily supported by the influential Project ENSI (Environment and School Initiatives). The later sections of the chapter present the current situation along with future trends and issues. Current developments are supported by Agenda 2030 and especially SDG 4. New Austria-wide projects and platforms are emerging on the basis of established initiatives, structures and networks. However, the Austrian education system continues to face challenges, including equal educational opportunities for all sectors of the population.

[1] University of Klagenfurt, Austria.
[2] PPH Augustinum Graz, Austria.
[3] University of Education in Upper Austria, Austria.
Emails: christiana.glettler@pph-augustinum.at; regina.steiner@ph-ooe.at; mira.dulle@aau.at
* Corresponding author: franz.rauch@aau.at

The Austrian education system and its political, economic, cultural and social context

In order to provide an understanding of the philosophy behind ESE and its development in Austria, we set out below a brief overview of the country's historical, socio-economic and cultural situation, and the development of its educational system.

The republic of Austria is a democratic state situated in Central Europe, with approximately 8.9 million inhabitants. Christianity (Roman Catholicism) is the predominant religion. Austria is a welfare state with a well-established health system and is one of the richest countries in the world in terms of gross domestic product per capita. The country has a well-developed social market economy and a high standard of living. The employment rate in 2020 was 71% (Statistik Austria 2020a). Austria has a very high rate of literacy, at 99%. Over recent years, the educational level of the population has increased: its 5,685 schools educate over one million pupils (Statistik Austria 2020a).

Hofstede (2011) defines six cultural dimensions that indicate the values of the world's cultures. In accordance with these, Austria can be described as an independent country with decentralized power structures and a focus on equal rights for all inhabitants (power distance dimension). With an individualistic (individualism dimension), masculine and highly success-oriented society (masculinity dimension), Austrians tend to adopt rigid rules and norms and have an inner urge to be busy and work hard (uncertainty dimension). They also have a pragmatic culture that easily adapts traditions to changed circumstances and a strong propensity to save, invest and achieve results (long term orientation dimension). Austrians put emphasis on realizing their desires and enjoying their life and leisure time; in addition, they take a positive attitude and have a tendency towards optimism (indulgence dimension) (Hofstede Centre 2020).

The cultural values described above developed over time and influenced Austrian's educational philosophy. There was a shift from a strong military orientation ('masculine') to a more humanistic and 'bel esprit' oriented ('feminine') system. The roots of compulsory schooling lie in the reign of Empress Maria Theresa, who initiated many reforms that still shape Austria today. In 1774, she introduced general compulsory education. During that time, the purpose of education was to strengthen the Austrian Empire and its military power and technology, as well as the economy and administration. Although traces of this can still be seen today, the focus is now more on the development of '*moral and social values, and the values of beauty, goodness and truth*' (School Organisation Act, or SchOG, Article 2). As also set out in the SchOG, education aims to equip young people with the knowledge and skills they need for their life and their future profession (ibid.).

Compulsory education in Austria consists of nine school years, with public schools being free of charge. Pupils' performance is assessed on a scale of grades running from 1 (very good) to 5 (not sufficient). In contrast to other OECD countries, pupils have to repeat a school year if they fail (Pongratz 2010). After four years of common primary education, pupils attend either a lower secondary school or a secondary academic school (another four years). To complete the nine years of compulsory education, pupils can attend a pre-vocational school. If they wish to take the Matura, Austria's equivalent to A-levels, they have to complete between four and five years at an upper secondary school (SchOG, Article 3). The structure of the Austrian educational system remains strongly hierarchical and closely interwoven with politics. The school system is regulated at the federal level by the Ministry of Education, Science and Research (Bundesministerium für Bildung, Wissenschaft und Forschung – BMBWF) and the provincial school authorities are responsible for the administration and supervision of schools.

In recent decades, many efforts have been made to reform the education system (Posch and Altrichter 1993). Some of the latest central reforms include the introduction of all-day schools alongside the traditional half-day establishments, a focus on competence-oriented instruction, enhanced autonomy for schools, and updates to curricula. All in all, access to education is excellent as a result of its having been compulsory for centuries; even access to Austrian universities is free for most fields of study, and currently no tuition fees are payable. However, entrance exams have been introduced in recent years for certain subjects.

Nevertheless, Austria's school system is currently facing a number of problems. Austria has the highest expenditure on education of all EU countries. Despite of this, Austrian students' performance in the PISA (Programme for International Student Assessment) and TIMSS (Trends in International Mathematics and Science Study) international studies is average (Bruneforth et al. 2012).

The latest reports, such as the Statistics Austria indicator report (Statistik Austria 2020b), point to positive developments in high-quality education in Austria over recent years. The broadening of educational provision in recent decades has resulted a general increase in the educational achievement of the Austrian population. Overall, the national employability index has been consistently at the relatively high level of 88.6% recently—also a result of the vocational school system in Austria—and thus above the EU-28 value of 81.7%. However, it should be noted that there is still considerable development potential in numerous areas, which is outlined briefly below and described in more detail in Section Six.

As the National Education Report 2018 states, equality of opportunity and participation are not guaranteed in Austria. There is a clear need for action, for example, to enable systematic access to pre-school support, which is seen as the cornerstone for later educational success. The Austrian educational landscape is characterized by high educational disparities (Schober 2019). Compared with other countries, Austria's social reproduction of educational inequality is relatively high. Accordingly, socio-economic status has a particularly strong effect on participation in education and learning as well as on economic and social outcomes. Young people who do not come from an academic household are significantly less likely to gain a degree (BMBWF and bifeb 2020). Similarly, the Austrian National Education Report 2018, Volume 1, indicates that only one third of the social inequality in the choice of secondary school level can be explained by differences in performance. Children of academics transfer to upper secondary school more often than children of others, on the basis of the same performance (Oberwimmer et al. 2019).

Overall, the environmental situation in Austria can be described as mixed. While the water quality is very high, the level of soil sealing and the climate situation are worrying. Temperatures in Austria have risen at more than twice as the global average in the last few years and 2019 was the third warmest year in the 252-year history of temperature measurements. Climate models predict that Austria and the alpine region will continue to experience greater warming than the global average in the future (Umweltbundesamt 2020). Political initiatives are now being put in place, especially with regard to travel, alternative energies and CO_2 taxation.

Historical background and the development of environmental and sustainability education in Austria

We will start with a brief historical overview of the concept of environmental education as discussed in the debate on education for sustainable development. The term environmental education (EE) has been in use in German and language literature since the 1960s. From its outset, EE was a policy instrument that aimed to deliver long-term solutions to existing environmental problems (Klenk 1987). The prevalent issues at the time were waste separation, recycling, water conservation and energy, and the use of burlap and glass instead of plastics. The 1990s added new frames of reference such as environmental ethics, ecological time, art and the environment, environmental education and the media and perceptions of landscape (Rauch and Steiner 2006).

With regard to legislative requirements, the 1979 decree on EE provided the impetus to embed EE in the Austrian system of education. It became a principle of instruction in all subjects at all secondary schools (catering to the 10–19-year age bracket). Together with the decree on political education, this created a relatively innovative legal framework. It aimed to develop competences to take eco-political action (Breiting and Mogensen 1999) and to integrate expertise, reflection and action in school-based teaching and learning. In 1992 another component was added, namely the decree on project teaching, which required teachers to undertake larger-scale, action-based, participatory projects in all schools and subjects. Environmental education topics were more than suited to this form of teaching, so the two decrees were mutually supportive (Rauch and Steiner 2006).

Environmental issues gradually transformed into environmental projects, conducted partly on an interdisciplinary, action-oriented basis and in cooperation with external organisations. Experience of the natural world, inspired mainly by the writings of Joseph Cornell (1979), became a popular antidote to ecological doomsday pedagogy. The aim was for the term 'environment' to have positive associations, and the enjoyment of nature was thought to encourage students to protect the environment without moralizing at them. Environmental education projects, however, frequently remained activity-based and failed to reflect on impacts and underlying causes, with the result that learning remained superficial (Lieschke 1993).

From development policy education to global learning and global citizenship education

While EE is one major strand of education for sustainable development in Austria, global learning is the other. The development of global learning was triggered by the public relations activities on 'third-world issues' in the 1960s and 1970s; these were largely non-political at that point, and involved the churches raising funds for aid projects. The 1987 decree on Political Education in Schools mentioned above also boosted development policy work, promoting awareness of what it means to be Austrian along with an openness to the world and a readiness to stand up for human rights, overcome prejudices, and champion the cause of the disadvantaged. This prompted organisations such as the Austrian Latin-America Institute, the Austrian Commission for UNESCO, the Youth Council for Development Aid working party on schools (which was later to form the core of the Austrian Information Service for Development Policy, ÖIE) to get involved with the in-service education of teachers. The development policy education of the 1980s developed out of a cognitive ideological critique in the 1970s, focusing on practical educational solutions in which affective elements started to play an ever-increasing role (Hartmeyer 2001).

In the 1990s, development policy education was gradually replaced by the concept of global learning. With the globalisation of all walks of life and the emergence of the idea of global citizenship, education was faced with new challenges. "The interfaces between world-wide equity, multicultural societies, global environmental issues, the peace issue, and the limits of growth in industrialized countries are all at the centre of global learning" (Hartmeyer 2001, 37). The concept of global learning in Austria took on another role: rather than simply conveying factual knowledge, its primary focus was a critical approach to concerns, interests, and experiences. The idea was that global learning cannot in and of itself create a better world, but encourages self-determination in a global context (Grobbauer and Wintersteiner 2019).

Education for sustainable development

When the concept of Education for Sustainable Development (ESD) emerged in the 1990s, proponents of both concepts (EE and global learning) claimed to be the legitimate pioneers and originators of this new approach. This led to some tensions, and intense discussions on leadership. However, the concept of the Sustainable Development Goals (SDGs) brought a welcome relaxation, integrating both approaches, especially in SDG 4.7, which explicitly mentions ESD, sustainable lifestyles, human rights and global citizenship.

Since the 1992 Earth Summit in Rio, the policy discourse has shifted from EE to a broader vision: ESD. As a result, apart from ecological issues (consumption of resources, environmental pollution, population explosion, etc.) EE, at a normative level, was from that point characterised by the idea of fair global distribution, which involved a new mix of ecological, economic, social, political and ethical dimensions (de Haan and Harenberg 1999).

Table 1 gives an overview of ESE developments in Austria since the mid-1990s. Some of the projects are described in more detail below.

Table 1. Overview of ESE developments in Austria (expanded from Rauch and Pfaffenwimmer 2020).

	ESD implementation and legal developments	ECO schools network (ECOLOG)	Teacher Education	Higher Education	International initiatives
1995	Education Support Fund	Launch of ECO-schools concept			ENSI decision on focus topics: ECO-schools, teacher education, (IT)-networking and quality assurance.
1996		Start of ECO-school pilot phase (1996–1998)			
1997			ENITE research project (Environmental Education in Teacher Education)		
1998					
1999		ECO-schools network concept			
2000					
2001		Start of ECO-schools network	ENITE network		
2002	ESD platform in Ministry of Education (2002–2008)	*National Environmental Performance Award* for Schools and Teacher Training Universities			EU-ENSI-SEED network project (2002–2005)
2003					
2004			First national university-level teacher training course 'Innovation in Teacher Education – Education for Sustainable Development' (BINE).		EU-ENSI-CSCT project (2004–2007)

Environmental and Sustainability Education in Austria 99

Year					
2005	Vilnius declaration; ESD Strategy process (2005–2007); Education Map			International conference 'Committing Universities to Sustainable Development'	UNECE Vilnius declaration
2006	ESD Strategy process; EU-ESD Conference		'Competences for Education of Sustainable Development' research project (KOM-BINE) (2006–2008).		UNECE evaluation; EU Project: FORM-IT - Take Part in Research
2007	Sustainability Award by UNESCO Austria			Sustainability Award established	EU-ENSI-SUPPORT network project (2007–2011)
2008	ESD Strategy decision; Austrian Agency for Education for Sustainable Development ('Dekadenbuero')		Second national BINE university course for trainee teachers		
2009					UNECE evaluation
2010		300 ECO-schools	ECO-schools network with teacher training universities		
2011	Legislation on quality management in schools				EU-ENSI-CoDeS-network project (2011–2014)
2012		400 ECO-schools	Third national BINE university course for trainee teachers	Alliance of Sustainable Universities	
2013				Launch: Future lectures	

Table 1 contd. ...

...*Table 1 contd.*

	ESD implementation and legal developments	ECO schools network (ECOLOG)	Teacher Education	Higher Education	International initiatives
2014	ESD Best of Austria Award				
2015	Global Action Programme and Sustainable Development Goals (International initiatives)				
2016			Fourth national BINE university course for trainee teachers		
2017	SDGwatch Austria	500 ECO-schools (ECO-schools network ECOLOG)			
2018			UniNEtZ Project (Higher Education)		
2019					
2020	Launch of joint ESD and Global Education webpage: www.bildung2030	600 ECO-schools			

The ENSI Project as an influential factor

Compared with other German-speaking countries, ESE in Austria bears more of the imprint of the ENSI project, in which school development and practical/action research play a key role.[1] In this regard, the link between action research and ESD is seen in an approach to learning that is autonomous and cross-linked, and reflects on research as it engages with the world. The principles of action research are thus very well aligned with the concept of ESD, because both aim to shape society through reflection (Rauch and Pfaffenwimmer 2019). In Austria, action research has been a firm fixture of quality development and assessment for schools since the late 1980s. The main reasons for this are the increasing autonomy of schools and the concomitant issues of on-site quality assessment, which meant that research and development projects and theoretical publications on action research fell on fertile ground (Altrichter and Posch 2009, Feldmann et al. 2018). These favourable conditions enabled ENSI to flourish and contribute to general school development (e.g., through studies and training schemes).

In 1986, a team of experienced Austrian teachers from different regions and school types were chosen to form the ENSI teacher team. The team was co-ordinated by staff at the Ministry of Education and supported and facilitated by university academics. The teacher team was trained in action research to enable it to document and publish their innovative work as case studies. The ENSI team bridged the gap between practice, policy and research for many years until its conclusion in 2017. It had a significant influence on the development of ESE in Austria (Affolter and Varga 2018).

In 2002, Austria submitted and coordinated the ENSI-EU project 'School Development through Environmental Education SEED' (2002–05) (www.ensi.org/projects). Its most influential publication is Quality Criteria for ESD Schools (Breiting et al. 2005), which has been translated into many languages (Lechner and Rauch 2014). Collaborations between schools and the communities in which they are based are regarded as crucial for real social development and change. The latest ENSI project, CoDeS (School and Community Cooperation for Sustainable Development) therefore focused on such collaborations, gathering together 29 experts in both fields from 17 countries.[2]

FORUM Environmental Education (FORUM Umweltbildung)

Another major role in ESE in Austria is played by FORUM Environmental Education (FORUM EE). Following the 1983 international symposium on *The Long-term Development of Environmental Policy and Environmental Education in Europe* in Vienna, the participants adopted the *Vienna Declaration* (Katzmann 1986). This triggered the launch of the ARGE Umwelterziehung (Working Group on Environmental Education), which acted as a central point of contact for environmental

---

[1] http://www.ensi.org.
[2] https://www.ensi.org/Projects/Our_Projects/CoDeS/.

education in Austria. The ARGE Umwelterziehung produced media and materials for teaching and organized topic-specific seminars, exhibitions and conferences. A regular magazine was published, providing teachers with theoretical and practical articles and was delivered to all German-speaking countries. The magazine was replaced in 2012 by an ESD yearbook.

In 1999 the ARGE Umwelterziehung was renamed FORUM Umweltbildung (FORUM Environmental Education – FORUM EE). The aim of this was to reflect the new focus on sustainable development and an openness towards social challenges and work on visions, learning processes, learning situations and learning methods. The focus shifted from conveying sustainable development content to participation in—and promotion of—a new 'educational culture.'

FORUM EE is part of an environmental NGO but is financed by the Ministries of Environment and Education, which also determine most of the content and outsource programme coordination to FORUM EE. It coordinates the eco-label for schools and the Environmental Education Fund, as well as the education map, and localization and presentation of ESD initiatives throughout Austria and beyond. The UN Decade Office (the coordinating office for the UN Decade of ESD in Austria (2005–2014)) was also part of FORUM EE. Between 2016 and 2019, FORUM EE coordinated the annual awards for projects that were particularly committed to the implementation of the World Action Programme on ESD, the *ESD—best of Austria* award. It works with ministries to define annual themes, in relation to which educational materials are then produced and symposia and conferences held. One of the tasks of FORUM EE was and remains to identify, pilot and evaluate current developments in ESE and progress with Education 2030. For this reason, FORUM EE is also repeatedly involved in international education and research projects.[3]

Understanding of environmental and sustainability education in Austria

Current discussions on ESE in Austria centre on the notions of sustainable development, environmental education, global citizenship education (GCED), international peace and civic education; they have also sparked debates on the nature of education in general (Rauch and Steiner 2013). Recent United Nations programmes such as the Sustainable Development Goals (especially goal 4 *Quality Education*) (United Nations 2015) and the 2015 UNESCO Global Action Programme on Education for Sustainable Development (UNESCO 2015) are in line with the discourse in Austria.

In Austria, most proponents of ESE regard sustainable development as a regulatory idea (Kant 1787/1956) that is similar to the concept of human rights. Such ideas do not determine the nature of an object but serve as structures for heuristic reflection. They give direction to research and learning processes. In terms of sustainability, this implies that the contradictions, dilemmas, and conflicting goals inherent in this vision need to be constantly re-negotiated in a process of discourse between the

---

[3] https://www.umweltbildung.at/.

participants in each and every concrete situation. This presents a significant challenge but also has considerable potential to enhance learning and innovation in education (Rauch 2015). A central goal is the transformation of individuals, organisations and the society. Learning in this sense is understood as transformative 'when the learners, integrate and reinterpret knowledge into their own frames and put it into practice in their own lives. Learning is also one mechanism for changing the society and for transforming the society' (Reardon 2010).

Within the UniNEtZ project, which links Austrian Higher Education institutions with the SDGs, the SDG 4 Position Paper sets out current conceptualisations of ESE in Austria. This paper related the concepts of GCED and ESD to each other in a participatory process that aimed to foster the implementation of SDG 4 (*Good Education for All*). The position paper emphasises the responsibility of research and education for finding solutions that foster an all-encompassing transformation toward a sustainable future for all. In the age of climate crisis, global inequalities and austerity measures, it is seen as crucial to empower people not only to change their thinking but also to drastically change their actions (UniNEtZ SDG 4 2019).

A recent UNESCO Austria document (Austrian Commission for UNESCO 2019) states that for education to become a real engine of social change, it must be transformative. This kind of pedagogy is characterised by three aspects:

- Transformative learning aims to transform individual perspectives
- Transformative learning is designed as a collective process of awareness development and emancipation (as defined by Paulo Freire)
- Transformative learning also strives to change cultures and structures (of the educational system) that hinder emancipatory learning (Singer-Bodrowski 2016).

Implementation of the ESD Global Action Program (GAP) and the follow-up programme 'ESD for 2030' in Austria

The previous sections have already presented several of Austria's ESE projects. The UNESCO GAP and ESD for 2030 programmes provide a framework and support in the field of education, both in the formal sector (e.g., curricula and decrees; the ECOLOG programme) and the informal sector (e.g., NGOs and the UNESCO Commission Austria). In this section, we will go into more detail, focussing on the first three priority action areas of the GAP.

Advancing policy

As described above, the UniNetZ project involves major Austrian higher education institutions as well as other stakeholders. Its aim is to provide the Austrian government with options for implementing the SDGs in Austrian society. The project team working on SDG 4 are likely have a particular impact on Austrian educational policy, as will be explained in the following section.

SDG Watch Austria, a platform representing and networking more than 200 civil society and non-profit organizations, increases the visibility of the implementation and delivery of Agenda 2030 and the UN sustainable development goals, highlighting their importance to all civil society organizations. A prominent theme is the *Thematic Initiative on Education,* which promotes SDG 4. The meetings under this initiative enable the exchange of information on political processes, activities and materials. In spring 2019, for instance, the initiative developed arguments for embedding Agenda 2030 more thoroughly in the curricula that were being drawn up, and sent them to the Federal Minister of Education.[4]

Transforming learning and training environments

Since it was founded in the 1990s, the ECOLOG network (Ecologisation of Schools) has played a major role in helping educators and school leaders make schools' practices more sustainable. In the summer of 1995, the Minister of Education commissioned the ENSI teacher team to design the schools' network, which after a two-year pilot phase became a wider network coordinated by FORUM EE until 2015 and by the Institute of Instructional and School Development at University of Klagenfurt after that. ECOLOG is a national support system that aims to promote and integrate ecological approaches into individual schools' development and attempts were being made to embed the programme in Austria's federal states through regional networks (Rauch and Steiner 2006). In order to provide support, a network structure involving ECOLOG regional teams in the nine Austrian provinces has been developed; a scientific advisory board has also been established. Central support is provided by the Ministry of Education and by the Institute of Instructional and School Development at the University of Klagenfurt. Additional support measures are provided by FORUM EE (an NGO) and via seminars for heads and coordinators of ECOLOG network schools, the Education Support Fund for Health Education and Education for Sustainable Development, and the National Environmental Performance Award for Schools and University Colleges of Teacher Education (Rauch and Pfaffenwimmer 2020).

The ECOLOG schools network contributed to the development of pedagogical criteria for The Austrian Eco-label for Schools and Teacher Training Colleges (www.umweltzeichen.at), which has been awarded by the government since 2002. Currently, the network consists of more than 700 institutions, including teacher education colleges and regional school boards.[5] ECOLOG has provided a reference point for other thematic networks in Austria focussing on ESD, for instance climate alliance schools (https://www.klimabuendnis.at/english), climate schools[6] (see below), nature park schools,[7] UNESCO schools[8] and healthy schools,[9] since the 1990s. Between 2013 and 2018 the Austrian Ministry of Education Department of Environmental Education compiled a list of all Austrian schools that are active

4 https://www.sdgwatch.at/de/.
5 https://www.oekolog.at/.
6 https://klimaschulen.at.
7 https://www.naturparke.at/schulen-kindergaerten/schulen/.
8 https://www.unesco.at/bildung/unesco-schulen/.
9 https://www.give.or.at/gesundeschule/.

members of these thematic networks. It lists 1000 schools, some active in a variety of networks. As there are 5,712 schools in Austria, we can state that every sixth school in Austria has an ongoing engagement with ESD (Rauch and Pfaffenwimmer 2020).

Another important driver for the transformation and ecologisation of schools also dates back to an ENSI project—the *National Environmental Performance Award* for Schools and University Colleges of Teacher Education. This is a national government award that recognises outstanding performance and has been awarded since 2002. About half of the 120 criteria relate to EE and ESD, the school curriculum and school development. The other half refer to technical aspects such as energy saving. The award is valid for four years, after which the compulsory external evaluation has to be renewed (Rauch and Pfaffenwimmer 2015). To date, over 100 schools have been awarded the Environmental Performance Award, some of them for the fourth time.

Throughout the history of ESE in Austria, and especially since the passing of the decree for Environmental Education for Sustainable Development 2014 (Austrian Federal Ministry for Education and Women's Affairs 2014), engagement with local educational activities has been a central focus. Partnerships with external agencies and stakeholders have proved a valuable approach (Lukesch et al. 2009). In the years 2012–2014 *School-Community Collaboration* was a focal topic for the ECOLOG programme, as was the ENSI-EU CODES Project (2011–2014) (Rauch and Pfaffenwimmer 2020).

In 2006, the Austrian UNESCO Commission decided to recognise projects within the UN Decade Education for Sustainable Development (DESD) that met the international ESD criteria. Between 2007 and 2014, 201 projects by 168 organisations were recognised and documented in four UNESCO commission publications, and the FORUM EE's Bildungslandkarte (Education Landscape) search tool.

Building the capacities of educators and trainers

As all the projects and initiatives described above and also in Section 6 below depend on teachers not only being willing to include ESE in their daily educational practices but also having the skills to do so, another important factor in transforming learning environments was, and remains, teacher education. Between 1997 and 2004, the ENITE project (Environmental Education and ESD in Teacher Education) by the University of Klagenfurt developed a research and development network that supported initiatives in teacher education and was inspired by ECOLOG, especially at universities of teacher education (Posch et al. 2000, Kyburz-Graber et al. 2003). The main outcome of the ENITE network so far is the National Teacher Training Course *Innovation in Teacher Education – Education for Sustainable Development* (German acronym BINE), offered by the Institute of Instructional and School Development at the University of Klagenfurt in cooperation with universities of teacher education. The four-semester in-service course has since been run successfully four times (Rauch and Steiner 2015), and the fifth course is currently in preparation. Since 2006–07, teacher education has been involved in a dynamic reform process based on new teacher training legislation. A positive result of the ENITE network and the BINE courses is that communication, collaboration and even participation between universities of teacher education and the ECOLOG network has been enhanced (Rauch and Pfaffenwimmer 2015).

In 2019, a new platform for Education 2030 was launched as a joint endeavour between ESE organisations and Organisations for Global Learning and Global Citizenship.[10] The platform brings together information on Agenda 2030 topics and provides a wide range of ideas on how global challenges can be addressed for all age groups and in different teaching and learning environments. It collates Austria-wide events and materials for schools and other educational institutions and provides an overview of current training and further education for stakeholders. The platform also presents the concepts GCED and ESD and contributes to critical and sustainable education. The preparation of quality criteria provided a particularly useful opportunity for intensive and fruitful discussions between the participating organizations. The platform is financed by the Austrian Development Cooperation and the Federal Ministry for Climate Protection, Environment, Energy, Mobility, Innovation and Technology.

Environmental and sustainability education in Austria: Emerging issues and trends, and current and future needs

We begin this section by describing issues that still need to be addressed as Austria works towards achieving the SDGs. Moving to a more positive outlook we then present impending policy changes and conclude with a description of current trends in ESE in Austria.

Unresolved issues

The Austrian Commission for UNESCO (2019) has identified several factors that make it difficult to implement the SDGs in Austria, thus highlighting issues that still need to be addressed:

- In the Austrian education system, informal learning processes are strongly linked to the experience of inequality, discrimination and exclusion. This is reflected in and intensified by educational approaches, and thus passed on from generation to generation.
- In the context of a complex education system, the interests of different stakeholders are at odds with each other. There does not appear to be a basic consensus in Austrian society that all students should have access to the best education.
- The formal education system provides the opportunities necessary to developing the critical thinking and political awareness that teachers and students need to be able to recognize problems, analyse them in a responsible manner and participate in identifying solutions in only a basic form.

---

[10] https://bildung2030.at/.

- The lack of integrated all-day schools and the lack of support in the free public education institutions for dealing with problems run counter to the achievement of appropriate and effective learning outcomes, especially with regard to children with a migrant background, a first language other than German, and/or living in poverty.

- There is a lack of nationwide childcare systems for children under the age of 3. The qualifications of the personnel in this sector, which do not involve academic training, do not meet international standards.

- Gender stereotypes remain a strong influence on children and young people. Austria has particularly large gender gaps when it comes to the competency development and choice of profession, and this has a negative impact on future opportunities for participation.

- There is an absence of assistance for people with disabilities in tertiary education. This impedes access to academic degrees and makes labour market opportunities very limited.

Policy and ESE

Despite the unresolved issues described above, on a more positive note, Austria is about to enter a new phase in the implementation of ESE in schools. A new curriculum for primary and lower secondary schools is currently under development (BMBWF 2020). This will include ESE both in general educational principles and selected subjects such as Biology, Geography and Mathematics, for the first time. As such, it is a milestone development and shows growing awareness of the importance of ESE on the part of Austrian policy-makers.

In similar spirit, the UniNEtZ project (see Chapters 4 and 5) is designed to further the efforts to advance sustainable development in Austria. As the SDGs have mutual impact on each other and can only be delivered through inter- and trans-disciplinary thinking, UniNEtZ represents a broad spectrum of expertise from fields seemingly as far apart as social sciences, STEM (Science, Technology, Engineering, Mathematics), and the arts. The project is likely to result in a number of synergies, including the implementation of sustainability in research and teaching, and stronger collaboration between universities and society. Networking between higher education institutions and the implementation of the SDGs in research, teaching and society are at the heart of UniNEtZ (UniNEtZ 2020).

The University of Klagenfurt and the University of Innsbruck are coordinating SDG 4 within UniNEtZ. Participating SDG 4 members come from 14 universities, and also include representatives of the Austrian Ministry of Education, university colleges of teacher education, Hochschulbildung Global [Higher Education Global], the Geological Survey of Austria, and Forum n. The SDG 4 Unit has recently published the Options Report—the final report with policy recommendations on how to achieve the SDGs, spanning the whole educational system from early childhood to university (UniNEtZ 2022).

Our understanding is that this project is novel in its form and structure. Since Austria is a relatively small country, this project can be considered a pilot study that could be adapted and rolled out to larger settings. We would thus like to share our experiences with the global sustainability community and discuss the potential and challenges of the communication and development process. We hope to launch further projects at the global level in the years to come.

Climate action as a central theme

While policy documents focus on the SDGs and ESE more generally, ongoing projects and stakeholder priorities reveal an increasing focus on climate change and thus have SDG 13 as a central theme. ECOLOG has identified *Climate.Action. Time* as the special theme that will guide network partners between 2020 and 2023. Accordingly, the homepage includes project suggestions and teaching materials on the topic for partners, in-service teacher education workshops on climate action will be held and schools are being encouraged to focus their initiatives on combating climate change.[11]

However, ECOLOG is not stand-alone with this focus. Forum EE and UBZ (the Environmental Education Centre, based in Graz) have a strong focus on SDG 13, providing links, materials and workshops for educators in order to support pro-environmental initiatives and action on climate change. The new Education 2030 platform[12]—a collaborative venture combining global learning and education for sustainable development also deserves a mention in this context (see also above).

The Tyrolean initiative k.i.d.Z. 21 (the German acronym for 'competent for the future') also focuses on climate change as one of the major challenges facing future generations. The project aims to

1) Increase young people's awareness of climate change and its consequences

2) Foster young people's adaptability and ability to take action, and

3) Prepare them for the social, ecological and economical challenges of the 21st century (Oberrauch et al. 2015)

The project combines inter- and trans-disciplinary approaches and focusses on the creation of direct partnerships, teaming up researchers with young people and their teachers.

Similarly, the 'makingAchange' project[13] aims to bring research and education together. The project targets secondary schools, helping them contribute to societal transformation by taking climate action and promoting sustainability.

As a bottom-up, social development, the Fridays-for-Future movement in Austria has also generated a positive dynamic in recent years, with active groups in schools and universities all over the country. Inspired by the uprising of young people, Teachers-for-Future and Scientists-for-Future have also been established. Fridays-for-Future has had a positive influence on the prioritisation of ESE in the

---

[11] https://www.oekolog.at/f%C3%BCr-den-unterricht/klima_wandel_zeit/.
[12] https://bildung2030.at/.
[13] https://makingachange.ccca.ac.at/

new curriculum that is currently under development (see above). Student groups are also actively involved in the UniNEtZ project.

Conclusion

In summary, it can be said that Agenda 2030 has given ESE increased momentum. New projects (such as UniNEtZ, makingAchange) and collaborations (like Platform Education 2030) have emerged and policy-makers have been positively influenced by bottom-up initiatives such as Fridays-for-Future. Furthermore, existing initiatives and networks, such as ECOLOG continue to develop and to adapt to new global programmes and challenges. The launch of the *Education for Sustainable Development: Towards achieving the SDGs* (*ESD for 2030*) programme in 2021 has also given new impetus to ESE in Austria (UNESCO 2020). The strong emphasis on transformative action and critical thinking should slowly move from the above-mentioned pioneering organizations and activities into the education mainstream.

As far as the weak spots of the Austrian education system are concerned, it remains to be seen whether the problems highlighted can really be overcome. In order to do so, a major paradigm shift would have to take place, with stronger detachment of educational policy from party politics. Additionally, the Covid-19 pandemic has also challenged the country and the education system in particular—as it has everyone else. It remains to be seen whether it will be possible to promote solidarity and a sustainable approach to the future. However, there is still hope.

References

Affolter, C. and A. Varga (eds.). 2018. Environment and School Initiatives. Lessons from the ENSI Network - Past, Present and Future. Wien, Environment and School Initiatives and Budapest, Eszterhazy Karoly University.

Altrichter, H. and P. Posch. 2009. Action research, professional development and systemic reform. pp. 213–225. *In*: Noffke, S.E. and B. Somekh (eds.). Educational Action Research. Los Angeles, SAGE Publications Ltd.

Austrian Commission for UNESCO. 2019. Positionspapier zur Umsetzung von SDG 4 in Österreich. https://www.unesco.at/fileadmin/Redaktion/Publikationen/Publikations-Dokumente/2019_Positionspapier_OEUK_Fachbeirat_Transformative_Bildung.pdf.

Austrian Federal Ministry for Education and Women's Affairs. 2014. Basic decree on environmental education for sustainable development. Wien. https://bildung.bmbwf.gv.at/ministerium/rs/2014_20_ge_umwelt_en.pdf.

BMBWF. 2020. Lehrplan für die Volksschule (under revision). www.bmbwf.gv.at/Themen/schule/schulpraxis/lp/lp_vs.html.

Breiting, S. and F. Mogensen. 1999. Action Competence and Environmental Education. Cambridge Journal of Education 29/3: 349–353.

Bruneforth, M., B. Herzog-Punzenberger and L. Lassnig (eds.). 2012. Nationaler Bildungsbericht Österreich 2012. Indikatoren und Themen im Überblick [National Education Report Austria 2012. Indicators and Topics: An overview]. Wien, BMUKK, Bifie, Leykam.

Cornell, J.B. 1979. Mit Kindern die Natur erleben. Soyen, Verlag an der Ruhr.

De Haan, G. and D. Harenberg. 1999. Expertise, Förderprogramm Bildung für nachhaltige Entwicklung. Berlin, FU Berlin.

Feldmann, A., H. Altrichter, P. Posch and B. Somekh. 2018. Teachers Investigate their Work. Milton Park, Routledge (3rd Edition).

Grobbauer, H. and W. Wintersteiner. 2019. Global Citizenship Education. Concepts, Efforts, Perspectives—an Austrian experience. Salzburg, Klagenfurt, Komment.

Hartmeyer, H. 2001. Globales Lernen in Österreich – Erfahrungen, Erwartungen, Perspektiven. *In*: Halbartschlager, F. (ed.). Eine Welt. Beiträge zu Globalem Lernen. Symposion Globales Lernen. Wien, SüdwindAgentur.

Hofstede, G. 2011. Dimensionalizing Cultures: The Hofstede Model in Context. Online readings in psychology and culture. International association for cross-cultural psychology. Unit 2, Subunit 1. https://academic. microsoft.com/paper/2133535394/citedby/search?q=Dimensionalizing%20Cultures%3A%20 The%20Hofstede%20Model%20in%20Context&qe=RId%253D2133535394&f=&orderBy=0.

Hofstede Centre. 2020. Cultural dimensions of Austria, Retrieved November 20, 2020, from https://www. hofstede-insights.com/country/austria/.

Kant, I. 1787/1956. Kritik der reinen Vernunft. Hamburg, Felix Meiner Verlag.

Katzmann, W. 1986. 'Die Wiener Deklaration' und die ARGE. *In*: ARGE Umwelterziehung (ed.). Eine Zwischenbilanz. Wien, ARGE Umwelterziehung.

Klenk, G. 1987. Umwelterziehung in allgemeinbildenden Schulen. Entwicklung, Stand, Probleme. Frankfurt, Haag und Herschen.

Kyburz-Graber, R., P. Posch and U. Peter. 2003. Challenges in Teacher Education—Interdisciplinarity and Environmental Education. Innsbruck-Wien-München-Bozen, StudienVerlag.

Lechner, C. and F. Rauch. 2014. Quality criteria for schools focussing on education for sustainable development (ESD). pp. 65–76. *In*: Rauch, F., A. Schuster, T. Stern, M. Pribila and A. Townsend (eds.). Promoting Change Through Action Research. International Case Studies in Education, Social Work, Health Care and Community Development. Rotterdam, Sense.

Lieschke, M. 1993. Ziel erreicht? Umwelterziehung 4/5: 6–9.

Lukesch, R., H. Payer, G. Pfaffenwimmer and P. Posch. 2009. Emerging partnerships between school and community—Results from a pilot study in Austria. *In*: Czippán, K., A. Varga and F. Benedict (eds.). Collaboration and Education for Sustainable Development. Case Studies Collected by SUPPORT Members 2008–2010. Oslo, Norwegian Directorate of Education and Training.

Oberrauch, A., L. Keller, M. Riede, S. Mark, A. Kuthe, A. Körfgen and J. Stötter. 2015. 'k.i.d.Z.21 – kompetent in die Zukunft' – Grundlagen und Konzept einer Forschungs-Bildungs-Kooperation zur Bewältigung der Herausforderungen des Klimawandels im 21. Jahrhundert. GW-Unterricht. 139: 19–31.

Oberwimmer, K., D. Baumegger and S. Vogtenhuber. 2019. Indikatoren A: Kontext des Schul-und Bildungswesens. pp. 25–48. *In*: Oberwimmer, K., S. Vogtenhuber, L. Lassnig and C. Schreiner (eds.). Nationaler Bildungsbericht Österreich 2018. Graz, Leykam.

Pfaffenwimmer, G. 2004. ENSI Study: Quality criteria for ECO school development. Austrian Report. Austrian Federal Ministry for Education, Science and Culture. *In*: Mayer, M. and F. Mogensen (eds.). Environmental Education ECO-schools: Trends and Divergences, A Comparative Study on ECO-school Development Processes in 13 Countries. Wien, BMUKK.

Posch, P., F. Rauch and I. Kreis (eds.). 2000. Bildung für Nachhaltigkeit. Studien zur Vernetzung von Lehrerbildung, Schule und Umwelt. Innsbruck, StudienVerlag.

Rauch, F. and R. Steiner. 2006. School development through education for sustainable development in Austria. Environmental Education Research 12(1): 115–127.

Rauch, F. and M. Dulle. 2012. Auf dem Weg zu einer nachhaltigen Schulkultur—15. Jahre ÖKOLOG-Programm, 10 Jahre Netzwerk ÖKOLOG. Wien, BMUKK.

Rauch, F. and R. Steiner. 2013. Competences for education for sustainable development in teacher education. CEPS-Journal (Centre for Educational Policy Studies Journal) 3(1): 9–24.

Rauch, F. and G. Pfaffenwimmer. 2015. Education for sustainable development in Austria. Networking for education. pp. 157–176. *In*: Mathar, R. and R. Jucker (eds.). Schooling for Sustainable Development in Europe. Dortrecht, Springer.

Rauch, F. and R. Steiner. 2015. BINE: Professional development ESD course for higher education teachers, Austria. pp. 114–119. *In*: Kapitulcinova, D., J. Dlouha, A. Ryan, J. Dlouhy, A. Barton, M. Mader, D. Tilbury, I. Mula, J. Benayas, D. Alba, C. Mader, G. Michelsen and K. Vintar Mally (eds.). Leading Practice Publication: Professional Development of University Educators on Education for Sustainable Development in European Countries. Prague, Charles University in Prague.

Rauch, F. 2016. Networking for education for sustainable development in Austria: The Austrian ECOLOG-schools programme. Educational Action Research 24(1): 34–45. (DOI: 10.1080/09650792.2015.1132000).

Rauch, F. and G. Pfaffenwimmer. 2018. ENSI pillars—Action research and dynamic qualities. pp. 40–45. *In*: Affolter, C. and A. Varga (eds.). Environment and School Initiatives. Lessons from the ENSI Network—Past, Present and Future. Wien, Environment and School Initiatives and Budapest, Eszterhazy Karoly University.

Rauch, F. and G. Pfaffenwimmer. 2020. The Austrian ECOLOG-schools programme—networking for environmental and sustainability education. pp. 85–102. *In*: Gough, A., J. Chi Kin Lee and E. Po Keung Tsang (eds.). Green Schools Globally: Stories of Impact for Sustainable Development. Dortrecht, Springer.

Reardon, B.A. 2010. Human Rights and Human Rights Learning as a Vehicle for the Renewal of the University. Klagenfurt, MS.

Schober, B. 2019. Bildung in der Welt von übermorgen: Herausforderungen und Chancen aus der Perspektive der Bildungssoziologie. pp. 498–505. *In*: Breit, S., F. Eder, K. Krainer, C. Schreiner, A. Seel and C. Spiel (eds.). Nationaler Bildungsbericht Österreich, Band 2. Graz, Leykam.

Singer-Brodowski, M. 2016. Transformatives Lernen als neue Theorie-Perspektive in der BNE. pp. 130–139. *In*: Umweltdachverband GmbH (ed.). Jahrbuch Bildung für nachhaltige Entwicklung – Im Wandel. Wien, Forum Umweltbildung im Umweltdachverband.

Statistik Austria. 2020a. http://www.statistik.at/.

Statistik Austria. 2020b. Agenda 2030 für nachhaltige Entwicklung in Österreich – SDGIndikatorenbericht. http://statistik.at/wcm/idc/idcplg?IdcService=GET_NATIVE_FILE&Revision SelectionMethod=LatestReleased&dDocName=122802.

Umweltbundesamt (ed.). 2020. Umweltschutzbericht 2020. Zusammenfassung. Vienna. https://www. umweltbundesamt.at/fileadmin/site/publikationen/rep0738bfz.pdf.

UniNEtZ SDG 4. 2019. Positionspapier. https://www.uninetz.at/beitraege/sdg-4-positionspapier-zum-bildungszusammenhang.

UniNEtZ. 2020. Perspektivenbericht Executive Summary. https://www.uninetz.at/beitraege/perspektivenbericht-executive-summary.

UniNEtZ. 2022. UniNEtZ–Optionenbericht: Österreichs Handlungsoptionen zur Umsetzung der UN-Agenda 2030 für eine lebenswerte Zukunft. https://www.uninetz.at/optionenbericht.

United Nations. 2015. Transforming our world: the 2030 agenda for sustainable development. https://www.un.org/ga/search/view_doc.asp?symbol=A/RES/70/1&Lang=E.

UNESCO. 2014. Roadmap for implementing the global action programme on education for sustainable development. http://unesdoc.unesco.org/images/0023/002305/230514e.pdf.

UNESCO. 2020. Education for Sustainable Development. A Roadmap. ESD for 2030. Paris, UNESCO.

Chapter 7

Environmental and Sustainability Education in the Czech Republic
Between Theory and Practice

Jan Činčera

Introduction

Environmental and sustainability education (ESE) in the Czech Republic has been shaped by the dramatic historical events of the 20th century. After having been founded as Czechoslovakia (in 1993, the country was split into the Czech Republic and the Slovak Republic) in 1918, the country went through a 20-year period of democracy, followed by more than four decades of oppression. And since 1989, it has successfully transformed into a democracy again. These events led to some aspects of ESE being promoted and other compromised. In addition to dealing with the specifics of the Czech experience, this chapter raises several broader questions that can be easily transferred to the situation in other countries: How do we develop ESE research when starting from the very beginning? How do we facilitate communication between theory and practice? How do we promote an international transfer of experience between countries with different cultural and educational contexts?

Masaryk University, Faculty of Social Studies, Department of Environmental Studies, Brno, Czech Republic.
Email: cincera@mail.muni.cz

The Czech Republic's political, economic, cultural, and social contexts and circumstances

Socio-demographic issues

The Czech Republic lies in Central Europe, bordering Austria, Slovakia, Poland, and Germany. With ten million inhabitants and covering an area of less than 79,000 square kilometers, it is a relatively small but economically developed country (Prague Castle 2021). Historically, the Czech territories were, at various points in time, a medieval kingdom, part of a multi-ethnic empire, an independent democratic republic (1918–1938), an occupied territory during WW II, a totalitarian Communist state (1948–1989), and, finally, since 1989, a democracy again.

Historically, the Czech Republic has been the subject of cultural influence from both the East and the West, and the Czech people tend to see their country as being 'in the middle,' not exactly in the West, but also not in the East. The narrative characterizing the country as a 'bridge' has emerged repeatedly throughout Czech history. Nevertheless, in the early 21st century, the country successfully joined NATO and became a member of the European Union. Despite its pro-Western political orientation, the Czech Republic has maintained certain nationalist policies, such as a strong anti-immigration policy and a preference for preserving the Czech national currency.

The Czech Republic is a green country with a temperate climate. Approximately 34% of its territory is forested. The country has a well-developed nature conservation system, with four national parks, 24 protected landscape areas, and 208 national nature reserves and national nature monuments (Nature Conservation Agency of the Czech Republic 2020). The severe pollution and environmental degradation in the final years of the Communist regime were among the reasons for the rising social dissatisfaction at the time, and environmentalism was one of the themes of the revolution in 1989 (Vaněk 1996). Thus, nature conservation was promoted in the early 1990s, and consequently the quality of the environment improved rapidly.

People and the environment

For the most part, Czech people enjoy outdoor activities and support environmental protection. They also generally take a responsible approach to recycling and are in favor of applying economic tools to better protect the environment. Interestingly, they are more concerned about global environmental problems than local issues. In comparison with other Europeans, Czech people are more skeptical about their ability to help the environment, and they are not willing to pay much money for environmental protection (Krajhanzl et al. 2018).

Czech people have also been largely skeptical about sustainability challenges, particularly human influence on climate change. However, the recurrent droughts in the final years of the second decade of the 21st century have led to a gradual change in public attitudes. In 2015, the Czech population was almost equally divided, with about half the people considering themselves 'skeptics' or 'passive' with regard to climate change (Chabada 2018). Today, most of the population agrees that human behavior needs to be modified in order to protect the climate (Akademie věd České Republiky 2020).

ESE in the Czech Republic: A historical background

ESE origins in the outdoor movement during the first republic

The roots of ESE in the Czech Republic can be traced back to the first democratic period in 1918–1938. Two trends were notable at that time. The first was a growing interest in outdoor learning. Since the beginning of the 20th century, outdoor movements such as scouting, woodcraft and others had become increasingly popular (Pecha 1999). Most young Czechs were involved in a variety of youth clubs; many of these operated on the basis of principles of self-organization inspired by popular youth journals. Jaroslav Foglar, a highly influential non-formal educator, published books and cartoons with adventure stories about groups of young boys, which encouraged them to do the 'right thing' and learn from nature. Foglar's work became an inspiration for similar youth clubs throughout the 20th century (Jirásek and Turčová 2017).

Formal education also included a range of approaches that were relevant to the ESE of the future. Eduard Štorch, a popular author of adventure and historical youth fiction, for example, founded the 'school farm,' an outdoor school on one of the river islands in Prague. Here, students 'learned by doing'—that is, by exploring nature or participating in practical projects (Štorch 1929). While Štorch associated his approach with an attempt to develop students' intra- and inter-personal competence, his work also encouraged students' interest in nature and in engaging with the natural world.

At the same time, the ideas of progressivist pedagogy (known as pragmatic pedagogy) were being discussed and promoted by some Czech educators in an attempt to challenge the prevailing, rather top-down and theoretical traditions of teaching. However, despite this approach being tested in selected schools, it was abandoned before WW II and never revived in the 20th century (Kasper and Kasperová 2007).

ESE in the Communist period

In the Communist period (1948–1989), formal education took a top-down, centralized approach in the context of which community-based activities and any kind of critical approach to environmental issues were largely impossible. However, non-formal education retained some of the principles of outdoor experiential learning, albeit in a controlled and partially re-oriented form. This led to the emergence of the outdoor education movement at the Czech Experiential Learning Centre (Prázdninová škola Lipnice) in the 1970s. The movement designed a new approach to learning that made intensive use of educational games and outdoor activities to further inter- and intra-personal development. This approach then spread to the other segments of non-formal education in the country (Jirásek and Turčová 2020).

The on-going degradation of the natural environment caused by severe industrial pollution of air and water led to the foundation of the Czech conservation movement. In the 1980s, two outdoor conservation organizations, the Czech Union for Nature

Conservation and the Brontosaurus Movement, organized leisure-based conservation events, often accompanied by non-formal educational activities. As open criticism of environmental policies was prohibited at that time, these organizations mostly focused on developing an affinity with nature, understanding ecological concepts, and promoting simple conservationist behavior (Vaněk 1996, Máchal 2000).

The spread of ESE in the post-1989 democratic period

The democratic revolution in 1989 provided the opportunity for rapid growth in non-formal environmental education (EE) in the Czech Republic. In the early 1990s, a new type of organization, environmental education centers or EECs, emerged from the outdoor and conservation education background. Apart from non-formal activities (such as voluntary conservation clubs, etc.), EECs began to offer EE programs to schools (Máchal 2000, Kulich 2006, Bureš 2006). As there was almost no tradition of EE except for teaching about ecology, EECs were the major locus of discourse on EE in the Czech Republic in the final decade of the 20th century. Drawing on established Czech tradition, this discourse focused on promoting affinity with nature, learning about ecological concepts, and encouraging simple behaviors to protect the environment. EECs used the game-based methodology and were influenced by the Czech tradition and approaches deployed in some other countries.

While EE was supported by the Czech Ministry of the Environment, it was not adequately represented within national educational institutions. There was no original research and universities were unable to prepare pre-service teachers to do anything beyond science education. This is where Czech EE found itself at the start of the 21st century.

How ESE has been understood in the Czech Republic

The initial understanding of ESE as fact-based and science-oriented

For the above-mentioned historical reasons, concepts of ESE in the Czech Republic were rather limited in the 1990s and for most of the first decade of the 21st century. EE was largely interpreted as 'ecology education,' meaning 'learning about ecology.' As such, it was often understood as a subset of science teaching.

This understanding was further supported by the description of EE in the newly-launched Framework for Education Programs, the national curricula for all types of schools. The Framework included EE as a cross-cutting theme. However, it was defined as focusing mostly on learning about ecological concepts, with a vague promotion of environmental awareness and responsible environmental behavior (Národní ústav pro vzdělávání 2020).

The relationship between ecology education and environmental education was intensively debated, until these debates resolved into a discussion of the differences between environmental education and 'education for sustainable development' (ESD) at the beginning of the 21st century (Machal 2000).

The limited effect of these discussions reflected the lack of academic involvement in this field in the Czech Republic at that time. The discourse was focused mainly on the emerging concepts of EE as they evolved within EECs' practice. However, this approach did not support in-depth theoretical investigation of the field or any great critical reflection. As a result, a few ESE practices spread into schools but, in general, these programs did not align with international trends. Often, the programs were interpreted as EE whilst consisting mainly of transmitting knowledge about plants or animals, with no relationship to environmental concepts. Most of the initial ESE programs were highly instrumental, assigned a strong role to teachers, and had vaguely defined objectives (Cincera 2006a, b).

Attempts to re-conceptualize ESE in the 21st century

ESE in the Czech Republic started to change at the end of the first decade of the 21st century. As part of the effort to re-define EE, two new documents, The Recommended Objectives for the Cross-cutting Theme of Environmental Education (Cincera et al. 2011) and Goals and Indicators for Environmental Education (Broukalova et al. 2012) were launched at the national level. Both documents were strongly influenced by the Guidelines for Excellence published by the North American Association for Environmental Education (NAAEE 2004). Responsible environmental behavior was adopted as the main aim of EE (Hungerford and Volk 1990). Further, both documents were based on international research (Hungerford et al. 1980, Jensen and Schnack 2006) and divided the specific aims and objectives into a number of categories representing the main themes or competences.'

According to these two documents, the main mission of EE was to develop environmental sensitivity, understanding of ecological processes, inquiry skills, sense of place, competence for environmental problems and issues, and students' action competence (Cincera et al. 2011, Broukalova et al. 2012). The authors intentionally preferred the American tradition to the competence-oriented approach presented by some European authors at that time (Jensen and Schnack 2006, Östman and Öhman 2007, Schnack 2009), though the documents still supported the competence-oriented approach to a limited degree.

The reason for this strategy was mainly pragmatic. The prevailing pedagogical tradition in the Czech Republic was rooted in top-down teaching and the prevailing approaches to ESE reflected fact-based and normative traditions. As a result, there was a high risk that if the new documents were based mainly on a pluralistic approach and emancipatory learning (Wals 2012), they would not be accepted in practice. Because of this, the documents took a pragmatic approach, supporting gradual transformation. Although the documents radically challenged the dominant, fact-based, science-focused tradition of EE, they still allowed schools to find ways to provide ESE that best suited their particular circumstance.

Today, the relationship between EE and ESD in the Czech Republic remains uncertain. According to the prevailing, if not fully accepted, understanding, ESD is a broader term than EE. However, the differences between ESD and EE are often difficult to identify in practice. The current focus on promoting civic competence in EE, the need to develop climate-change education, and the growing cooperation of organizations interested in global development education within the EE/ESD community call for further re-conceptualization of the national interpretation of the field. One of the options is to adopt the concept of ESE as an umbrella term for these two linked fields.

Implementing ESE in the Czech Republic

Although originally rooted in non-formal education, EECs, as we can see, were crucial to the development of ESE in the Czech Republic. However, as ESE has become more widespread, the current situation is being shaped by the interaction of a number of key players.

EECs and non-formal education: the main players

EECs have remained the leading stakeholders when it comes to delivering ESE in the Czech Republic. The group of Czech EECs mainly comprises non-profit organizations, with a few centers managed directly by their municipality, their region, or other institutions such as zoological gardens. In 1992, the most active centers founded the Pavučina network (The Web), an umbrella organization providing support and facilitating cooperation between centers. In 2020, this network consisted of 44 members and two observers (SSEV Pavučina 2020). Together, network members offer approximately 5,000 programs that are under eight hours and are delivered to around 100,000 students, and roughly 100 longer, residential programs delivered to about 3,000 students every year (SSEV Pavučina 2015). More than 60% of schools in the Czech Republic use some of the programs offered by these EECs (SSEV Pavučina 2009). However, it is estimated that at least 150 EECs provide ESE programs in the Czech Republic (Kulich et al. 2009).

Of the other non-formal education organizations, Junák—the association of the Czech Scouts and one of the largest non-formal youth movements—is highly active with regard to ESE. In 2008, Junák thoroughly re-worked its educational objectives, adding a strong emphasis on EE (Klapste 2008). In 2019, the organization launched a discussion about how to support climate change education in its activities (Junák 2015). Since then, Junák has organized climate change events for its members and for the wider public.

Apart from implementing ESE in practice, Czech EECs actively cooperate with the Ministry of the Environment which provides them with support for ESE and helps them organize an annual national EE conference.

The development of ESE in Czech universities

The role of Czech universities in ESE has grown from very limited to reasonably significant. In 2006, the first ESE-related article affiliated with a Czech university was published in an international journal with impact factor. That same year, the Charles University Environment Center launched the first Czech peer-reviewed ESE journal, Envigogika.

Today, Czech research on ESE is relatively strong in the Central European region, but remains proportionally smaller in scope when compared with ESE research in Austria or Germany (see Figure 1).

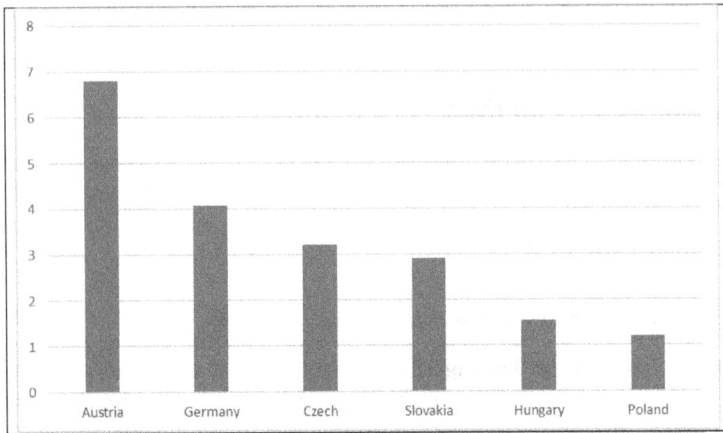

Figure 1. Number of published articles in journals with impact factor per 1 million inhabitants in 2017–2021 (January 03, 2022). Source: Web of Knowledge.[1]

Czech ESE research has focused mainly on the evaluation of formal and non-formal ESE programs and also the evaluation of ESD at universities. While several universities have become involved in this research, the greatest number of published papers are affiliated with Masaryk University (Brno), the Environment Center of Charles University (Prague), and the University of J. E. Purkyně (Ústí nad Labem). These universities are also the main academic leaders in the Czech Republic with regard to ESE. Since 2017, these institutions have held an annual conference to promote cooperation among ESE-oriented Czech universities and other stakeholders. The CeSFER network (Czech and Slovak Forum for Environmental and Sustainability Education) was also founded in 2017, in order to facilitate communication between stakeholders.

Despite these successes, the impact of the Czech ESE community is still lagging behind expectations. The reasons for this slower development include the limited number of Czech ESE researchers and the institutional barriers to changing curricula for trainee teachers. However, the situation has been gradually improving, and the universities are increasingly playing a social role in shaping ESE in the Czech Republic.

Formal education: what changes are needed most

EE is a mandatory part of the curricula for all types of Czech schools. However, its definition in the national curricula document reflects the older, science-education-based approach. This situation is reflected in the way ESE is being implemented in Czech schools. Most elementary schools engage well with the teaching of ecological concepts, but only a small number of schools enable

---

[1] TS = "environmental education" OR TS = "education for sustainable development" OR TS = "environmental and sustainability education" OR TS = "education for sustainability" and Articles (Document Types) and Articles (Document Types).

students to learn about these concepts outdoors. Further, while most students participate regularly in outdoor courses, fewer than 20% of schools encourage them to investigate sustainability issues and participate in sustainability activities in the community. Similarly, even though almost all elementary schools introduce students to the idea of their personal impact on the environment, such as through recycling or energy-saving, fewer than 10% of schools teach students about public participation in decision-making, such as public hearings or local referendums. Moreover, the participative approach is not widely used in teaching and the top-down approach remains the most widespread form of formal ESE (Cincera et al. 2016).

Currently, ESE practices in elementary and secondary schools are very diverse. Although a minority of schools provide exceptional ESE, the majority retain the outdated approach. Some whole-school programs offered by ESE, such as EcoSchool, GLOBE, or the place-based School for Sustainable Development program, are playing a very positive role in the transformation of formal education. However, their implementation brings specific challenges, notably with regard to the interpretation of the participative or community-based approaches such programs assume (Cincera et al. 2019a, b, c).

Government agencies and municipalities: the supporters of ESE

Czech government agencies and municipalities are mostly supportive of ESE in cases when the Ministry of the Environment plays the leading role. The regular support of the Ministry is one of the main reasons for the increase in EECs and for maintaining their services over the last thirty years. Besides financial support, the Ministry initiates research into ESE and facilitates the Council for ESD which is made up of members of EECs and university teachers.

At the same time, the Ministry of Education, Youth, and Sport is considerably less supportive of ESE, but does not directly discourage it. The need to redefine the strategic goals for education in the Czech Republic led to the initiation of a national discussion on the Strategy for Educational Policy 2030+; its draft version proclaims the importance of meeting contemporary global challenges through education, and promotes the participative approach and the development of students' citizenship competences. Nevertheless, the role of ESE in future national curricula remains unclear, although it has been elaborated to a certain extent.

Emerging issues and trends

Interaction between the main ESE players brings a multitude opportunities but it also brings challenges for ESE. Several issues have emerged from this interaction and remain topical, including quality and evaluation.

The issue of quality

The prevailing practice-oriented approach to designing and implementing ESE in the decade following 1989 led many educational programs to espouse active learning methods, but it also left unresolved questions regarding the quality of such programs (Cincera et al. 2009, Cincera 2006a, b). These questions began to be discussed in the early years of the 21st century. With the growing number of EECs and limited (if long-term) support from the public sector, EECs soon felt the need to lay down some quality criteria for their programs. The aim was both to stimulate the professional growth of the EEC community and to defend its services from the criticism that was being levelled at it by a segment of the broader educational community at that time.

EECs therefore gradually published a set of internal documents defining and assessing the quality of their programs, titled the Eight Points of the Eco-educator and the Assessment Table for Environmental Education Programs (published 2006–2008). While both these documents were designed by the EECs, they reflected a somewhat limited conceptual understanding of the field and indirectly supported the prevailing practice of providing short programs lasting only several hours rather than, for example, facilitating community-based projects. Today, both these documents are considered outdated, and new guidelines setting out quality standards for ESE programs are being prepared.

At the same time, the large number of EECs meant there was a need to further differentiate their quality in and of themselves. The solution adopted in 2013 was a national system of EEC certification, based on external evaluation of centers, which is conducted by certified auditors. The auditors evaluate the quality of EECs' programs, environmental management and human resources, and then decide whether to certify them. The criteria for certification were developed by the EECs, and the auditors are experienced ESE leaders (SSEV Pavučina 2020b).

The issue of evaluation

The gradual development of an influential group of ESE scholars in the Czech Republic has generated a fruitful but challenging process of cooperation with the EECs. While EECs have dealt with the issue of quality since the early 21st century, they have tended to stick to their own definitions of it, which are based on their own understanding of the field. Thus, EECs were at first rather skeptical regarding the role of ESE academics, criticizing what they saw as outdated methods of pre-service teacher training at universities.

When Czech academics started conducting research to evaluate the ESE programs offered by EECs, the EECs were ambivalent. Some EECs welcomed the research, which included external critical analyses of their programs, and saw the confrontation of their practice with ESE theory as an opportunity for professional growth, adopting new approaches and differentiating their services from those of other centers. Periodic evaluation of the EcoSchool program, for example, helped to transform the program's main approach from an instrumental, top-down one to a participative one in which teachers do not direct but facilitate the activities of their students. The positive impact of this transformation was reflected both in students' increased satisfaction with the program and the program's promotion of student empowerment (Cincera et al. 2019).

However, other EECs remained skeptical of theory that was not rooted in their practice. Some experienced program leaders expressed concern about what they interpreted as an attempt to 'measure the unmeasurable' or the risk of destroying spontaneous and authentic ESE practices by the forced adoption of theory-driven methods (Bartos 2010, Cincera 2013). Others were satisfied with their current practices and the status quo and felt threatened by critical reflection from outside of the EEC community (Cincera 2013).

As the need for evaluation gained support from the Ministry of the Environment, most EECs developed more advanced methods for analyzing their programs based on either internal or external evaluation. Nevertheless, even though program evaluation has now become common practice, the field still struggles with a shortage of experienced external evaluators. Cooperation among EECs remains open and positive, but is limited by the scarcity of academics.

Other challenges: extending existing networks, promoting new approaches

As discussed above, the Czech ESE community originated in a non-formal outdoor movement that developed into a well-functioning network of EECs. While the Pavučina network (The Web) played a crucial role in shaping the national discourse on ESE and assisted communication and cooperation among its members, the prevailing practice-based approach was also a limiting factor for further qualitative development of the ESE field. One way of overcoming this limitation is to develop and link the existing networks of the country's various ESE stakeholders. Although the initial steps have already been taken (founding the CESFeR; organizing a regular national conference), this task remains a challenge for the future.

Simultaneously, the Czech ESE community has been gradually opening up to the international ESE community and to international networks. Good examples of this are the membership of some Czech EECs in international organizations (e.g., The Foundation for Environmental Education), EECs' participation in international projects (the Real World Learning project is an example of international cooperation among EECs and universities), and their involvement in international ESE research and in preparing an international joint Masters degree in ESE.[2] The Czech Republic is also organizing and hosting the 11th World Environmental Education Congress (WEEC).

At the same time, the transfer of some ESE approaches developed in different socio-cultural environments has proved challenging. As already mentioned, Czech elementary schools are still not making widespread use of the participative/emancipatory approach or involving students in the process of investigating and developing solutions to sustainability issues in their communities, but the situation is slowly improving. The main barriers are the outdated national ESE curriculum and the inadequacy of pre-service teacher training at most universities.

Finally, the urgency of climate change makes it necessary to dramatically challenge the ways ESE is implemented in formal education. In response, a new working group has been initiated, consisting of university teachers, activists, EEC members, administrators, and teachers. This initiative aims to promote cooperation

---

[2] The program is being prepared by a consortium comprising Masaryk University (Czech Republic), the University of Vechta (Germany), the University of Klagenfurt (Austria), and Karlstad University (Sweden).

among the various stakeholders, facilitate the exchange of experience with other countries, and transform schools' approaches towards sustainability.

Meeting these challenges is the main task for the Czech ESE community in the years to come.

Conclusion

The story of ESE in the Czech Republic is a story of transition from an enthusiastic but limited movement to a well-established field. The main factor driving this process has been the dynamic relationship between EECs, the strongest ESE players in the country and the creators, along with other stakeholders, of the first national ESE discourse after 1989. In many respects, this relationship has shaped some of the country's specific ESE features, such as the strong emphasis on outdoor education, the predominance of the pragmatic approach in Czech research, the repeated efforts to define the quality of ESE, and the primary theme of the interrelationship between theory and practice.

While we cannot predict the future, the process of shaping these issues will likely define ESE in the Czech Republic for years to come. The emerging answers may lead Czech ESE to make an original contribution to ESE in other countries.

Acknowledgments

This chapter has been supported by the project MUNI/A/1201/2019, 'Cesty k udržitelné a spravedlivější společnosti – environmentalní přístupy' ['Paths toward Sustainable and More Just Society – Environmental Approaches'].

References

Akademie věd České republiky. 2020. Postoje české veřejnosti ke změnám klimatu na Zemi – tisková zpráva. https://cvvm.soc.cas.cz/media/com_form2content/documents/c2/a5249/f9/oe200805.pdf.

Bartoš, M. 2010. O ekologické výchově s láskou. Sluňákov, Olomouc. http://www.slunakov.cz/index.php?1-1004x461.

Broukalová, L., Broukal, V., Činčera, J. Daniš, P., Kažmierski, T., Kulich, J., Lupač, M., Medek, M. and Novák, M. 2012. Cíle a indikátory pro environmentální vzdělávání, výchovu a osvětu v České republice. Ministerstvo životního prostředí České republiky, Praha. http://www.mzp.cz/cz/cile_indikatory_evvo_dokument.

Bureš, J. 2006. Historie pavoučí sítě. Bedrník. 4(2): 15–16.

Chabada, T. 2018. Postoje a chování ke změně klimatu. Jak je rozdělena česká společnost? Konference Sociální rozměry klimatické změny: globální problémy v lokálních souvislostech.

Činčera, J. 2006a. Osm problémů environmentální výchovy. Bedrník - časopis pro ekogramotnost. 4(1): 18–20.

Činčera, J. 2006b. Problémy a příležitosti environmentální výchovy v České republice [online]. Envigogika 1: 1. http://www.czp.cuni.cz/envigogika.

Činčera, J., J. Kulich and D. Gollová. 2009. Efektivita, evaluace a podpora programů environmentální výchovy. Envigogika: Charles University E-journal for Environmental Education 4: 2. http://envigogika.cuni.cz/index.php/Envigogika/article/view/39/pdf_39.

Činčera, J., K. Jančaříková, P. Šimonová and K. Volfová. 2011. Environmentální výchova. In: Pastorová, M. (ed.). Doporučené očekávané výstupy. Metodická podpora pro výuku průřezových témat na základních školách. Výzkumný ústav pedagogický, Praha. http://www.vuppraha.cz/nova-publikace-divize-vup-%E2%80%93-doporucene-ocekavane-vystupy-pro-zakladni-skoly.

Činčera, J. 2013. Střediska ekologické výchovy mezi teorií a praxí. BEZK, Agentura Koniklec a Masarykova univerzita, Brno.

Činčera, J., K. Jančaříková, T. Matějček, P. Šimonová, J. Bartoš, M. Lupač and L. Broukalová. 2016. Environmentální výchova z pohledu učitelů. BEZK, Agentura Koniklec a Masarykova univerzita, Brno.

Cincera, J., J. Boeve-de Pauw, D. Goldman and P. Simonova. 2019a. Emancipatory or instrumental? Students' and teachers' perception of the EcoSchool program. Environmental Education Research 25(7): 1083–1104.

Činčera, J., R. Kroufek, K. Marková, Š. Křepelková and P. Šimonová. 2019b. The GLOBE program: what factors influence students' and teachers' satisfaction with science education. Research in Science & Technological Education, Latest Articles.

Činčera, J., B. Valešová, Š. Doležalová Křepelková, P. Šimonová and R. Kroufek. 2019. Place-based education from three perspectives. Environmental Education Research 25(10): 1510–1523.

Hungerford, H.R., B.R. Peyton and R.J. Wilke. 1980. Goals for curriculum development in environmental education. The Journal of Environmental Education 11(3): 42–47.

Jensen, B.B. and K. Schnack. 2006. The action competence approach in environmental education. Environmental Education Research 12(3): 163–178.

Jirásek, I. and I. Turčová. 2017. The Czech Approach to Outdoor Adventure and Experiential Education: The Influence of Jaroslav Foglar's Work. The Czech approach to outdoor adventure and experiential. Journal of Adventure Education and Outdoor Learning 17(4): 321–337.

Jirásek, I. and I. Turčová. 2020. Experiential pedagogy in the Czech Republic. pp. 8–18. In: Parry, J. and P. Allison (eds.). Experiential Learning and Outdoor Education. Traditions of Practice and Philosophical Perspectives. Routledge, New York.

Junák – český skaut. 2015. Otevřený dopis a výzva náčelnictvu a výkonné radě Junáka ve věci klimatické krize. https://krizovatka.skaut.cz/clanky-na-webu/4689-otevreny-dopis-a-vyzva-nacelnictvu-a-vykonne-rade-junaka-ve-veci-klimaticke-krize?autologin=1.

Kasper, T. and D. Kasperová. 2007. Dějiny pedagogiky. Grada, Praha.

Klápště, P. 2008. Příroda kolem nás. Metodika ke skautské stezce. Junák, Praha.

Krajhanzl, J., T. Chabada and R. Svobodová. 2018. Vztah české veřejnosti k přírodě a životnímu prostředí. Reprezentativní studie veřejného mínění. Masarykova univerzita, Brno.

Kulich, J. 2006. Co jsou a kde se vzala střediska ekologické výchovy, ekocentra, ekologické poradny. Bedrník 4(2): 12–14.

Kulich, J., Činčera, J., Kunssberger, D., Lupač, M., Činčera, P. and Gollová, D. 2009. Analýza potřebnosti a využití environmentálních vzdělávacích center na území České republiky. Hlavní zjištění 2009. SEVER, Agentura Koniklec and BEZK, Horní Maršov.

Máchal, A. 2000. Praktická ekologická výchova. Rezekvítek, Brno.

NAAEE. 2004. Excellence in Environmental Education - Guidelines for Learning (PRE K-12). North American Association for Environmental Education, Washington. http://www.naaee.org/npeee/learner_guidelines.php.

Národní ústav pro vzdělávání. 2020. Rámcové vzdělávací programy. NUV, Praha.

Nature Conservation Agency of the Czech Republic. 2020. https://www.ochranaprirody.cz/en/.

Öhman, J. 2009. Sigtuna think piece 4 climate change education in relation to selective traditions in environmental education. Southern African Journal of Environmental Education 26: 49–57.

Östman, L. and J. Öhman. 2007. Selective Traditions within Environmental Education. WEEC, Durban.

Pecha, D. 1999. Woodcraft. Lesni moudrost a lesni bratrstvo. Votobia, Olomouc.

Prague Castle. 2021. About the Czech Republic. https://www.hrad.cz/en/czech-republic/about-the-czech-republic.

Schnack, K. 2009. Action Competence, Conflicting interests and Environmental Education. Aarhus University, Copehagen. http://www.MUVIN.net.

SSEV Pavučina. 2009. Analýza stavu environmentálního vzdělávání, výchovy a osvěty. SSEV Pavučina, Praha.

SSEV Pavučina. 2015. Činnost členských organizací. http://www.pavucina-sev.cz/?idm=14.

SSEV Pavucina. 2020a. Kdo jsme. http://www.pavucina-sev.cz/rubrika/58-kdo-jsme/index.htm.

SSEV Pavucina. 2020b. Cíl certifikace. https://www.certifikace-sev.cz/certifikace/.

Štorch, E. 1929. Dětská farma. http://www.archiv.prirodniskola.cz/knihy-publikace/Detska%20farma%20
    -%20Eduard%20Storch.pdf.
Vaněk, M. 1996. Nedalo se tady dýchat: ekologie v českých zemích v letech 1968–1989. Maxdorf, Praha.
Wals, A. 2012. Learning our way out of unsustainability: the role of environmental education.
    pp. 628–644. *In*: Clayton, S. (ed.). The Oxford Handbook of Environment and Conservation. Oxford
    University Press, Oxford.

Chapter 8
Education for Sustainable Development in Germany

Marco Rieckmann[1,]* and *Mandy Singer-Brodowski*[2]

Introduction

Education for Sustainable Development (ESD) has a long history in Germany. Theoretical examination of ESD and its pilot implementation in the German education system began in the 1990s. But even in the decades before that, environmental education and development education were already playing an increasingly important role in educational research and practice. After an initial pilot implementation of ESD, the UN Decade and the subsequent Global Action Programme on ESD have led to more extensive integration of ESD into the German education system. In educational research, too, ESD has increasingly established itself as a field of research in its own right.

The chapter first presents the contexts and circumstances of the education system in Germany and the institutional and organizational principles of this education system before going on to describe the historical background and development of ESD and the understanding of ESD in the country. Against this background, the main part of the chapter then describes and discusses the implementation of ESD in the different sectors of the German education system. Finally, the chapter highlights and discusses some emerging issues and trends, and current and future needs with regard to ESD in Germany.

---
[1] University of Vechta, Germany.
[2] Freie Universität Berlin, Germany.
Email: s-brodowski@institutfutur.de
* Corresponding author: marco.rieckmann@uni-vechta.de

The education system in Germany: Contexts and circumstances

Currently, there are 83.7 million people living in Germany,[1] which makes it the country with the highest population in the European Union. Education is compulsory for young people from the age of 6 to (usually) the age of 18. Based on the International Standard Classification of Education (ISCED), it includes voluntary early childhood education (0–6), compulsory primary education (6–10 or 6–12), compulsory lower and upper secondary education (10–18 or 12–18), which may comprise some vocational activities in the later years and tertiary education, which is divided into academic education and vocational education and training.[2]

In political terms, the structure of the educational system is federal, whereby the 16 federal states (Länder) have prime responsibility for financing, administering and developing educational activities. The federal states also author and revise school curricula. The key organisation for formal education issues (school education, and to some extent vocational education and training and higher education) is therefore the Standing Conference of the Ministers of Education and Cultural Affairs of the Länder in the Federal Republic of Germany (Kultusministerkonferenz, KMK), where the 16 federal states negotiate and align their educational policies. Local authorities, in particular, are accountable for the buildings of educational organisations, and for managing their material resources.

In addition to the state organisations, a variety of non-governmental organisations (NGOs) are involved in the German educational system. This is particularly the case for early childhood education, where NGOs such as church associations are responsible for day care facilities and are funded by the state to supply kindergarten sites for young children. NGOs are also involved in non-formal learning— here, environmental or social organisations run a range of educational activities and programmes. Their funding comes from participant fees (environmental or developmental), ministries and foundations. The Federal Ministry of Education and Research (BMBF) does not usually fund individual educational organisations, but supports research and development projects that contribute to quality improvements in the education system more generally. In this context, it funds the Monitoring of ESD[3] research project, which is establishing a broad knowledge base about the implementation of ESD in Germany.

Educational reports in Germany show an overall trend towards participation in education across the whole lifespan of people, although this trend has slowed in recent years for a number of reasons (Autorengruppe Bildungsberichterstattung 2022). Though the boundaries between the different educational institutions are becoming more permeable, the economic and social background of the family continues to influence the chances of young people gaining any given qualification

---

[1] https://www.destatis.de/EN/Themes/Society-Environment/Population/Current-Population/_node. html;jsessionid=2FBF0C95AAF8D88894AB4E62A1788DB9.live732.

[2] http://uis.unesco.org/en/country/de and https://www.datenportal.bmbf.de/portal/en/G294.html.

[3] https://www.ewi-psy.fu-berlin.de/einrichtungen/weitere/institut-futur/Projekte/ESD_for_2030---English/index.html.

(*ibid.*). With regard to families who have migrated to Germany in recent years and decades, the most pressing issue for participation in education is the achievement of a certain level of competence in the German language.

Historical background of education for sustainable development in Germany

From environmental education to education for sustainable development

Germany has a long tradition of educational approaches involving the environment and the natural world, such as the progressive education movement, for instance. However, environmental education did not develop as a standalone concept in Germany until the early 1970s. Three phases of environmental education can be distinguished (Gräsel 2010, based on Michelsen 1998):

1. *Initial phase*: In the 1970s, the first consideration was given to environmental education.

2. *Initial realisation and differentiation*: In the 1980s, environmental education began to be implemented in educational institutions, model projects were implemented and different currents and approaches to environmental education began to develop (such as nature education, instrumental environmental education, eco-pedagogy, ecological learning, Table 1).

3. *Development of ESD*: In the 1990s, following the Rio Conference, and on the basis of Chapter 36 of Agenda 21, there was reflection on ESD, which has gradually replaced environmental education as the dominant approach, in line with the international efforts of UNESCO. While environmental education in Germany had mainly dealt with topics of environmental protection and nature conservation, ESD has been understood as an integrative concept that draws links between ecological, social and economic issues. The concept of ESD was also accompanied by a paradigm shift from a threat-based scenario to a modernisation-based scenario.

Table 1. Different approaches of environmental education in the 1980s.

Pedagogical approach	Nature education	Instrumental environmental education	Eco-pedagogy	Ecological learning
Topics	Nature	Environmental protection in everyday life	Societal causes of environmental degradation	Grassroots ecology
Learning objective	Experience of nature	Environmental behaviour	Critical reflection on society	Learning outside educational institutions

Source: Michelsen (1998).

From development education to global learning

In parallel to the concept of environmental education, Germany has also had a discourse on development education for several decades. Development education emerged alongside the institution of a state development policy in the country at the beginning of the 1960s, its task initially being primarily 'to increase public understanding and support for the commitment to development of the state, the church and non-governmental aid organisations' (Seitz 2022: 37, own translation). In the subsequent years and decades, different approaches to development education developed and four main currents emerged; these can be distinguished primarily by their underlying understanding of development (Table 2).

The pedagogy of instruction, which focused on imparting knowledge, was aligned with the modernisation theories that dominated the discourse on international development in the 1960s. The pedagogy of ideological critique emerged in the context of the theories of imperialism that evolved in the first half of the 1970s and had an emancipatory agenda, aiming to raise awareness. Against the background of the dependency theory of the later 1970s, action-focused pedagogical approaches aimed to ensure learners had positive experiences of exercising political influence through action in their own lifeworld (Seitz 2022).

Globalisation and changing perceptions of global development challenges finally led to the emergence of the concept of Global Learning in the early 1990s (*ibid.*, see also Lang-Wojtasik 2022). 'Whereas development education was still focused on the problem of the unfair distribution of global wealth between North and South and the provision of fair compensation' (Overwien and Rathenow 2009: 12, own translation), Global Learning integrates development, human rights, environmental and peace education (Asbrand and Scheunpflug 2005, Overwien and Rathenow 2009, Scheunpflug 2001).

Table 2. Concepts of development education and underlying development theories.

Development problems are seen as...	Backlog	Result of exploitation	Dependent development	Global maldevelopment
Underlying development theory	Modernisation theories	Imperialism theories	Dependency theories	Globalisation/ world society theories
Pedagogical approach	Instruction	Ideological critique	Action focus	Global learning
Learning topics	Third World development deficit	International structures	Lifeworld	Local-global nexus, sustainable development
Learning objective	Knowledge/ Conscience	Critical awareness	Motivation to act	Competence development

Source: Seitz (2022: 38).

Establishment of education for sustainable development

Since the end of the 1990s, education science discourse and educational practice in Germany has seen increasing reference to sustainable development. One important step towards ESD has been the development of the BLK 21 programme. The BLK framework, *Education for Sustainable Development* (BLK 1998), provided important guidelines on the implications of sustainability for educational policy and practice. The theoretical framework of Education for Sustainable Development was developed in the report of the same name (de Haan and Harenberg 1999), which also formed the basis for the BLK 21 programme. This report represented the completion of the paradigm shift from environmental education to ESD. The BLK 21 programme (1999–2004) and the follow-up Transfer-21 programme (2004–2008) began to integrate ESD into school education in Germany (Programm Transfer-21 2008, Rode 2005). BLK 21 and Transfer-21 have both produced results that are fruitful for other areas of education, with a variety of initiatives promoting ESD being introduced in the non-school sector, for example (Dieckmann et al. 2006).

A significant contribution to the introduction of Global Learning in schools was the publication of the first edition of the *Curriculum Framework Education for Sustainable Development* (BMZ and KMK 2007). This document ensured that Global Learning, which had previously been primarily been promoted by NGOs, was more strongly recognized in education policy from 2007 onwards.

Understanding of education for sustainable development in Germany

The concept of Education for Sustainable Development (ESD) combines environmental education, development education, peace education, health education and political education (BLK 1998: 25), using the lens of sustainable development to link the content and focus of all these educational approaches. ESD thus attempts to understand complex interrelationships that cannot be tackled by environmental education or development education alone.

ESD aims to enable all people to contribute to sustainable development, through participation in social learning, communication and design processes. Its goal is therefore to develop key competences that are particularly relevant to sustainable development, but which most people still lack (de Haan et al. 2008, Rieckmann 2021a). It should enable individuals to act in the interests of sustainable development 'if they have corresponding goals, purposes or intentions' (de Haan et al. 2008: 117, own translation). It is therefore about 'opening up possibilities' (*ibid.*: 123, own translation) and not about educating people to behave in a certain way that is supposedly in conformity with sustainability. Learners—as 'sustainability citizens' (Schank and Rieckmann 2019)—should be enabled to think about sustainable development issues and find their own answers. This emancipatory understanding of ESD thus meets Klafki's requirements for general education (Rieckmann 2021b).

As already pointed out, the concept of Global Learning also plays an important role in the current German discourse on ESD. Global learning is understood 'as the pedagogical reaction to the development of the world society' and thus 'reacts to the learning challenges arising from the increasing globalisation of the world' (Scheunpflug 2001: 87, own translation, see also Overwien 2009). The development of the world society results in learning requirements that have a factual dimension (dealing with the simultaneity of knowledge and non-knowledge), a temporal dimension (acceleration and lack of time), a spatial dimension (dissolution of boundaries and interconnection) and a social dimension (familiarity and strangeness) (Lang-Wojtasik 2019, Scheunpflug 2011). In this sense, Global Learning is concerned with 'the formation of personality in the global context' (Seitz 2009: 44, own translation) and 'enabling learners to acquire the competences for life in the world society' (Asbrand and Scheunpflug 2005: 469, own translation).

There are very different views on the relationship between ESD and Global Learning: Parts of the German discourse see ESD as an over-arching concept that includes Global Learning; other parts view ESD as the current focus of Global Learning; and elsewhere ESD and Global Learning are often described as concepts of equal rank that both reference sustainable development. Indeed, even though ESD and Global Learning are rooted in different traditions—environmental education and development education respectively—and to some extent take different approaches to sustainability-related themes and issues, they both seek to promote sustainable development and to provide individuals with the capacity to take action in an increasingly complex world. The aim of both ESD and Global Learning is thus the acquisition of key competences that will enable people to shape (world) society in a sustainable way. This chapter principally uses the term ESD, but also draws on approaches and literature from the field of Global Learning.

The ESD discourse has for some years seen intensive discussion on the key competences individuals should have in order to be able to actively build sustainable development into their own lives and social environments (Rieckmann 2018). In Germany, the most widely used concept in ESD is *Gestaltungskompetenz* (shaping competence), which refers to the ability to 'recognise problems relating to unsustainable development and to successfully apply knowledge about sustainable development' (de Haan et al. 2008: 12, own translation, see also Barth et al. 2007). The concept is made up of a series of sub-competences that have been repeatedly modified and supplemented in recent years. It currently distinguishes between twelve sub-competences (Figure 1).

The concept of *Gestaltungskompetenz* thus covers competences 'that enable people to take forward-looking and autonomous action to co-create sustainable development' (Michelsen 2009: 84, own translation). A further concept that forms part of the discussion and application of ESD in Germany is the competence to make decisions (*Bewertungskompetenz*) (Eggert and Bögeholz 2010, Gausmann et al. 2010).

In Germany, the Curriculum Framework Education for Sustainable Development (KMK and BMZ 2016) is widely used in ESD, but more particularly in the context of Global Learning. It sets down core competences for Global Development Education/ESD under the three headings of recognising, assessing and acting (Table 3).

1. Gathering knowledge in a spirit of openness to the world, integrating new perspectives
2. Forward-looking thinking and acting
3. Acquiring interdisciplinary knowledge and acting accordingly
4. Dealing with incomplete and overly complex information
5. Working with others on decisions
6. Coping with problems and issues arising with regard to individual decision-making
7. Participating in collective decision-making
8. Motivating oneself and others to take action
9. Reflecting upon one's own principles and those of others
10. Espousing the idea of equity in decision-making and planning actions
11. Planning and acting autonomously
12. Showing empathy for and solidarity with the disadvantaged

Figure 1. *Gestaltungskompetenz* sub-competences (de Haan 2010: 320).

Table 3. Core competences for Global Development Education/ESD.

Recognising	Assessing	Acting
• Acquisition and processing of information • Recognising diversity • Analysis of global change • Differentiation between levels of action	• Change of perspectives and empathy • Critical reflection and comment • Evaluation of development projects	• Solidarity and shared responsibility • Understanding and conflict resolution • Ability to act in times of global change • Participation and active involvement

Source: KMK and BMZ (2016: 102)

However, in the German discourse, ESD is not limited to the development of competences; as transformative education, it is also concerned with the 'transformation of the relationship between the individual and the world in a global context' (Scheunpflug 2019: 66, own translation), and thus with conceptual changes, i.e., changes to fundamental orientations (attitudes, values, paradigms and worldviews) (*ibid.*). ESD is thus also expected to contribute to critical discourse on values. It can, and should, provide suggestions to encourage learners to reflect on their own values and take a position in the debate on values en route to sustainable development (Schank and Rieckmann 2019). ESD establishes a direct link between individual and social change and can thus be considered to be transformative education (Scheunpflug 2019).

Even if ESD—as emancipatory education (Rieckmann 2021b)—is not about imparting certain predetermined attitudes, values, paradigms and world-views, it nevertheless takes its cue from the ideas of intra- and inter-generational justice (de Haan et al. 2008). Against this background, it always pursues the value-related goal of facilitating 'sensitisation to the responsibility to survive' (Mokrosch 2009: 38, own translation). At the same time, one of ESD's pedagogical goals is to promote discussion and the 'clarification of values' (*ibid.*: 36, own translation) with regard to sustainable development, in particular the preservation of the natural basis of life, human dignity and justice (Stoltenberg 2009). Understood in this way, ESD can help effect a 'change of values towards sustainability' (WBGU 2011, own translation) without overwhelming learners[4] and indeed by supporting the development of reflective competences.

Competences (and related values) cannot simply be taught but must be developed by learners themselves (Weinert 2001). ESD therefore requires an action-oriented, transformative pedagogy (Rieckmann 2018), characterised by pedagogical principles such as a learner-centred approach, action-oriented learning, reflection, participation, systemic learning, future orientation, and transformative learning (Künzli David 2007, Rieckmann 2018).

Implementation of education for sustainable development in the German education system

The UN Decade of ESD (2005–2014) created remarkable momentum in Germany, strengthening ESD in the education system, particularly by building up professional networks that functioned as centres for disseminating ESD (Rode and Michelsen 2012: 36). Nearly 2000 projects, networks and local authorities were designated as exemplifying good practice at the end of the UN Decade. The subsequent UNESCO Global Action Programme (GAP) (2015–2019) aimed to focus on 'accomplishing the transition from project to structure' (National Committee 2013: 4). In 2015, the Federal Ministry of Research and Education (BMBF) took over the coordination of the political process for implementing ESD, in particular developing a strategy for involving new stakeholders. This set out a structure for implementation, with a National ESD Platform chaired by the BMBF Secretary of State, six expert forums on the different educational areas, and a youth panel (Figure 2). Additionally, the former UN Decade working groups continued as partner networks. The expert forums had particular responsibility for developing a National Action Plan on ESD, which was adopted in 2017 (National Platform on Education for Sustainable Development 2017). Germany's long history of initiatives and political efforts with regard to ESD let to it hosting UNESCO's World Conference on ESD in 2021, which led to the adoption of the Berlin Declaration on ESD (UNESCO 2021) and which formed the official starting point for the GAP follow-up programme, ESD for 2030.

With the start of the GAP in 2015, the BMBF decided to fund a research project to monitor and track progress with ESD in the various areas of education. This

---

[4] Of crucial importance to the German ESD discourse is the idea of the *Überwältigungsverbot* (which translates as a 'ban on overwhelming', meaning that students should not be overwhelmed or indoctrinated), which is one of the main principles of the idea of civic education in Germany. This means that (school) education should not manipulate learners, or force them to think or behave in a particular way or to adopt specific values. Against this backdrop, de Haan and others advocate the development of *Gestaltungskompetenz*, but emphasise that this has to be through self-directed learning.

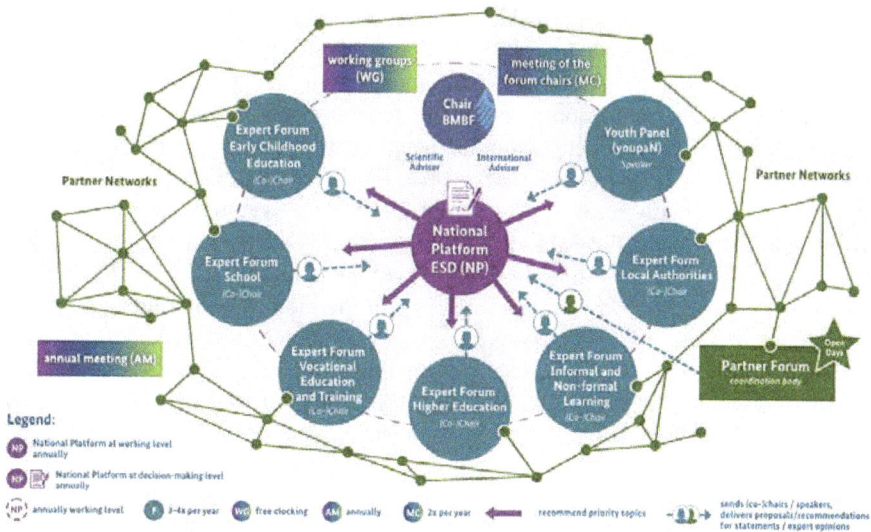

Figure 2. GAP implementation structure in Germany.[5]

project is led by the working group run by the National ESD Platform's scientific advisor (Prof. Dr. Gerhard de Haan, Freie Universität Berlin). Over recent years the monitoring team has conducted a number of studies (document analysis, expert interviews, focus group discussions, a large online survey of young people and educators and online surveys of non-formal learning organisations). The multi-methods design was adopted following suggestions that different methods should be used to monitor ESD, in order to gain not only a broad overview of implementation trends, but also a contextualised and deep understanding of ESD in one country (Stepanek Lockhart 2018).

To date, the monitoring studies have generated a systematic overview of the implementation of ESD, Global Learning and Environmental Education in Germany in the various educational sectors over recent years. As the particular conditions and contexts of each educational sector have an impact on the efforts and successes of ESD (Singer-Brodowski et al. 2019a), the following sections report results separately for each sector.

Early childhood education

Early childhood education has only started to integrate ESD systematically in recent years (including internationally) (Somerville and Williams 2015). At the same time, educators in this sector see themselves as providing learner-centred, democratic education with an emphasis on the natural world and where the values of solidarity and participation are accorded high importance. This makes this educational sector particularly well aligned with efforts to strengthen ESD (Singer-Brodowski et al.

---

[5] https://www.bne-portal.de/bne/en/german-national-action-plan/gap-implementation-structures-in-germany/gap-implementation-structures-in-germany_node.html.

2019a). It has therefore been possible to integrate ESD in early childhood education in Germany 'en passant,' alongside other efforts to develop quality and transforming the sector in the context of "Endogenous Capacity Building" (*ibid.*: 6).

Analysis of the education plans of the federal states in particular reveals an ongoing trend towards structural implementation: In 2020, 9 out of 16 education plans had integrated ESD—some of them quite comprehensively—and a few of them were revised only recently (Holst et al. 2020). However, there were significant differences between the federal states as regards curricula for pre-service courses for early childhood educators (non-university level, generally provided in the form of vocational education and training, which is still the most common preparation for educators in this sector). In summary, the level of implementation proved insufficient to enable high quality ESD practices (Singer-Brodowski et al. 2019b). The module descriptions of academic programmes on early childhood education developed by higher education institutions in recent years have more systematic mentions of ESD and related concepts, but this was also a result of their strong focus on intercultural education (*ibid.*). This analysis shows that the ongoing general strengthening of early childhood education professionalization processes in Germany in recent years has facilitated the structural implementation of ESD.

School education

School education policies in Germany tend to be hierarchical and top-down, although new stakeholders are also gaining importance (Altrichter 2015). Expert interviews characterise the transformation process as 'waiting for ESD policy and its legitimation in schools' (Singer-Brodowski et al. 2019a: 7). For this reason, the implementation of ESD in all curricula is a high priority for teachers (Grund and Brock 2020).

Nearly all federal states at least mention ESD in their year four (primary education) and year nine (secondary education) curricula (Holst et al. 2020). Holst et al. (*ibid*: 8) point out that 'recently revised documents show a generally high quantitative abundance and qualitative integration of ESD.' This embedding of ESD highlights state-specific priorities as well as centres of gravity within known ESD-supporting subjects, such as geography and biology, 'and a rather marginal integration of ESD in the central subjects that are either obligatory or take up a lot of time in the curricula' (Singer-Brodowski et al. 2019b: 501). The overall picture of ESD implementation within the school curricula is therefore characterised by huge differences between federal states and school subjects. Furthermore, 'there is evidence of a structural gap in the delivery of ESD in teacher education higher education institutions' (Holst et al. 2020, Rieckmann 2019).

In addition to this view of the implementation of ESD in the German school system, the online survey of 525 teachers and 2564 young people (between 14 and 24 years old) showed that although many students are aware of sustainability issues, they do not have hope that the transformation towards sustainability will be successfully achieved and therefore do not behave more sustainably themselves (Grund and Brock 2019). This emphasizes the necessity of focusing on ESD approaches that are more emotionally-appealing and sensitive.

Vocational education and training

Vocational education and training (VET) in Germany has a dual structure, with vocational schools on one side and in-service training for apprentices on the other. There are currently 325 state-approved training occupations.[6] When curricula or programmes are revised, the new content for the specific professional documents in question is negotiated between all VET partners (state officials, employers' associations, trade unions, and academia).

The interview study of monitoring revealed that VET has a tendency to aim for a concept of sustainability that brings all three dimensions (social, ecological and economical) into harmony, ignoring conflicting aims and ambiguities (Singer-Brodowski et al. 2019a). In this context the VET path toward ESD can be described as a 'consensual concretisation of sustainability' where sustainability has to be spelled out for each individual profession in the context of the professional practice in question (*ibid.*). For this reason, practical guidelines and recommendations have to be developed for each single training profession.

With regard to the implementation of ESD within documents, the Federal Institute for VET (Bundesinstitut für Berufsbildung, BIBB) has had a broad impact on VET in Germany, fostering innovation. In 1991, it adopted a resolution on environmental protection, which drove the integration of environmental protection in all training regulations (Holst et al. 2020) and thereby had a significant influence on the discourse about environmental education in VET. While sustainability and sustainable development were not integrated into the guidelines for VET in Germany at that time, in 2020 the BIBB adopted a new resolution on the integration of digitalization and sustainability into every profession,[7] which can be considered a major achievement by the experts concerned.

Higher education

Higher education institutions in Germany (as in many other democratic countries) have a long tradition of subject-specific focus and autonomy from societal expectations such as the need for a stronger focus on sustainability. This is both an advantage and a disadvantage for the implementation of inter- and transdisciplinary forms of ESD in higher education (Singer-Brodowski et al. 2019a).

There has been progress on political documents, for example, with sustainability being addressed in the higher education regulations of the various federal states and in the target agreements between the federal states and higher education institutions (Holst et al. 2020). With regard to higher education's supra-regional self-governing bodies and the documents stemming from higher education institutions themselves, such as module descriptions, internationalization strategies and university profiles, progress with the implementation of ESD appears slow and there are significant differences between disciplines. 'While considerable momentum can be observed within the political and mostly federal state-controlled laws, plans, and agreements,

---

[6] https://www.bibb.de/veroeffentlichungen/de/publication/show/16754.
[7] https://www.bibb.de/en/134898.php.

the analysed documents that are specific to and written by higher education institutions prove to be less ambitious at present' (Holst et al. 2020: 11).

Nevertheless, a few early adopters (especially among the small and medium-sized institutions) have taken a Whole-Institution Approach (Bauer et al. 2021, Niedlich et al. 2020, Singer-Brodowski et al. 2019a), developing a sustainable profile and professionalising the sustainability of their research, learning, teaching and organisational management. Against the background of academic freedom and universities self-determination, these institutions serve as good examples of the holistic integration of ESD and, in terms of their progress with transformation, are thereby 'beacon strategy[ies] in Higher Education Institutions' (Singer-Brodowski et al. 2019a). In recent years, the BMBF-funded project, Sustainability at Higher Education Institutions (HOCH-N) (2018–2021), has significantly increased visibility and fostered further implementation of ESD in the higher education system (e.g., by establishing a network, holding workshops and publishing guidelines).[8]

Non-formal learning organizations

Historically, ESD was driven by a range of non-formal learning organizations, which are characterized by a high degree of diversity with regard to their structure, size, profile and funding strategies. Although the thematic and organisational backgrounds of these non-formal organisations are diverse, they share 'a specific understanding of non-formal learning as voluntary, needs-oriented, learner-centred, experimental and participatory' (Singer-Brodowski et al. 2019a: 8) and thereby serve as an important source of high quality ESD practices (Brock and Grund 2020). While the efforts of the last decades to advocate for and advance ESD practices and materials have seen some success and state stakeholders are increasingly taking on responsibility for implementing ESD, new challenges are arising for non-state, non-formal learning organizations.

One challenge is that the non-formal learning organizations have to some extent ceased to set the agenda and are increasingly being asked to cooperate on a longer-term basis with partners, 'in tandem with the formal education sector' (Hopkins et al. 2002: 13). this is characteristic of a transformation that involves the 'interwoven efficacy of non-formal learning organizations' (Singer-Brodowski et al. 2019a: 8). The intensified cooperation between formal and non-formal educational organisations, which are pivotal in enabling more ESD (*ibid.*: 12), also carries the risk of 'formalising' non-formal learning environments. Another big challenge is the precarious financial situation of most non-formal learning organisations (Brock and Grund 2020). This also impacts on the relationship between non-formal civil society organisations and state organisations: there is a 'permanent demarcation and simultaneous symbiosis between non-state and state actors' (von Seggern and Singer-Brodowski 2020).

---

[8] https://www.hochn.uni-hamburg.de/en.html.

Overall results of the implementation of ESD in Germany

Synthesizing and summarizing the overall findings about implementation in Germany, an analysis of nearly 4500 documents up to 2020 revealed that ESD and related concepts are increasingly forming an integral part of the German educational system's documents (laws, curricula, module descriptions, etc.), though with some differences between the federal states and certain subjects, professions or disciplines. While this illustrates that ESD has had some success in moving 'from projects to structure,' it also shows that implementation has been slower with regard to documents and practices for educators in all educational sectors.

The qualitative studies of monitoring looking at how ESD is delivered in the different educational sectors show, on the one hand, that the social innovation of ESD (Bormann 2013) could transform the different educational sectors in specific ways because of its conceptual vagueness. On the other hand, 'the institutional characteristics, structural pre-conditions, and cultural aspects within the different innovation systems of the educational areas influence the success of the ESD diffusion process decisively' (Singer-Brodowski 2019a: 13). This circular transfer or scaling process with regard to ESD has led to different 'transformation paths.' The dynamics of scaling up ESD in specific educational contexts are also mirrored in the communication between the multiple stakeholders and change agents involved in the BMBF process and thereby in ESD governance (Singer-Brodowski et al. 2020).

The quantitative studies revealed that although a high percentage of young people have an affinity for sustainability, many of them have low expectations with regard to the success of the process of transformation towards sustainability. When ESD is delivered in young people's education institutions, a higher percentage of them report that they engage in sustainable behaviours (Grund and Brock 2020).

Education for sustainable development in Germany: Emerging issues and trends, and current and future needs

From some of the overarching trends regarding the implementation of ESD in the German educational system, it can be said that the move from projects to structure requires huge effort and time and, for this reason, has not yet been completed. Nevertheless, the structure for systematic political implementation, with a national platform, expert forums and partner networks, has generated additional momentum, and an increasing number of federal states are adopting ESD as an overall and compulsory approach across all levels and types of education. Baden-Wuerttemberg, for example, has introduced ESD as one of six guiding principles in its curricula for all school subjects[9] and thereby significantly strengthened ESD. Lower Saxony has adopted a decree on 'Education for Sustainable Development in public general and vocational schools and private schools,'[10] which should support a wide variety of

---

[9] http://www.bildungsplaene-bw.de/,Lde/Startseite/BP2016BW_ALLG/BP2016BW_ALLG_LP_BNE.
[10] https://www.mk.niedersachsen.de/startseite/aktuelles/aktuelle_erlasse_und_gesetze/erlass-bildung-fur-nachhaltige-entwicklung-bne-an-offentlichen-allgemein-bildenden-und-berufsbildenden-schulen-sowie-schulen-in-freier-tragerschaft-199018.html

ESD-related educational activities. Such developments are also having an influence on higher education institutions, especially in the context of teacher education. While the higher education system generally is demonstrating an increased appetite to address sustainable development issues, the sector is also displaying some reaction against political initiatives to strengthen sustainable development (e.g., Strohschneider 2014).

In terms of teacher education, there is evidence from some federal states that teachers are demonstrating increasing knowledge of ESD and a more positive attitude towards sustainable development; there are also reports of a greater number of ESD-based activities over the last 12 years (Waltner et al. 2020). 'However, still half of the teachers (49.5%) (surveyed) indicated that the new curriculum was of little or no importance for their decision to teach ESD or not' (ibid: 9). This raises the question of how international, national or even regional policy documents or new curricula can be supported and translated into specific teaching and learning practices.

For young learners, the emotional aspects (such as feeling connected to nature, or the experience of ecological fear and anxiety) are more likely to drive sustainable behaviour than sustainability-related teaching and learning activities in educational institutions (Grund and Brock 2020). Nevertheless, ESD activities are the third-strongest predictor for self-reporting of sustainable activity, and for this reason ESD can be described as successful in terms of behavioural change (*ibid.*)—although this should not be the main aim of (emancipatory) ESD (see section on the understanding of ESD above). Questions arise in pluralistic approaches to ESD (such as the facilitation of contentious discussions in the classroom) that are considered crucial to high-quality ESD, but these were not predictors of more sustainable behaviour (*ibid.*). Future policies and initiatives aiming to improve the quality of ESD should focus on re-inventing and 'updating' ESD, i.e., to address how best to cope with sustainability-related emotions, to encourage contentious discussions and to foster the search for (non-simplistic) solutions to complex sustainability problems (*ibid.*).

In addition to ESD, other cross-cutting issues are also currently given high priority in the German education system, such as inclusion and digitisation. Teachers and other educators sometimes perceive it as a challenge—and time-consuming—to deal with all these cross-cutting issues at once. Against this background, Germany has recently made efforts, both in educational practice and in academia, to relate ESD and inclusive education (Vierbuchen and Rieckmann 2020) or ESD and digitisation (Ketter 2021, Muheim 2021) to each other, or even to combine all three concepts (Schluchter 2021). This may provide further stimulus for quality improvements in ESD.

Conclusion

This chapter has shown how, against the backdrop of the traditions of environmental and development education, ESD has developed and established itself as an independent educational concept in Germany—in line with UNESCO's international proposals. Whereas initially it was more a matter of individual projects in individual educational institutions, ESD now seems to be becoming part of the structure of the German education system. However, there remain significant differences between the various educational sectors and the German federal states.

More education policy initiatives and pressure from civil society will be required before ESD is comprehensively integrated into the education system and all educational institutions have adopted the Whole-Institution Approach. If this succeeds, the education system in Germany (and elsewhere) will indeed fulfil its task of contributing to the sustainable development of the society in the world as a whole.

References

Altrichter, H. 2015. Theory and evidence on governance: conceptual and empirical strategies of research on governance in education. pp. 25–43. *In*: Schrader, J. (ed.). Governance von Bildung im Wandel. Interdisziplinäre Zugänge. Springer VS. Wiesbaden.

Asbrand, B. and A. Scheunpflug. 2005. Globales Lernen. pp. 469–484. *In*: Sander, W. (ed.). Handbuch politische Bildung. 3rd edition. Bundeszentrale für Politische Bildung. Bonn.

Autor:innengruppe Bildungsberichterstattung (ed.). 2022. Bildung in Deutschland 2022. Ein indikatorengestützter Bericht mit einer Analyse zum Bildungspersonal. Wbv. Bielefeld. https://www.bildungsbericht.de/de/bildungsberichte-seit-2006/bildungsbericht-2022.

Barth, M., J. Godemann, M. Rieckmann and U. Stoltenberg. 2007. Developing key competencies for sustainable development in higher education. International Journal of Sustainability in Higher Education 8(4): 416–430. https://doi.org/10.1108/14676370710823582.

Bauer, M., M. Rieckmann, S. Niedlich and I. Bormann. 2021. Sustainability governance at higher education institutions: equipped to transform? Frontiers in Sustainability 2: 640458. https://doi.org/10.3389/frsus.2021.640458.

BLK – Bund-Länder-Kommission für Bildungsplanung und Forschungsförderung. 1998. Bildung für eine nachhaltige Entwicklung. Orientierungsrahmen (Materialien zur Bildungsplanung und zur Forschungsförderung, Heft 69). Bonn.

BMZ – Bundesministerium für wirtschaftliche Entwicklung und Zusammenarbeit and KMK – Kultusministerkonferenz. 2007. Orientierungsrahmen für den Lernbereich Globale Entwicklung im Rahmen einer Bildung für nachhaltige Entwicklung. https://www.kmk.org/fileadmin/veroeffentlichungen_beschluesse/2007/2007_06_00_Orientierungsrahmen_Globale_Entwicklung.pdf.

Brock, A. and J. Grund. 2020. Non-formale Bildung für nachhaltige Entwicklung: Divers, volatil und dabei feste Säulen der Nachhaltigkeitstransformation. Berlin. https://www.ewi-psy.fu-berlin.de/einrichtungen/weitere/institut-futur/Projekte/Dateien/Brock_-A__-Grund_-J__2020__Non-formale_BNE_Divers_volatil_und_ dabei_feste1.pdf.

de Haan, G. and D. Harenberg. 1999. Gutachten zum Programm Bildung für eine nachhaltige Entwicklung (Materialien zur Bildungsplanung und zur Forschungsförderung, Heft 72). Bonn.

de Haan, G., G. Kamp, A. Lerch, L. Martignon, G. Müller-Christ and H.G. Nutzinger (eds.). 2008. Nachhaltigkeit und Gerechtigkeit: Grundlagen und schulpraktische Konsequenzen. Springer. Berlin, Heidelberg.

de Haan, G. 2010. The development of ESD-related competencies in supportive institutional frameworks. International Review of Education 56(2-3): 315–328. https://doi.org/10.1007/s11159-010-9157-9.

Dieckmann, A., K. Hübner and B. Paulsen. 2006. Bestandsaufnahme der Aktivitäten der Umweltverbände zur Bildung für eine nachhaltige Entwicklung (2001–2005). http://www.leuphana.de/fileadmin/user_upload/Forschungseinrichtungen/infu/files/pdf/infu-reihe/32_06.pdf.

Eggert, S. and S. Bögeholz. 2010. Students' use of decision-making strategies with regard to socioscientific issues: An application of the Rasch partial credit model. Science Education 94(2): 230–258.

Gausmann, E., S. Eggert, M. Hasselhorn, R. Watermann and S. Bögeholz. 2010. Projekt Bewertungskompetenz: Wie verarbeiten Schüler/-innen Sachinformationen in Problem- und Entscheidungssituationen Nachhaltiger Entwicklung – Ein Beitrag zur Bewertungskompetenz. Zeitschrift für Pädagogik 56: 204–215.

Gräsel, C. 2010. Umweltbildung. pp. 845–859. *In*: Tippelt, R. and B. Schmidt (eds.). Handbuch Bildungsforschung. VS Verlag für Sozialwissenschaften. Wiesbaden. https://doi.org/10.1007/978-3-531-92015-3_44.

Grund, J. and A. Brock. 2019. Why we should empty pandora's box to create a sustainable future: hope, sustainability and its implications for education. Sustainability 11(3): 893. https://doi.org/10.3390/su11030893.

Grund, J. and A. Brock. 2020. Education for sustainable development in Germany: not just desired but also effective for transformative action. Sustainability 12(7): 2838. https://doi.org/10.3390/su12072838.

Holst, J., A. Brock, M. Singer-Brodowski and G. de Haan. 2020. Monitoring progress of change: implementation of Education for Sustainable Development (ESD) within documents of the German education system. Sustainability 12(10): 4306. https://doi.org/10.3390/su12104306.

Hopkins, C. and R. Mckeown. 2002. Education for sustainable development: an international perspective. pp. 13–24. *In*: Tilbury, D. (ed.). Education and Sustainability: Responding to the Global Challenge. Cambridge. IUCN Commission on Education and Communication.

Ketter, V. 2021. Digital geprägte Bildung für nachhaltige Entwicklung: Eine systematische Betrachtung medienpädagogischer Praxis. merz | medien + erziehung (4): 46–55.

KMK – Standing Conference of the German Ministers of Education and Culture and BMZ – German Federal Ministry of Economic Cooperation and Development. 2016. Curriculum Framework 'Education for Sustainable Development'. Engagement Global gGmbH. https://www.globaleslernen.de/sites/default/files/files/link-elements/curriculum_framework_education_for_sustainable_development_barrierefrei.pdf.

Künzli David, C. 2007. Zukunft mitgestalten: Bildung für eine nachhaltige Entwicklung – Didaktisches Konzept und Umsetzung in der Grundschule. Haupt Verlag.

Lang-Wojtasik, G. 2019. Große Transformation, Bildung und Lernen – Chancen und Grenzen einer Global Citizenship Education. pp. 33–50. *In*: Lang-Wojtasik, G. (ed.). Bildung für eine Welt in Transformation: Global Citizenship Education als Chance für die Weltgesellschaft. Verlag Barbara Budrich. Leverkusen.

Lang-Wojtasik, G. 2022. Globales Lernen für nachhaltige Entwicklung. Einleitende Überlegungen und zukunftsgewandte Perspektiven. pp. 11–30. *In*: Lang-Wojtasik, G. (ed.). Globales Lernen für nachhaltige Entwicklung: Ein Studienbuch. Waxmann. Münster, New York.

Michelsen, G. 1998. Theoretische Diskussionsstränge der Umweltbildung. pp. 61–65. *In*: Beyersdorf, M., G. Michelsen and H. Siebert (eds.). Umweltbildung: Theoretische Konzepte – empirische Erkenntnisse – praktische Erfahrungen. Luchterhand. Neuwied.

Michelsen, G. 2009. Kompetenzen und Bildung für nachhaltige Entwicklung. pp. 75–86. *In*: Overwien, B. and H.-F. Rathenow (eds.). Globalisierung fordert politische Bildung: Politisches Lernen im globalen Kontext. Verlag Barbara Budrich. Opladen, Farmington Hills.

Mokrosch, R. 2009. Zum Verständnis von Werte-Erziehung: Aktuelle Modelle für die Schule. pp. 32–40. *In*: Mokrosch, R. and A. Regenbogen (eds.). Werte-Erziehung und Schule: Ein Handbuch für Unterrichtende. Vandenhoeck & Ruprecht. Göttingen.

Muheim, V. 2021. Bildung für nachhaltige Entwicklung und Medienpädagogik. Desiderate und Denkanstöße. merz | medien + erziehung (4): 38–45.

National Committee (The German National Committee for the UN Decade of Education for Sustainable Development). 2013. Position paper 'Strategy for ESD 2015+'. German UNESCO Commission. Bonn. https://www.bne-portal.de/files/ESD-Position-paper-2015plus_english.pdf.

National Platform on Education for Sustainable Development. 2017. National Action Plan on Education for Sustainable Development: The German contribution to the UNESCO Global Action Programme. https://www.bne-portal.de/bne/shareddocs/downloads/files/bmbf_nap_bne_en_screen_2.pdf?__blob=publicationFile.

Niedlich, S., B. Kummer, M. Bauer, M. Rieckmann and I. Bormann. 2020. Cultures of sustainability governance in higher education institutions: A multi-case study of dimensions and implications. Higher Educ Q 74(4): 373–390. https://doi.org/10.1111/hequ.12237.

Overwien, B. 2009. Globalisierung und Globales Lernen. pp. 63–80. *In*: Lucker, T. and O. Kölsch (eds.). Naturschutz und Bildung für nachhaltige Entwicklung: Fokus: Globales Lernen. Bundesamt für Naturschutz. Bonn-Bad Godesberg.

Overwien, B. and H.-F. Rathenow. 2009. Globalisierung als Gegenstand der politischen Bildung - eine Einleitung. pp. 9–21. *In*: Overwien, B. and H.-F. Rathenow (eds.). Globalisierung fordert politische Bildung: Politisches Lernen im globalen Kontext. Verlag Barbara Budrich. Opladen, Farmington Hills.

Programm Transfer-21. 2008. Programm Transfer-21: Bildung für eine nachhaltige Entwicklung. Abschlussbericht des Programmträgers 1. August 2004 bis 31. Juli 2008. http://www.transfer-21. de/daten/T21_Abschluss.pdf.

Rieckmann, M. 2018. Chapter 2 – Learning to transform the world: key competencies in ESD. pp. 39–59. *In*: Leicht, A., J. Heiss and W.J. Byun (eds.). Education on the move. Issues and trends in education for sustainable development. United Nations Educational, Scientific and Cultural Organization. Paris.

Rieckmann, M. 2019. Education for sustainable development in teacher education. An international perspective. pp. 33–48. *In*: Lahiri, S. (ed.). Environmental Education. Studera Press. New Delhi.

Rieckmann, M. 2021a. Bildung für nachhaltige Entwicklung. Ziele, didaktische Prinzipien und Methoden. merz - Zeitschrift für Medienpädagogik 65(04): 12–19.

Rieckmann, M. 2021b. Reflexion einer Bildung für nachhaltige Entwicklung aus bildungstheoretischer Perspektive. Religionspädagogische Beiträge 44(2): 5–16. https://doi.org/10.20377/rpb-153.

Rode, H. 2005. Motivation, Transfer und Gestaltungskompetenz. Ergebnisse der Abschlussevaluation des BLK-Programms "21" 1999–2004. Forschungsgruppe Umweltbildung/Working Group Environmental Education. http://www.transfer-21.de/daten/evaluation/Abschlusserhebung.pdf.

Rode, H. and G. Michelsen. 2012. Der Beitrag der UN-Dekade 2005–2014 zur Verbreitung und Verankerung der Bildung für Nachhaltige Entwicklung. Deutsche UNESCO-Kommission (DUK): Bonn. https://www.bne-portal.de/publikationen/1204/downloads/DUK%20-%20UN-Dekade%20 Verbreiterung%20&%20Verankerung.pdf.

Schank, C. and M. Rieckmann, M. 2019. Socio-economically substantiated education for sustainable development: development of competencies and value orientations between individual responsibility and structural transformation. Journal of Education for Sustainable Development 13(1): 67–91. https://doi.org/10.1177/0973408219844849.

Scheunpflug, A. 2001. Die globale Perspektive einer Bildung für nachhaltige Entwicklung. pp. 87–99. *In*: Herz, O., H. Seybold and G. Strobl (eds.). Bildung für nachhaltige Entwicklung. Globale Perspektiven und neue Kommunikationsmedien. Leske + Budrich. Opladen.

Scheunpflug, A. 2011. Lehren angesichts der Entwicklung zur Weltgesellschaft. pp. 204–215. *In*: Sander, W. and A. Scheunpflug (eds.). Politische Bildung in der Weltgesellschaft: Herausforderungen, Positionen, Kontroversen; Perspektiven politischer Bildung. Bundeszentrale für Politische Bildung. Bonn.

Schluchter, J.-R. 2021. Medienbildung, Bildung nachhaltige Entwicklung und Inklusion/inklusive Bildung. Eine Annäherung. merz | medien + erziehung. Online First. https://doi.org/10.25969/ mediarep/16847.

Seitz, K. 2009. Globales Lernen in weltbürgerlicher Absicht: zur Erneuerung weltbürgerlicher Bildung in der postnationalen Konstellation. pp. 37–48. *In*: Overwien, B. and H.-F. Rathenow (eds.). Globalisierung fordert politische Bildung: Politisches Lernen im globalen Kontext. Verlag Barbara Budrich. Opladen, Farmington Hills.

Seitz, K. 2022. Herkunft und Zukunft Globalen Lernens. Vorgeschichte und Werdegang eines pädagogischen Arbeitsfeldes. pp. 33–45. *In*: Lang-Wojtasik, G. (eds.). Globales Lernen für nachhaltige Entwicklung: Ein Studienbuch. Waxmann. Münster, New York.

Singer-Brodowski, M., N. Etzkorn and J. von Seggern. 2019a. One transformation path does not fit all—insights into the diffusion processes of education for sustainable development in different educational areas in Germany. Sustainability 11(1): 269. https://doi.org/10.3390/su11010269.

Singer-Brodowski, M., A. Brock, N. Etzkorn and I. Otte. 2019b. Monitoring of education for sustainable development in Germany—insights from early childhood education, school and higher education. Environmental Education Research 25(4): 492–507. https://doi.org/10.1080/13504622.2018.1440380.

Singer-Brodowski, M., J. von Seggern, A. Duveneck and N. Etzkorn. 2020. Moving (Reflexively within) Structures. The Governance of Education for Sustainable Development in Germany. Sustainability 12(7): 2778. https://doi.org/10.3390/su12072778.

Somerville, M. and C. Williams. 2015. Sustainability education in early childhood: an updated review of research in the field. Contemp. Issues Early Child. 16: 102–117.

Stepanek Lockhart, A. 2018. Monitoring ESD: Lessons learned and ways forward. pp. 215–231. *In*: Leicht, A., J. Heiss and W.J. Byun (eds.). Education on the Move. Issues and Trends in Education for Sustainable Development. United Nations Educational, Scientific and Cultural Organization. Paris.

Stoltenberg, U. 2009. Mensch und Wald: Theorie und Praxis einer Bildung für nachhaltige Entwicklung am Beispiel des Themenfeldes Wald. oekom verlag. München.

Strohschneider, P. 2014. Zur Politik der Transformativen Wissenschaft. pp. 175–192. *In*: Brodocz, A. (ed.). Die Verfassung des Politischen. Festschrift für Hans Vorländer. Springer VS. Wiesbaden.

UNESCO. 2021. Berlin Declaration on Education for Sustainable Development. https://en.unesco.org/sites/default/files/esdfor2030-berlin-declaration-en.pdf.

Vierbuchen, M.-C. and M. Rieckmann. 2020. Bildung für nachhaltige Entwicklung und inklusive Bildung – Grundlagen, Konzepte und Potenziale. ZEP – Zeitschrift für internationale Bildungsforschung und Entwicklungspädagogik 43(1): 4–10. https://doi.org/10.31244/zep.2020.01.02.

von Seggern, J. and M. Singer-Brodowski. 2020. Why context matters for educational policy. Analysing interactive practice in the governance of education for sustainable development in Germany. Zeitschrift für internationale Bildungsforschung und Entwicklungspädagogik 43(4): 25–29.

Waltner, E.-M., K. Scharenberg, C. Hörsch and W. Rieß. 2020. What teachers think and know about education for sustainable development and how they implement it in class. Sustainability 12(4): 1690. https://doi.org/10.3390/su12041690.

WBGU – German Advisory Council on Global Change. 2011. World in Transition – A Social Contract for Sustainability. WBGU. https://www.wbgu.de/en/publications/publication/world-in-transition-a-social-contract-for-sustainability.

Chapter 9

Swedish Approaches to and Engagement with ESD in Education Policy and Formal Education During the Global Action Programme on ESD

*Stefan Bengtsson** and *Paul Plummer*

Introduction

This chapter aims to provide an overview of the historical background, development and current situation with regard to environmental and sustainability education (ESE) in the Swedish formal education system, focusing on education for sustainable development (ESD). As the chapter will argue, a number of historical trajectories have influenced ESE in the Swedish context. However, given that this chapter focuses on linkages with the Global Action Programme (GAP) on ESD, we have limited our discussion to providing an account of the discourse on ESD policy and curricula and on the general uptake of the concept in formal education. This chapter thus reflects Sweden's transition from working with ESD as part of the *Global Action Programme on ESD* (2014–2019) and the introduction of the follow up programme of *Education for Sustainable Development: Towards achieving the SDGs* (ESD for 2030). It argues that current engagement with ESD in formal education is aligned with—and dependent on—broader governmental responses to the policy framework

Uppsala University.
Email: paul.plummer@angstrom.uu.se
* Corresponding author: stefan.bengtsson@edu.uu.se

of the Sustainable Development Goals (SDGs), partially delaying responses to the GAP but also integrating ESD into wider discourses about education and its linkages to sustainable development.

Sweden's political, economic, cultural and social circumstances, including its education system

During both halves of the 20th century, Sweden saw dramatic social and economic changes. The country industrialized late compared to other central European countries, with its second industrialization starting in the 1890s (Jörberg 1965). This second industrialization led to significant societal changes, uprooting traditional ways of life as well as the political system, particularly with the organization of workers and the emergence of workers movements (Olsson and Ekdahl 2002). The Social Democratic Party-led government in 1920 initiated a substantial reform of the state, the economy and society, focusing on the national vision of *folkhemmet* (people's home) and the nation's approach to socialism (Stråht 1993). With regard to the latter, Sweden's socialist reform programme combined the principles of market economics with strong intervention where workers' rights and influence were concerned. Sweden tends to perform well on comparative environmental sustainability indexes, being ranked 8th out of 180 countries in the 2020 environmental performance index (EPI 2020) and second out of all 193 UN Member States for SDG achievement as measured by the Sustainable Development Report (Sachs et al. 2021).

With regard to education, the post Second World War period saw substantial education reforms replacing previous structures, with the introduction of differentiated secondary education tracks similar to those in Germany. During this period, education policy focused on the principle of *En skola för alla* (one school for all), with the principles of equality and egalitarianism being central (Lidensjö and Lundgren 2000). The guiding ideas for these reforms were partially imported from the pedagogical reforms in the US, with John Dewey's work being recast against the background of the Swedish social democratic *folkhemmet* (cf. Olsson and Pettersson 2005). During the 1990s, the Swedish idea of a single 'school for all' was challenged and partially deconstructed through the introduction of neoliberal reforms to the state and the economy (Lundahl 2005). One key reform was the decentralization of the education system, with responsibility for—and ownership of—its operation transferring from the state to the *kommunerna* (municipalities). A second key reform was to the legal system, which allowed for private ownership and provision of state funded education. Based on the principle of freedom of choice, competitiveness and market forces, private provision of schooling once more drastically changed education in Sweden.

While education policy and curricula remain under the governance of the state, with the Ministry of Education responsible for education policy and the National Agency of Education (*Skolverket*) responsible for curricula, responsibility for delivery has been transferred to municipalities. It is important to mention that when it comes to the overarching objectives of education, there is no clear-cut separation between education policy and curricula. Sweden has different *läroplan* (curricula) for different levels and forms of education (e.g., pre-school, year 1–9 basic education and upper secondary education). These curricula are divided into two main parts. The first part lays out the foundational values (*värdegrunden*) and central tasks of schooling and the overarching goals of education, namely the basic values and principles that education should strive to live up to and which represent the Swedish democratic ideal. The second part of curricula consists of *kursplaner* (syllabi) for the specific subject being taught.

Historical background and development of environmental and sustainability education in Sweden

Three historical trends can be seen as having influenced environmental and sustainability education (ESE) in Sweden. These trends have been verified through empirical analyses of ESE teaching in Sweden (e.g., The National Agency of Education 2001, Öhman 2006). The first trend is formal science education, which began to address environmental issues at an early stage, particularly through concepts arising from biology, including ecology (Östman 1995). Science education can be seen as having resulted in what Sandell, Öhman and Östman (2005) describe as a *fact-based tradition*, where environmental education, as part of science education, is seen to provide conceptual methods, often borrowed from ecology, and scientific facts that in turn create more knowledgeable students.

The second trend is education as a form of environmental activism. During the late 1960s and early 1970s, socialist activism drifted off into environmental movements, such as the anti-nuclear movement and later the Green Party. The environmental activist trend sees education as a primarily political endeavour, meaning that instead of focusing on scientifically derived facts, education promotes norms and values that are associated with environmental protection. This is not to say that the environmental activist trend opposes the idea of scientific facts and knowledge, but that it sees them as supporting political and normative imperatives.

The third influential trend in Swedish ESE is the outdoor pedagogy (*utomhus/friluftspedagogik*) movement, which is separate from, and has different historical roots from, environmental activist approaches to education. The main difference from the second trend is that *utomhus/frilufts* education focuses on personal, authentic, and physical experiences of place (cf. Szczepanski 2013). Accordingly, it has similarities with the international outdoor education movement, and yet it also emerges out of and feeds back into Swedish historical narratives and the Swedish imagination, often being associated with the idea of identity being formed by 'being outside in nature.'

With regard to the influence of these three trends on formal education, the engagement of science education with environmental issues has had the most significant impact. In the context of NGOs' engagement with environmental issues, however, environmental activist approaches, which often take a clearly normative approach to education, can also be seen as having influenced teaching practices. There is a clear crossover between outdoor education and formal education, particularly in formal preschool education, with a strong outdoor movement and an alliance of outdoor preschools.

Sweden's engagement with *Education for Sustainable Development* (ESD) was mainly the result of the Social Democrats' strong policy focus during the Göran Persson government (1996–2006). National policy engagement with ESD was also strongly influenced by, and aimed at promoting, cooperation on ESD with the regions (Baltic 21) and the United Nations (UN). Much of the national engagement with ESD took place during the period leading up to the Unit ed Nations Decade of Education for Sustainable Development (UNDESD) in particular. One of the key milestones in Sweden's engagement with ESD, both before and during the UNDESD, was the report of a committee that was formed to author a white paper on ESD in Sweden. This white paper, 'Learning for Sustainable Development' (*Att lära för hållbar utveckling*) (SOU 2004, 104), came to act as a guiding policy framework and remains influential in defining how ESD is interpreted in the Swedish context. The committee identified (ibid., 22) the following essential aspects of ESD:

1. The many and multifaceted aspects of economic, social and environmental circumstances and processes should be dealt with in an integrated manner through the use of interdisciplinary working methods.

2. Conflicting objectives and synergies between different needs and interests should be clarified.

3. Content should take a long-term perspective, encompassing the past and the future, the global and the local.

4. Democratic working methods should be used so that students can influence the design and content of educational programmes.

5. Learning should be reality-based with close and frequent interaction with nature and society.

6. Learning should focus on problem-solving and promote critical thinking and readiness to take action.

7. Both the process and the product of education are important.

When it comes to the terminology and conceptual framework used in Sweden, the two main concepts are environmental education (EE) and learning/education for sustainable development (utbildning för hållbar utveckling/UHU). EE, or *miljöundervisning*, is primarily used in certain conceptions of science education and in the third trend of outdoor education mentioned above. It was in the context of the political engagement of the Persson administration that formal education came to embrace the concept of ESD. Critiques of the idea of ESD have been primarily directed at terminology implying the idea of 'educating for' rather than 'learning for' (*lärande för hållbar utveckling* or *LHU*). LHU is however often understood in terms of the key principles of ESD as characterized above and is seen as embracing the idea of democratic citizenship education. UHU (ESD) tends to be more prominent in policy discourse, while LHU is often more commonly used among educators. Given the confluence of the meanings of UHU and LHU as regards underlying principles, we will utilize the term ESD in the Swedish context and the discussion below will explain the understanding of ESD in formal education, where it has been the predominant concept in relation to EE since the 2010s.

Understanding of environmental and sustainability education in Sweden

Having looked at the historical aspects of ESE/ESD in Sweden, this section will describe three interconnected elements in the country's current understanding of ESD, in order to outline some general patterns in how it is interpreted and applied. We will start with a brief overview of the presentation of ESD in Swedish policy and legislature, moving on to describe how ESD is dealt with in the compulsory curriculum and syllabi (primary education and lower secondary education). Finally, this section will outline the ways in which ESD is applied within certain teaching environments.

ESD in policy and legislature

A sense of the understanding of ESD in Sweden can be gained by looking at key policy documents such as The Education Act (2010: 800) and The Higher Education Act (1992: 1434). In recent decades, both these documents have been amended to take account of EE and ESD: the wording of the Education Act, for example, was amended in 1990 to include the consideration of environmental issues, while in 2006, the Higher Education Act was changed to include the aim of promoting sustainable development (Östman and Östman 2013). Although there are no direct references to sustainable development within the Education Act, the initial regulations outlined in Chapter 1 feature certain ESD related concepts. For example, it is stated that education should promote respect for human rights and that there should be no discrimination on the basis of gender or ethnicity, which limits access to education, both of which can be connected to the social dimensions of SD (The Swedish Government 2010).

The Higher Education Act on the other hand (1992: 1434) does explicitly mention SD, stating that 'higher education institutions shall promote sustainable development to assure for present and future generations a sound and healthy environment, economic and social welfare, and justice.' Thus, at the level of policy and regulations, ESD appears largely to be understood as the fostering of responsible democratic citizens, as suggested by the emphasis in the Education Act on respect for human rights and non-discrimination. A more detailed sense of the intended sustainability outcomes of fostering such citizens is evident in The Higher Education Act: as well as a healthy environment, this refers to social and economic welfare, which links back to Sweden's social democratic political history.

ESD in the curricula and syllabi

A more detailed picture of the interpretation and application of ESD in Sweden can be gained by looking at steering documents such as national curricula. In the compulsory curriculum (The National Agency of Education 2018), the section on 'Fundamental values and tasks of schools' describes four perspectives which schools should provide to pupils, namely historical, environmental, international and ethical. These perspectives are related to qualities such as 'an understanding of the present, and a preparedness for the future,' the 'ability to think dynamically,' the 'ability to develop personal standpoints and to act responsibly towards oneself and others,' and 'responsibility for the environment in areas where they themselves can exercise direct influence.' These perspectives can thus be seen as establishing an ESD focus within the curriculum, which ties in with many of the recommendations of the Committee on ESD. Insofar as these perspectives are intended to permeate all aspects of education, it is also clear that ESD represents a holistic framework guiding the curriculum, and again tying in with the Committee on ESD's recommendation of an integrated and multi-subject approach.

In the second part of the compulsory curriculum, which addresses the syllabi and goals of different subjects, many topics are explicitly linked to sustainable development (SD). For example: sustainable consumption in Home and Consumer Studies (ibid. 48) or the role and impact of technology in Physics and Technology (ibid. 176). Although an environmental perspective on SD is given the greatest emphasis here, social dimensions are also included, for example, in the Home and Consumer Studies and Civics curricula (ibid. 44, 227). The social dimension of SD is also represented in the curriculum's consistent emphasis on issues such as gender equality, democratic values and human rights. The economic dimension of SD arguably receives the least attention in the curriculum, although subjects such as Civics cover topics that can be seen as relevant to this (ibid., 232, 230). Thus, in terms of the balance between the three pillars of sustainability within the curriculum, ESD in Sweden appears to lean more towards environmental considerations at a surface level, while social issues relating to SD are also consistently featured, albeit in a less direct way.

Student-centred and active learning also feature in the curriculum, and are connected explicitly to SD: the Biology curriculum, for example, states that 'pupils should be given the ability to manage practical, ethical and aesthetic situations involving health, the use of natural resources and ecological sustainability' (ibid., 166). The curriculum can also be seen as encouraging inter/multi/trans disciplinary approaches. For example, it encourages 'teaching in different subject areas [to be] coordinated such that the pupils are provided with opportunities to understand wider areas of knowledge as a whole,' and 'teaching in different subjects [to] integrate cross-disciplinary areas of knowledge, such as the environment, traffic, gender equality, consumer issues...' (ibid., 17). These points suggest that students should be able to draw on multiple disciplinary perspectives to attain broader understandings of large-scale issues and phenomena. SD is linked to this through the examples of cross-disciplinary knowledge provided, such as 'environment, traffic, gender equality, consumer issues.' The fact that a number of individual subjects include disciplinary questions relating to SD also suggests that this is one of the 'wider areas of knowledge' which it is intended pupils should understand.

A number of the values and attitudes that are featured in the curriculum can also be related to an understanding of ESD, such as respect, compassion and empathy for others, as well as social justice, citizenship and stewardship (ibid., 227, 10). Here again we see a strong emphasis on the notion of fostering responsible democratic citizens. Competencies and skills relating to ESD are also featured, such as 'a holistic perspective' and the 'ability to critically review information, facts and relationships' (ibid., 198, 208, 9). Anticipatory competencies are also alluded to, with the aim of pupils 'develop[ing] an understanding of the present, and a preparedness for the future, and [...] their ability to think dynamically' (ibid. 8). The curriculum also states that students should be encouraged to connect real world actions and behaviours to SD, stimulating their 'desire to translate ideas into action' and enabling them to 'solve problems [...] in a creative and responsible way' (ibid., 8, 11). A sense of the type of student Swedish ESD seeks to foster thus becomes clear: specifically, ESD should promote concern for others and for the consequences of one's actions, as well as a desire and ability to contribute positively through one's actions towards society.

With regard to the role education plays in social learning and the development of society, the curriculum states that democratic forms of working should be used to 'prepare pupils to participate and take responsibility, and to exercise the rights and fulfil the obligations that characterise a democratic society' (ibid., 14). This is also connected explicitly to SD in the curriculum, which states for example that 'Teaching should illuminate how the functions of society and our ways of living and working can best be adapted to achieve sustainable development' (ibid., 8). Effort is also made to include both local and global perspectives. For instance, the Chemistry syllabus deals with 'People's use of energy and natural resources, locally and globally, and what this means in terms of sustainable development' (ibid., 191). With regard to the inclusion of local contexts within education, the curriculum states that each school's development should be pursued in close contact with the local community, and that a high-quality education 'presupposes close co-operation between work and the local community' (ibid., 15). Engagement with local contexts/environments is incorporated in the curricula for many individual subjects: the analysis of local images in Art; the study of local flora and fauna and ecosystems in Biology; and the study of local political institutions and decision-making processes in Civics.

Teaching

When it comes to the teaching of ESD in Sweden, it has to be acknowledged that practices are diverse. In the following section we will focus on teaching in formal education. In this context we can differentiate between two main approaches.

The first approach treats ESD primarily as content specific to the syllabi of particular subjects. In this context, teaching focuses on integrating ESD related content into existing teaching, subject areas and disciplines. This is similar to the fact-based tradition mentioned earlier. At the same time, the curriculum stipulates that the teaching of such topics should be 'holistic,' which creates incentives for collaboration across subject areas; this is achieved with varying degrees of success (cf. Borg et al. 2012).

The second teaching approach treats ESD as primarily related to the overarching objectives of the curriculum (*Fundamental values and tasks of the school*). From this perspective, ESD is related primarily to what the Swedish tradition might call 'citizenship education' (*medborgar/demokratifostran*), namely, the socialization and qualification of individual citizens through education. In this tradition, ESD is seen as going beyond educational content and relating primarily to the overarching ambition of realizing democratic ideals in and/or through education, which is seen as a form of socialization or subjectification (cf. Biesta 2009). The normative and pluralist traditions mentioned above are aligned with this perspective on ESD. In this second approach, there is a well-documented tension between the normative and pluralist traditions, reflecting two different understandings of the overarching objective of citizenship education (*demokratifostran*) (cf. Rudsberg and Öhman 2010, Sund and Wickman 2011, Hasslöf and Malmberg 2015, Lundegård and Wickman 2007). On the one hand, ESD is seen as a form of socialization where teaching inculcates predetermined societal values, norms and principles, and on the other, ESD is seen as a form of subjectification where teaching delivers democracy by encouraging learners to think critically and reflexively, engaging with content as well as norms and values.

Implementation of the Global Action Program (GAP) on ESD in Sweden

Sweden was well represented at the Nagoya conference that launched the Global Action Programme (GAP) on ESD in 2014, with the then Swedish minister calling for national action plans on ESD. However, the development of such action plans was put on hold while national engagement with ESD was aligned with the Sustainable Development Goals (SDGs), which were launched a year later in 2015. Accordingly, Swedish policy and action on the part of state agencies was aligned with work to gather baseline information from such agencies and other stakeholders and develop a strategy for working with the SDGs, referred to at the state level as *Agenda 2030*. The government response to Agenda 2030 primarily comprised engagement with two fora: first, an inter-ministerial working group was founded and, second, an Agenda 2030 commission was formed, composed of societal representatives and given the task of facilitating broader societal engagement and consultation. The Agenda 2030 commission presented its final report and recommendations to the Swedish government in 2019 (SOU 2019: 13) in the form of a white paper, highlighting a number of recommendations specific to ESD. These long-term recommendations included capacity development for educators and principals/education managers by the Swedish National Agency for Education (*Skolverket*), and monitoring and evaluation by the Swedish Schools Inspectorate (*Skolinspektionen*) of the delivery of ESD in the formal education sector. A national Agenda 2030 action plan (Fi 2018: 3) for the 2018–2020 period was launched prior to the release of the recommendations in 2018, stating that ESD was an issue requiring long-term perspectives with regard to action within both the formal education system and higher education.

Thus, when summarizing the intra-national governmental response to the GAP in Sweden, it might be concluded that action was delayed as ministries and governmental agencies decided on how to direct and align different sectors and areas of responsibility as part of national engagement with the SDGs. Yet, during that period, a number of parliamentary motions were put forward to strengthen and expand the work being done on ESD, for instance by requesting increased funding and assignments for the National Agency of Education in this area. The first three of these motions were rejected, with the government pointing to the work that had already been undertaken in this area (Carlsson and Jonsson 2013, Eriksson 2014, Bergstrom 2017, Wallentheim 2019).

The government also responded to the GAP by making Uppsala University and the Swedish International Centre of Education for Sustainable Development (SWEDESD) national focal points for GAP. National coordination was closely coordinated with the Swedish National Agency for Education (*Skolverket*) and the Swedish Council for Higher Education (*Universitets och Högskolerådet*) and its Global School (*Globala Skolan*) programme. The Swedish Council for Higher Education and its Global School provided in-service training and capacity development on ESD throughout the GAP period (2014–2019) and continues to do so. The Swedish National Agency for Education developed two training modules on SD and ESD for lower and upper secondary education during the GAP period. In its capacity as the national focal point for the GAP, SWEDESD participated in a number of networks and facilitated a range of working groups, aiming to create cross-sector collaboration and identify needs and trends.

SWEDESD organized two national conferences in 2016 and in 2018 to bring key stakeholders together and identify needs and challenges to the implementation of the GAP on ESD. The reports highlighted a mismatch between existing educational structures and governance since they did not support cross-sectoral work with ESD. Further, despite the 2004 white paper, it pointed out that there was no clear framework and a lack of a cohesive national strategy for local implementation. Existing frameworks did not promote an understanding of education for sustainable development at the municipal, school and classroom level. In order to support systematic implementation and development work, stakeholders recommended that ESD should be integrated into the education system and form a central part of a coherent education policy. The findings from the conferences highlighted that policies should be developed in dialogue with teachers and teacher trainers to ensure that existing frameworks and opportunities align with teaching practices and local needs. Furthermore, ESD goals, guidelines and criteria within policies and curricula should not set a restrictive framework, but should be designed and formulated so as to support and guide development processes and to provide common points of departure for teaching and learning.

Consultation and knowledge sharing with stakeholders, as facilitated by the national focal points, have highlighted four key areas for state engagement with ESD. First, *governance* in the form of policy-making, legislative changes and state agency directives; second, *monitoring and evaluation*, which was identified as a key mechanism for the successful coordination, development and upscaling of government directives, policy making and legislative regulations. Third, *upscaling* of existing initiatives was seen as an important potential intervention by the state through the provision of financial resources but also through communication platforms and networks for mobilizing and sharing knowledge. Fourth, despite Sweden having a strong research base on ESD, additional and ongoing *research* was identified as an important area for state support to enable national implementation to be consolidated and to develop existing initiatives during the latter half of the GAP on ESD, as part of initial national engagement with the follow up programme, ESD for 2030.

Environmental and sustainability education in Sweden: emerging issues and trends, and current and future needs

The launch of the SDGs framework has had a significant impact on engagement with ESD, highlighting the linkages between education and sustainable development to wider audiences. While, as stated above, the introduction of the SDGs framework can be seen to have delayed governmental engagement with the GAP, it opened up the concept of ESD to non-education focused governmental agencies as well as non-governmental organisations. Although uptake of the concept has increased in Sweden, it also needs to be pointed out that this broader engagement often did not entail awareness of the history of the concept or of the defining characteristics, as outlined in the white paper 'Learning for Sustainable Development' (*Att lära för hållbar utveckling*) (SOU 2004: 104), for example. This observation applies not only to non-governmental stakeholders but also to state agencies and ministries. Hence, one of the current trends that can be observed with regard to ESD is a form of *policy amnesia*, where the history and interpretation of a policy is forgotten or reinterpreted to the extent that its original meaning is erased or replaced.

Another trend is the return to what we have labelled the fact-based tradition in education policy discourse in Sweden. The revival of this focus is prevalent both in education policy discourse and in the broader public debate on education. It has also led to a return to subject-specific content and teaching. Accordingly, there has been a shift of focus in how teachers and school principals relate to the curriculum. Given the focus on subject-specific content and facts, the second part of the curriculum is now seen as providing primary directives on how to teach them. As a result, the overarching ambitions of education to create responsible citizens have received less attention. This move away from the holistic aspects of education towards specific syllabi has also been facilitated by increased national testing that is measuring students' knowledge in specific subjects. The concept of ESD is thus shifting as education is increasingly treated as a product rather than a process (cf. discussion of key aspects of ESD above). Educational outcomes are measured and such measurement is influencing teaching practices.

At the time of writing, the Swedish Schools Inspectorate (*Skolinspektionen*) is undertaking an assessment of the quality of schools at primary and lower secondary level with regard to ESD. Quality is here understood to refer to the holistic organization and premises of education as provided over different school years. The assessment hints at a return to a more holistic and citizenship-based understanding of ESD, as it is particularly interested in the potential of learners to influence their own schooling and learning.

Another emerging trend relating to ESD in Sweden is increasing alignment between ESD and the concept of entrepreneurial learning. Promoted by the EU as a competency to ensure competitiveness 'in the globalised 'knowledge-based society' of today,' entrepreneurial learning has become an increasingly mainstream concept in the Swedish education system, and was integrated into the compulsory curriculum in 2011 (Dahlstedt and Fejes 2019). In terms of its implicit emphasis on individual choice, innovation and competitiveness, the increased traction of the concept of entrepreneurial learning has apparent linkages to broader neoliberal reforms in the

Swedish political and educational systems. As well as having similar aims to ESD in that it aims to foster citizens who can respond to the challenges facing society, entrepreneurial learning also promotes similar qualities, such as 'creativity, curiosity, self-confidence, and willingness to test ideas and solve problems' (ibid. p,2). In view of these areas of overlap, there has been an increasing tendency to align the two policy concepts among institutional stakeholders. WWF Sweden, for example, argues that the goals and objectives of ESD and entrepreneurial learning should be united, as creative and innovative people are needed both to 'maintain the competitiveness of a globalized economy' and to 'tackle global ecological and social challenges.' (WWF Sweden 2012). Similarly, the environmental protection agency writes that '[t]he ongoing concentration on entrepreneurial learning should be complemented and related to ESD' (Naturvårdsverket 2015).

However, the alignment of these two concepts has been met with some resistance, particularly among Swedish educational scholars. The objections raised often relate to the fact that this would downplay the tensions between the promotion of economic expansion and the pursuit of ecological and social sustainability, and emphasise individual responsibility over collective political action. Thus, authors such as Knuttson (2013) argue that '[r]esponsible individual entrepreneurs with high moral standards are expected to deal with, or even solve, problems generated by the prevailing economic system,' while Norberg (2016) writes that 'fostering a democratic citizen is subordinate to fostering citizens with entrepreneurial abilities.' On one level, the fact that some see alignment and others misalignment between ESD and entrepreneurial learning speaks to the fact that 'the breadth of the discourse of sustainable development means that it lends itself to contrasting and sometimes contradictory interpretations' (Spaiser et al. 2017). This example is also a useful mirror of broader discursive struggles within the Swedish education system, and society more broadly, as the concept of the role of the citizen, and hence the student, is shifting with the reimagining of the Swedish social democratic model along increasingly neoliberal lines.

Conclusion

This chapter aimed to give an account of Sweden's comparatively long-standing history of engagement with ESD, which has evolved alongside wider political and educational reforms. As seen in the different historical trends, teaching approaches, and attitudes to entrepreneurial learning, ESD remains a locus of ongoing negotiation in Sweden at the level of both policy and delivery. The struggle for ESD can here be interpreted as inherently interwoven with the struggle to define societal development more generally. Our account also highlights that ESD in Sweden involves a notion of citizenship education that is closely aligned with the Swedish notion of democracy and citizenship education. Accordingly, there has been a strong tradition of understanding ESD in terms of socialization and subjectification in and through education, highlighting tensions between the idea of fostering particular citizens and seeing education as democracy in alignment with the idea of one school for all. Emphasis on education as a place that is open to political and democratic deliberation has, however, been a key aspect of the interpretation of ESD in Sweden. When we come, then, to key conceptions of ESD that have had an enduring influence on its interpretation at a policy level as well as in the context of teaching, the white paper SOU: 2004 can be seen as a cornerstone for aligning historical trajectories and informing policy engagement with the Global Action Programme on ESD as well as initial efforts to engage with the ESD for 2030 programme in Sweden.

References

Bergstrom, S. 2017. Motion till riksdagen: 2017/18:2396: Prioritera lärande om hållbar utveckling i skolan. Retrieved from: https://www.riksdagen.se/sv/dokument-lagar/dokument/motion/prioritera-larande-om-hallbar-utveckling-i-skolan_H5022396.

Biesta, G. 2009. Good education in an age of measurement: on the need to reconnect with the question of purpose in education. Educational Assessment, Evaluation and Accountability 21(1): 33–46.

Borg, C., N. Gericke, H.O. Höglund and E. Bergman. 2012. The barriers encountered by teachers implementing education for sustainable development: discipline bound differences and teaching traditions. Research in Science & Technological Education 30(2): 185–207.

Carlsson, G. and M. Jonsson. 2013. Motion till riksdagen 2013/14:Ub341 Lärande för hållbar utveckling i all utbildning. Retrieved from: https://data.riksdagen.se/fil/DEE12755-DBE5-4A4C-BAD3-499B859B2CA2.

Dahlstedt, M. and A. Fejes. 2019. Shaping entrepreneurial citizens: a genealogy of entrepreneurship education in Sweden. Critical Studies in Education 60(4): 462–476.

EPI (Environmental Performance Index). 2020. EPI Results. Yale University. Retrieved from: https://epi.yale.edu/epi-results/2020/component/epi.

Eriksson, S. 2014. Motion till riksdagen: 2014/15:442: Utbildning för hållbar utveckling. Retrieved from: https://data.riksdagen.se/fil/144DD193-C096-45E7-B19E-ECADC5FA5CB4.

FI (Finansdepartementet/Ministry of Finance). 2018. Handlingsplan Agenda 2030: 2018–2020. Retrieved from: https://www.regeringen.se/49e20a/contentassets/60a67ba0ec8a4f27b04cc4098fa6f9fa/handlingsplan-agenda-2030.pdf.

Hasslöf, H. and C. Malmberg. 2015. Critical thinking as room for subjectification in education for sustainable development. Environmental Education Research 21(2): 239–255.

Jörberg, L. 1965. Structural change and economic growth: Sweden in the 19th century. Economy and History 8(1): 3–46.

Knutsson, B. 2013. Swedish environmental and sustainability education research in the era of post-politics? Utbildning and Demokrati 22(2): 105–122.

Lindensjö, B. and U.P. Lundgren. 2000. Utbildningsreformer och politisk styrning. Stockholms Universitets Förlag, Stockholm.

Lindster-Norberg, E.L. 2016. Hur Ska Du Bli När Du Blir Stor? Umeå Unversitet, Umeå. Retrieved from: https://umu.diva-portal.org/smash/get/diva2:1046939/FULLTEXT01.pdf.

Lundahl, L. 2005: A Matter of self-governance and control. The reconstruction of Swedish education policy: 1980–2003. European Education 37(1): 10–25.

Lundegård, I. and P.O. Wickman. 2007. Conflicts of interest: an indispensable element of education for sustainable development. Environmental Education Research 13(1): 1–15.

Naturvårdsverket. 2015. Styr med sikte på miljömålen. Retrieved from: https://www.naturvardsverket.se/Documents/publikationer6400/978-91-620-6666-6.pdf?pid=16477.

Olsson, L. and L. Ekdahl. 2002. Klass i rörelse : arbetarrörelsen i svensk samhällsutveckling. Nacka: Inläsningstjänst. Retrieved from: http://urn.kb.se/resolve?urn=urn:nbn:se:lnu:diva-67592.

Olsson, U. and K. Petersson. 2005. Dewey as epistemic figure in the Swedish discourse of governing the self. pp. 39–60. *In*: Popkewitz, T. S. (ed.). Inventing the Modern Self and John Dewey: Modernities and the Traveling of Pragmatism in Education. Palgrave Macmillan, New York.

Öhman, J. 2006. Pluralism and criticism in environmental education and education for sustainable development: a practical understanding. Environmental Education Research 12(2): 149–163.

Östman, L. 1995. Socialisation och mening: No-utbildning som politiskt och miljömoraliskt problem. Acta Universitatis Upsaliensis, Uppsala.

Östman-Aaro, E. and L. Östman. 2013. Sweden, in National Journeys towards Education for Sustainable Development. UNESCO, Paris.

Rudsberg, K. and J. Öhman. 2010. Pluralism in practice—experiences from Swedish evaluation, school development and research. Environmental Education Research 16(1): 95–111.

Sachs, J., C. Kroll, G. Lafortune, G. Fuller and F. Woelm. 2021. Sustainable Development Report 2021. Sustainable Development Report 2021. Oxford University Press, Oxford. https://doi.org/10.1017/9781009106559.

Sandell, K., J. Öhman and L. Östman. 2005. Education for Sustainable Development. Studentlitteratur AB, Stockholm.

SOU (The government's official investigations) (2004: 104). Att lära för hållbar utveckling. Retrieved from: https://www.regeringen.se/49b71d/contentassets/09ac8f7b0f9d402395ff95af1f6eb7cf/att-lara-for-hallbar-utveckling-sou-2004104.

SOU (The government's official investigations) (2019: 13). Agenda 2030 och Sverige: Världens utmaning – världens möjlighet. Retrieved from: https://www.regeringen.se/493ab5/contentassets/a1d21f7c7c7c484e96c759f2b3c44638/agenda-2030-och-sverige-varldens-utmaning--varldens-mojlighet-sou-201913.pdf.

Spaiser, V., S. Ranganathan, R. Bali Swain and D.J.T. Sumpter. 2017. The sustainable development oxymoron: quantifying and modelling the incompatibility of sustainable development goals. International Journal of Sustainable Development and World Ecology 24(6): 457–70.

Stråth, B. 1993. Folkhemmet mot Europa: Ett historiskt perspektiv på 90-talet. Tiden. Retrieved from https://researchportal.helsinki.fi/en/publications/folkhemmet-mot-europa-ett-historiskt-perspektiv-på-90-talet.

Sund, P. and P.O. Wickman. 2011. Socialization content in schools and education for sustainable development—I. A study of teachers' selective traditions. Environmental Education Research 17(5): 599–624.

Szczepanski, A. 2013. Platsens Betydelse För Lärande Och Undervisning – Ett Utomhuspedagogiskt Perspektiv. Nordic Studies in Science Education 9 (1): 3–17.

The National Agency of Education. 2001. Miljöundervisning Och Utbildning För Hållbar Utveckling i Svensk Skola. Skolverket, Stockholm.

The National Agency of Education. 2018. Curriculum for the compulsory school, preschool class and school-age educare. Retrieved from: https://www.skolverket.se/download/18.31c292d516e7445866a218f/1576654682907/pdf3984.pdf.

Utbildningsdepartementet (The Ministry of Education and Research). 2010. Skollag (2010:800) (The Education Act). Retrieved from: https://www.riksdagen.se/sv/dokument-lagar/dokument/svensk-forfattningssamling/skollag-2010800_sfs-2010-800.

Utbildningsdepartementet (The Ministry of Education and Research). 2019. Högskoleförordning (1993:100) (The Higher Education Act). Retrieved from: https://www.riksdagen.se/sv/dokument-lagar/dokument/svensk-forfattningssamling/hogskoleforordning-1993100_sfs-1993-100.

Wallentheim, A. 2019. Motion till riksdagen 2019/20:933: Lärande för hållbar utveckling (LHU). Retrieved from: https://data.riksdagen.se/fil/B1806B17-C14F-45B7-93DD-2181097CC9BA.

WWF Sweden. 2012. Entreprenörskap och lärande för hållbar utveckling inom skola och högskola. Retrieved from: https://wwwwwfse.cdn.triggerfish.cloud/uploads/2019/01/utredning.pdf.

Chapter 10

The Case(s) of Environmental and Sustainability Education in the United Kingdom

Alun Morgan and *Paul Warwick**

Introduction

The United Kingdom (UK) has a well-established legacy of imperialism, industrialisation and urbanisation that means its reformative path towards environmental and sustainability education (ESE) is unique in the scale of its spatial and temporal dimensions (Carter and Lowe 2014). Arguably, it is this very context that has provided the incubation space to nurture an alternative history of other-wise; where radical grassroots movements for educational and societal change within the UK have not only been seeded but evolved into communities of enquiry and practice. With this hope of preferable sustainable futures that UK citizens advocate and mobilise for, often in partnership with global neighbours, this chapter begins by highlighting the complex geo-political-cultural context of the UK. From this contextual framing, it seeks to explain how the UK educational policy context is actually far from united; with ebbs and flows of reform progress and conservative retreat. But in the midst of these political tides, this chapter highlights the thriving resilience of the third sector, where charitable and non-governmental organisations have steadfastly advanced ESE in the UK. Consequently, this chapter emphasises the crucial role to be played by a wide variety of co-creation networks in taking forwards ESE in the current context of the Global Action Program (GAP) of ESD and the follow-up programme 'ESD for 2030' within the UK.

University of Plymouth, UK.
Email: alun.morgan@plymouth.ac.uk
* Corresponding author: paul.warwick@plymouth.ac.uk

A complex geo-political-cultural context in the United Kingdom informing educational policy

The United Kingdom (UK) is a Nation State composed of four relatively autonomous 'home nations' (England, Wales, Scotland and Northern Ireland). England is dominant in terms of geographical and population size, but also given a historical legacy of military and political ascendency over the other 'home nations.' Indeed, too often the UK is erroneously considered co-terminous with 'England.' However, there has been a gradual attenuation of England's dominance since the 1950s, as the so-called 'Celtic fringe' nations have (re)asserted their cultural distinctiveness. This led to devolution movements at the end of the twentieth century, and ultimately to the creation of semi-autonomous 'National Assemblies'. These 'identity politics' have taken on a new impetus in the context of a lengthy period of Anglocentric policy and most recently in the light of Brexit (or the UK's separation from the European Union).

A consideration of these separate constituent parts of the United Kingdom, and their inextricably linked histories, is helpful as it accounts for the contemporary diversity across the UK in terms of education generally, and ESE specifically. The different cultural and political contexts have given rise to more or less supportive or progressive policies and practices, with England arguably falling behind, or failing relative to, the other home nations in recent years, after a promising start.

Historical educational considerations

England and Wales were effectively governed as one political unit from Tudor times, resulting in a shared education system. This situation gradually changed in the latter part of the 20th century resulting in a partial divergence in educational systems. Northern Ireland was created in 1922 through Partition from the remainder of the island, tying it closely to the rest of the UK. Consequently, it too shares a similar education system to England and Wales, although it has a particular legacy of sectarianism to contend with. Scotland, on the other hand, has always had an independent education system, and is therefore significantly different from England, Wales and Northern Ireland. However, since devolution in 1999, each respective Assembly has pursued its own, often divergent, educational policies. A notable tension which partly accounts for these divergences is between so-called traditional vis-à-vis progressive educational ideologies. The latter tends to emphasise 'the value of play, the use of the outdoors, and learning by discovery with the child at the heart of the educational process' (Scott and Vare 2020, 105) and arguably this is currently more apparent in non-English contexts.

The legacy of a vibrant third sector in the UK

Within the UK context there has been a long history of the 'third sector' being vital in advancing democracy (Rochester 2013). Charities, voluntary and community organisations, social enterprises and cooperatives have played a leading role in civil societal development across all of the home nations as an advocate and protagonist for compassionate social and environmental change over recent centuries. This has given rise to a number of environmental and civic educational reform movements and alliances. These have spanned a spectrum from reformist to radical—and have been extremely influential, particularly in relation to the emergence of so-called 'adjectival educations' in the UK, namely: environmental education, development education, peace education, world studies, global education, multicultural, and 'anti-racist' education. Together, these have informed contemporary ESE practice across the UK. The list of charities contributing to these 'adjectival educations' has expanded in recent years, and there has been an increasing trend towards collaboration both within the sector (between charities) and with other sectors (such as formal education).

The contemporary situation of ESE and its early development in the UK

Despite the shared histories of the four home nations, there have been significant differences and divergences in policy since devolution. Each Assembly has tried to redress centuries of Anglocentrism, achieve self-determination, celebrate their cultural distinctiveness and carve out a place in the international arena, all while addressing issues particular to their context. This is expressed in emphases which are often quite different (and deliberately so) from England. This is particularly the case in terms of education, and by extension ESE, which fall within the remit of the devolved Assemblies. In recent years, each of the home nations (including England) has undertaken revision of their respective curricula, which, to a greater or lesser extent, support or constrain the ESE agenda. We attempt to outline these key distinctions below. However, before doing so, it is appropriate to consider the continuing commonalities across the UK, particularly in terms of the key policy players, influencers and stakeholders in each context. Whether in terms of Central Government of the Nation-State or devolved Assemblies, the key government departments are those with a remit for Education or the Environment. Then there are associated Quasi-Autonomous Government Organisations (Quangos) or Non-Departmental Public Bodies (NDPB) which influence policy related to formal education including official qualifications, accreditation and/or assessment agencies for school qualifications or further and higher education; and inspectorates. Examination boards also exert a significant influence on choices around content and pedagogy since their respective syllabi dictate what is taught for examination. Some examination boards in the UK have paid more attention than others to including relevant aspects of ESE into their 'examination specifications.' Also significant are the professional support networks which act on behalf of specific subject areas (Subject Associations and Learned Societies) and/or cross-curricular areas to provide teachers with support, advice and continuing professional development and which can also serve as pressure groups for further reform.

Finally, the higher education sector has been a significant player in promoting research informed innovation and development in ESE, for example the Centre for Research in Environmental Education (CREE) at Bath University; the Centre for Cross-Curricular Initiatives (CCSI) at London South Bank University; and the Development Education Research Centre (DERC) at the Institute of Education, UCL.

UK NGOs leading the way

A significant source of ESE advocacy and leadership in the UK has been non-governmental organisations (NGOs) including development and/or environmental charities and associated centres; outdoor learning organisations; and youth and inclusion organisations.

In terms of 'environmental education,' Scott and Vare suggest that there have been four major strands, namely: an environmental/nature studies tradition, a conservation tradition, an outdoor education tradition, and an urban studies tradition (Scott and Vare 2020, 173). Environmental/outdoor-oriented youth and educational movements have contributed greatly to these strands. These include the Scouts and Woodcraft Folk, the latter reacting against the overly imperialistic-militaristic culture of the former (Mills 2014). They also include outward bound, open air schools, and field studies; and educational wings of charities such as the Royal Society for the Protection of Birds (RSPB), The Word Wildlife Fund-UK, the Wildlife Trusts, Keep Britain Tidy and UK-wide or national landscape and/or heritage charities such as the National Trust, English Heritage/Cadw/Muir Trust, etc.

Historically speaking, a number of charitable organisations have been set up in the UK to act as umbrella organisations specifically helping to advance the ESE agenda. One such example is the Council for Environmental Education (CEE), a co-ordinating body in the UK first established in the 1970s that over a number of decades drew over 70 organisations together to respond to the educational priorities that conferences on the state of the countryside were helping to highlight. Similarly, the Development Education Association (DEA) and then its rebranded successor, Think Global, helped to co-ordinate a significant third sector movement for global learning and global citizenship education for 35 years until its closure in 2018. Vital aspects of this global learning education continue to be advanced by long-standing charitable organisations in the UK such as Christian Aid, Action Aid, and Practical Action as well as a host of locally based development education centres. Of particular relevance has been the work of Oxfam including its *Curriculum for Global Citizenship* (Oxfam 1997); and World Wide Fund for Nature's UK branch (WWF-UK) with its emphasis on supporting continuing teacher professional development (CPD). Oxfam and WWF-UK have worked collaboratively on numerous initiatives as both are firmly committed to wedding environmental and social justice issues.

Within this manifold context it is possible to highlight some of the key contemporary developments in ESE within each of the four home countries.

ESE in England

A conducive policy environment for ESE in England has ebbed and flowed through numerous political interventions into the work of schools and teachers with each change of government. The assertion of a 'New Right' agenda by the Conservative Government in the 1980s marginalised more progressive educational approaches such as the 'adjectival educations' (by way of illustration, the UK left UNESCO in 1985, not returning until 1997 in response to a change in government). Most significantly, the 1988 Education Reform Act (ERA) introduced a centrally-devised, content-driven and traditional subject-oriented National Curriculum for England and Wales (NC) which was in place from 1991. The earliest version did promote the idea of 'cross-curricular themes,' including Environmental Education (National Curriculum Council 1990). However, these themes were often overwhelmed by more pressing, core curriculum agendas. Since that time, the NC has undergone several revisions, most notably in 2000 (DfEE and QCA 1999) under the jurisdiction of the Labour Government which used it to promote a 'New Agenda' incorporating Personal, Social, Health Education (PSHE), Citizenship Education, and Education for Sustainable Development (ESD). A particular innovation was the inclusion of a statement on the values and purposes of education and the goal to "… secure their [learners'] commitment to sustainable development at the personal, local, national and global level" (DfEE and QCA 1999, 11). ESD was a statutory part of the Programmes of Study of four subjects: geography, science, design and technology, and citizenship; and 'opportunities to contribute to ESD' were explicitly highlighted in art, history, information and communications technology, and physical education (Ibid.). This provided the context for some innovative ESE work in England, which is discussed in more detail below.

The Department for Education and Skills (DfES) published a national framework in 2006 for every school to become a sustainable school by 2020 (DfES 2006a). This concept of a sustainable school was founded upon a notion of compassionate care and an active concern for well-being in the broadest sense—with sustainable development being defined as:

- care for oneself
- care for each other (across cultures, distances and generations), and
- care for the environment (both near and far).

The national framework for sustainable schools was built around eight interconnected 'sustainability doorways':

- food and drink
- energy and water
- travel and traffic
- purchasing and waste
- buildings and grounds
- inclusion and participation
- local well-being, and
- the global dimension.

For each doorway a recommendation was stated to be achieved by 2020. For example with regard to energy and water the 2020 recommendation was: 'We would like all schools to be models of energy efficiency, renewable energy and water conservation, showcasing opportunities such as wind, solar and biomass energy, insulation, rainwater harvesting and grey water recycling to everyone who uses the school' (DCSF 2009, 35).

This sustainable schools framework was important for articulating an integrated ESE approach where the eight doorway objectives for sustainability were to be achieved through a joined-up approach to curriculum (teaching provision and learning); campus (values and ways of working); and community (wider influences and partnerships). In support of this, schools in England were provided with a range of resources including a self-evaluation framework to monitor progress (DCSF 2009).

At the same time, the Qualifications and Curriculum Authority (QCA) established 'the global dimension and sustainable development' as one of its cross-curricular themes in the national curriculum. Citizenship Education was also introduced as a new curriculum subject at primary, secondary and Post 16 levels, again with clear links to the ESE agenda. For example, within the Citizenship Education directive, students were to be encouraged and supported to identify, investigate and think critically about controversial global issues that were of concern to them, and to be engaged with making a difference through experiential and participatory pedagogies that facilitated their learning through active citizenship in their communities (QCA 2004).

The ESE agenda in the early years of the new century in England importantly started to identify a broad set of competencies, capacities and skills to be at the heart of what ESE was essentially seeking to draw out from all learners: 'We do not know exactly what will be the skills needed for sustainable development, but we expect that they will include teamwork, flexibility, analysis of evidence, thinking critically, making informed choices and participating in decisions' (DCSF 2008, p.15).

At the same time, a review of the secondary phase of the curriculum further supported the ESE agenda in England (DCSF and QCA 2007). This iteration kept the subject-oriented structure in place, but introduced a number of innovations. First, there were three overarching goals to create: successful learners, confident individuals, and (most pertinent for this chapter), responsible citizens. This final goal included emphases on social justice and environmental sustainability. Second, seven cross-curricular dimensions were identified including community participation, and the 'global dimension and sustainable development.' Finally, a 'big picture' was presented that located the subjects within an overarching framework that foregrounded the purposes outlined above. Alongside these developments, in 2006 the Manifesto for Learning Outside the Classroom (LOtC) was published by the government (DfES 2006b) to promote outdoor learning and community-based learning. The Council for Learning Outside the Classroom (CLOtC) was created in 2008 to help advance this work.

However, in 2010, the Conservative party regained political power (first as the leader in a coalition government with the Liberal Democrats party 2010–2015; and then outright from 2015 to present) and led to an significant initial decline in ESE policy in England. A more traditional conception of education has reversed the emphasis on cross-curricular subjects, ESD and LOtC. Thus, all reference to the Sustainable Schools Framework was pointedly removed from the curriculum and associated government websites, and whilst the LOtC Manifesto is still in operation, it has received very little support. From this time on, the other devolved governments have taken on the progressive ESE mantle. Nevertheless, throughout this time of slippage in ESE across the formal education sector in England, the National Association for Environmental Education provided helpful curriculum guidance where the existing curriculum still allowed opportunities to deliver ESE (NAEE 2015, 2018). Additionally, some aspects of the ESE agenda have continued to flourish. For example, in England, as in the other home countries, there has been a significant rise in forest schools, and their adoption in a range of formal settings although this is not without criticism (Leather 2018, Morgan 2018).

At the time of writing, there is hope that ESE policy in England is about to ebb forwards once again. Coinciding with the UK hosting the 26th UN Climate Change Conference of the Parties (COP26) in 2021, and in direct response to UNESCO's 'ESD for 2030,' the Department for Education (DfE) in England has established a Sustainability and Climate Change Unit. In April 2022 the DfE issued a policy paper 'Sustainability and climate change: A strategy for the education and children's services systems' (DfE 2022). This strategy makes explicit mention of operating within the context of the SDGs as well as UK government legislation to meet net zero carbon targets by 2050. It seeks to support ESE through learners re-connecting with nature and through initiatives such as a virtual National Education Nature Park to engage young people with the natural world and directly involve them in increasing biodiversity. It also proposes a Climate Leaders Award as a structured route for children and young people to participate in practical positive action for sustainable futures and acknowledges the need for structures that support the development of sustainability leadership (DfE 2022).

ESE in Wales

As noted previously, Wales and England have followed similar paths, up to at least Devolution in 1999, although Wales arguably had a stronger internationalist and social-democratic political milieu. But where the Welsh formal education system has led the way in comparison to England has been in its more consistent inclusion of ESE considerations over the last decade, and therefore more established positioning to respond to the educational imperative of GAP and 'ESD for 2030.' At Devolution, the Welsh Assembly began to set itself the ambitious goal of 'Creating a sustainable, inclusive and equal Wales' (NAW 2001). The national curriculum in Wales underwent two reviews during the New Labour years, first in 2000 (NAW and ACCAC 2000) and then 2008 (WAG 2008). Neither was radically different from that of England, although a strong focus was placed on how to promote 'Education for Sustainable Development and Global Citizenship' (ESDGC). This identified seven ESDGC themes that have some clear points of resonance with the sustainable schools doorways outlined above but makes much clearer reference to the climate change priority. The seven themes of ESDC are as follows (Welsh Assessembly Government 2008a):

1. *Consumption and waste*: Understand that resources are finite and that this has implications for people's lifestyles, industry and future generations.

2. *Choices and decisions*: Understand that choices and decisions have consequences and that conflicts are a barrier to development and a risk to us all and why there is a need for their resolution and the promotion of harmony.

3. *Health*: Understand the importance of a healthy lifestyle and acknowledge that basic needs must be met universally.

4. *Climate change*: Recognise the importance of taking individual responsibility and action to make the world a better place and understand the range of alternative ways to both save and generate energy.

5. *Identity and culture*: Understand, respect and value human diversity, cultural, social and economic; and recognise the negative impact of discrimination and prejudice on individuals and groups.

6. *The natural environment*: Develop respect for all living things and acknowledge the relationship between people and the environment.

7. *Wealth and poverty*: Understand how people, the environment and the economy are inextricably linked at all levels, from local to global.

In comprehensive support of this educational policy in Wales, guidance has been provided specifically tailored for schools, teacher educators and the adult and community learning sector (Welsh Assembly Government 2008a, 2008b, 2012).

Building upon this momentum, a new curriculum has been created for implementation from 2022 (Education Wales 2020) which represents a significant departure, being reformulated as an 'aims-based education, where the aims of education are to enable people to lead fulfilling personal, civic and professional lives' (Gatley 2020, p.203). This is structured around two strands. First, there are Four Purposes, with the fourth being particularly pertinent: 'ethical, informed citizens who are ready to be citizens of Wales and the world,' and which incorporates a specific commitment to 'the sustainability of the planet' (Donaldson 2015, pp.31–32). The second strand is Areas of Learning and Experience (AoLE), which are: expressive arts; health and well-being; humanities; languages, literacy and communications; maths and numeracy; and science and technology. The curriculum guidance emphasises how schools and teachers must address the sustainability agenda with significant emphasis placed upon applied and experiential teaching and learning approaches (Education Wales 2020).

ESE in Northern Ireland

The Northern Ireland Curriculum (NIC) has similarly followed that of England and Wales. The political milieu is generally more Conservative and Unionist and more closely resembles the prevailing English system. However, Northern Ireland has had to deal with its ongoing legacy of sectarianism, which has been reflected in segregated schools and communities. Thus, much emphasis is placed on intercultural dialogue and reconciliation with some integrated schools. The current curriculum introduced in 2007 (CCEA 2007) is based largely on the National Curriculum implemented in England and Wales at that time, and remains in place despite the retrogressive changes in England discussed above. It is structured around a 'Big Picture' framework incorporating three key objectives related to skills and competencies: to develop the young person as … 'an individual,' 'as a contributor to society;' and 'as a contributor to the economy and environment,' this last objective expressly including 'education for sustainable development'. The curriculum promotes 'Areas of Learning' including 'Environment and Society' and moved Citizenship, Employability and Personal Development from the margins to the centre with subjects being subordinated to them. Within this educational policy, of particular relevance to ESE is the Department of Education policy document 'Schools for the Future: A Policy for Sustainable Schools' (DoE 2009). This has a broad notion of 'sustainability' including financial, but crucially acknowledges the need for sustainable development to be an important part of the education system through the curriculum, school facilities and learning space design. It highlights the importance of students learning sustainability through the day-to-day practices of the school and the high value placed on the well-being of its pupils and the school environment. In addition, an implementation plan for sustainable development was published in November 2006. This strategy committed the Department of Education to a wide-ranging set of actions to embed sustainable development in all aspects of education (DoE 2009).

Thus, the current curriculum still provides a more positive context for addressing ESE than has been the case in England in recent years, although there are particular challenges related to the specific 'post-conflict' and continuing sectarian context (Niens and Reilly 2012, Reilly and Niens 2014). Currently there are discussions around the desirability of further revising the Northern Ireland Curriculum. However, this as yet does not appear to represent a significant departure from the earlier curriculum framework.

ESE in Scotland

From Devolution, Scotland initially followed the existing curriculum inherited from the earlier period. However, from 2002 a creative review of education was undertaken resulting in a radically new approach called 'Curriculum for Excellence' (CfE) (Scottish Executive 2004) which has been implemented since 2010. The CfE foregrounds 'experiences' and 'outcomes' expressed as four fundamental capacities: successful learners, confident individuals, effective contributors, and responsible citizens. The Scottish context uses the term 'Learning for Sustainability' (LfS) in an expansive sense to describe a process which weaves together sustainable development education (SDE), global citizenship and outdoor learning (Scottish Government 2012). LfS has been developed by the Scottish Government in partnership with the 'One Planet Schools Working Group' and the 'Learning for Sustainability National

Implementation Group' with the latter group publishing their 'Vision 2030+' report which revisited and expanded upon the original LfS recommendations that:

1. All learners should have an entitlement to Learning for Sustainability.

2. In line with the new General Teaching Council for Scotland Professional Standards for Teachers, every practitioner, school and education leader should demonstrate Learning for Sustainability in their practice.

3. Every school should have a 'whole-school' approach to Learning for Sustainability that is robust, demonstrable, evaluated and supported by leadership at all levels.

4. All school buildings, grounds and policies should support Learning for Sustainability.

5. A strategic national approach to support for Learning for Sustainability should be established (LfS National Implementation Group 2016: 6).

This understanding of LfS and these five strategic objectives have received governmental commitment (Scottish Government 2013, 2016). A Learning for Sustainability Action Plan was published in 2019 and sets out how the Scottish Government will, with others, implement the recommendations of the Vision 2030+ report. Within this strategic approach emphasis is placed on Outdoor Learning as an entitlement in terms of offering 'opportunities for all children and young people to enjoy first-hand experience outdoors, whether within the school grounds, in urban green spaces, in Scotland's countryside or in wilder environments. Such experiences motivate our children and young people to become successful learners and to develop as healthy, confident, enterprising and responsible citizens' (LTS 2010, p.7).

Emerging shared understandings of ESE practice across the UK

From the brief historical review of the unfolding and diverse context of the UK's engagement with ESE, it is possible to identify some common approaches. Within the numerous educational policies connected to the ESE agenda it is possible to see links made to responsible and active citizenship education agendas and to notions of personal well-being. It is also possible to highlight three unifying understandings of best practice in ESE, as summarised below.

A systemic and holistic approach that encompasses environmental care and social justice

The approach of 'Learning for Sustainability' adopted in Scotland, and 'Education for Sustainable Development and Global Citizenship' recognised in Wales, embody a common recognition amongst many ESE practitioners and theorists of the synergistic relationship between sustainable development education (SDE), global citizenship, global learning and outdoor learning. Key to this is the recognition of the need to bring together environmental and social justice agendas. Adopting a systems thinking educational approach that recognises the interdependence of the environmental and societal SDGs is seen as key. Similarly, ESE, Outdoor Learning, and Health and Well-being are being increasingly recognised as interconnected and synergistic in formal and informal learning contexts. Whilst not without their critiques, the UN Sustainable Development Goals (SDGs) are seemingly presenting a common framework for exploring these issues and their interconnectedness across the UK.

The importance of place-based approaches and the increasing emphasis on outdoor immersive learning

This interconnected systemic approach is also evident through increased attention to the learning context, a particular emphasis in the 'Place-Based Education' (PBE) movement which seeks to orient learning around the issues and characteristics presented in the home locality and lived experience of learners, while simultaneously relating them to broader regional and global issues (Gruenewald and Smith 2008). While initially different and often separate strands of place-based learning were in evidence associated either with rural (environmentally-oriented) vis-à-vis urban (social justice-oriented) contexts, effort is increasingly being made to synergistically combine the 'best of both worlds' (Gruenewald 2003). Today, Place-Based Learning or Place-Based Education represents an emerging focus in the UK as it does across the world, offering fertile ground for the intersection of formal and informal learning contexts (Morgan 2009, 2012, Smith and Sobel 2010). The devolved administrations of Wales, Northern Ireland and Scotland, with their flexible curricular structures, greater emphasis on progressive pedagogies *and* their strong focus on 'home nation' specific issues represent a potentially fruitful ground for developing this 'place-based' orientation. For example, What Works Scotland (Bynner 2016) and Children's Neighbourhoods Scotland (McBride 2018) have both advocated for 'place-based approaches' which are relevant to the Curriculum for Excellence. England with its traditional subject-based curriculum structure prevailing in the National Curriculum, has had more limited opportunities for promoting place-based learning. However, there is hope that in the recent DfE strategy for sustainability and climate change the emphasis on locality-based activities presents a far more place-based educational approach for ESE within England going forwards (DfE 2022).

The emphasis on participatory, dialogic and reflexive pedagogical approaches

ESE in the UK can also be associated with an emerging emphasis on inquiry-based approaches that seek to connect 'head, heart and hand' through participatory and experiential pedagogies. This supports problem-based learning pedagogies and collaborative, dialogic and reflexive approaches such as Philosophy for Children and the Open Space for Dialogue and Enquiry (OSDE) methodology. OSDE is another example of the voluntary sector and higher education in the UK collaborating with international partners in order to develop new educational practice. Led by educators at the Centre for the Study of Social and Global Justice at the University of Nottingham and the charity Global Education Derby, the OSDE initiative developed resources for teacher-educators as well as secondary and primary schools seeking to respond to the challenges for global citizenship education in an interdependent, diverse and unequal world (OSDE 2006). Drawing from critical literacy theory, it represents a valuable methodology for de-colonising the curriculum and considering within ESE how the voices of the marginalised, disadvantaged or excluded need to be considered when engaging with the complexity and contested nature of global sustainability challenges. In particular, OSDE serves to highlight the importance of learners engaging in ESE processes that raise critical consciousness and reflexivity—where learners are helped to become self-aware of their assumptions, bias, and responsibilities, and introduced to alternative perspectives on the SDGs, for example.

The UK's implementation of ESE in the light of GAP and 'ESD for 2030'

As has been highlighted above, at a formal educational policy level, England has lagged behind the other three countries in the UK with regard to its integration of both the priorities outlined in the Global Action Programme (GAP) for ESD and the re-stated ambitions of its follow-up programme 'ESD for 2030.'

Arguably, there has been a much more unified and comprehensive response across the UK within and across the voluntary and higher education sectors. This is particularly with regard to the priority action areas of building the capacities of educators and empowering and mobilizing youth (UNESCO 2020). Consequently, there are many associations and networks helping to co-ordinate and mobilise ESE at a grassroots level in the UK. Three such examples that are briefly highlighted here are the Environmental Association for Universities and Colleges (EAUC); the Teacher Education for Equity and Sustainability Network (TEESNet) and the charitable organisation Sustainability and Environmental Education (SEEd).

The EAUC was launched in 1996 and has been a registered charity since 2004. It serves as an environmental and sustainability champion within further and higher education in the UK and Ireland. Based at the University of Gloucestershire and Queen Margaret University, it has over 300 institutional members. With regard to its vital contribution to the 'ESD for 2030' agenda, it provides a forum for ESE educators to share experiences and information, disseminating good practices and research on all aspects of improving the sector's overall environmental performance. This it achieves through a community of practice and a regional networking approach, as well as offering an annual conferencing opportunity and promoting the Green Gown Awards as a means of championing ESE initiatives and leaders (EAUC 2021a). Alongside an extensive website of supporting resources, it offers continuing professional development opportunities including the Emerging Leaders Programme (EAUC 2021b). It has also been a leading organisation within the UK for facilitating the SDG Accord as the global further and higher education sector's collective response to the SDGs. As of early 2022, this accord had over 2000 signatories from over 100 countries; demonstrating a truly globally scaled initiative (SDG Accord 2022).

TEESNet is a UK network aligned with the GAP Priority Action Area 3 of seeking to build the capacities of educators and trainers by providing a variety of support to teacher education institutions. First established by South Bank University in London with a variety of NGOs and teacher trainers, TEESNet is now co-ordinated through Liverpool Hope University, the charity Liverpool World Centre, and a steering group of UK wide partners. Currently TEESNet has over 300 members and representation from across the four UK nations, including university-based teacher education, school teachers, NGOs and Development Education Centres. In 2020, a major work theme for TEESNet was encouraging teacher-educators to consider how training could support ESE being a pedagogy of hope and possibility, exploring how the current myriad of global challenges can be engaged with in education to cultivate a sense of agency and possibility, especially amongst young people. This recognises the need to shift from individual to collective responses, for people to engage critically with their responsibilities towards others and the planet, and to foster creativity in imagining and adapting to new possibilities (TEESNet 2020). TEESNet also helps UK teacher educators to be aware of and access specialised training opportunities that otherwise might only exist in one of the home countries. One example is the Leading Change in Learning for Equity and Sustainability in ITE programme that is led by Learning for Sustainability Scotland (LfS) and the University of Edinburgh. This course has responded to the growing recognition of the need to inspire new teachers' engagement with issues relevant to young peoples' development in local and global contexts. The aim is to support teacher educators across the UK in their collective journey towards implementing and leading Learning for Equity and Sustainability in practice.

SEEd has existed as a charitable organisation since 2009 and has provided a vital source of support to schools and educators across the UK in ESE. In 2015, SEEd was selected by UNESCO as one of the key partners on their GAP ESD programme and has played a significant role in generating and scaling up concrete action on ESD in the UK. SEEd's work in Priority Action Area 2 has focused on the transformation of learning and training environments to embed whole school/institution approaches to ESD (SEEd 2016). In doing so, they have been able to gather examples of best practices on whole school approaches from across the UK as well as serve as an advocate for research–informed and transformative learning approaches that are also pertinent beyond the UK. Led by the pioneering ESE educator Ann Finlayson, SEEd's school tailored programme of support includes key features such as:

- Supporting individual schools through workshops and resources to develop their own goal and action plan
- The adoption of an appreciative inquiry approach for leading change
- Linking ESE starting schools into a cohort network to provide additional support and peer to peer learning opportunities.

SEEd currently co-chairs a broad coalition advocating for SDG4.7 in England by 2030. Our Shared World is a coalition of more than 150 members including NGOs, Businesses, Educational Institutions, Educators and Youth Groups and is supporting an evidence-based action learning approach to advancing the ESE agenda (Our Shared World 2022).

Issues and trends: An emerging integrated learning model for ESE in the UK

As this chapter has sought to emphasise, there is no singular case of how ESE has been advanced in the UK context. Instead there are many cases of educational policy directives, pioneering educators and pedagogical reform movements that over a significant period of time have contributed to advancing ESE. But from this brief review and a temporal engagement with ideas of preferable futures for ESE in the UK, we offer here a representation of an emerging integrated ESE learning model. Outlined in Figure 1 is a conceptualisation of emerging ESE in the UK, that identifies four interacting dimensions towards a transformative approach to ESE. Rather than seeing these as being organised as distinct and sequential, our observation is that apt ESE embodies all four layers of activity in an intertwined ecology of learning process. Briefly considering each in turn, the 'Connect' layer uses the SDGs framework as a common cause meeting point for a compassionate solutions focused approach to learning. It signifies a clear move towards working with students as partners, co-enquirers and co-creators, harnessing and engaging with their existing environmental and sustainable issues of concern. This is in recognition of young people in the UK, as in many other countries of the world, being increasingly encountered as concerned and engaged with a wide range of environmental and sustainability issues, rather than apathetic or passive. The 'Participate' layer to our ESE model builds upon the spirit of the student voice approach and recognises the social dimensions to learning. Embracing a critical literacy and dialogic learning space approach, this aspect of ESE places importance on facilitating for all learners the space for inter-disciplinary, inter-generational and inter-cultural exchange and, in particular, the need to encounter alternative, marginalised or excluded voices and perspectives; to imagine otherwise. The 'Activate' layer of the ESE model recognises the competency agenda increasingly underpinning ESE in the UK and seeks to provide experiential and applied pedagogical opportunities for a creative

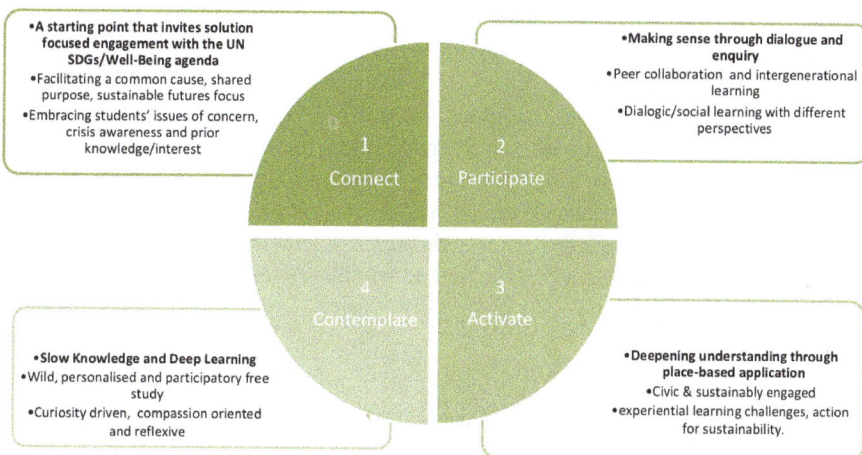

Figure 1. An emerging ESE learning model from the UK perspective (adapted from Warwick 2023).

engagement with processes of sustainable development and change. This very much requires transforming learning environments in line with 'ESD for 2030' where ESE partners with 'the local community as a valuable setting for interdisciplinary, project-based learning and action for sustainability' (UNESCO 2020: 28). The 'Contemplate' layer recognises the importance of reflective and reflexive elements of learners grappling with the contested and complex nature of the sustainability agenda. It seeks to facilitate 'deep and slow learning' through empowering students to articulate their own self-directed learning and independent study in ESE as well as reflexively consider their own points of view; its implications and assumptions, and their responsibilities in moving towards more sustainable futures. In order for such an integrated learning model of ESE to be made available for all, significant transformation of teacher education is required to enable comprehensive ESE capacity development across the UK schooling and non-formal education sector.

Conclusion

This chapter has highlighted how, within the context of an imperialist, industrial and urbanised legacy, the development of ESE in the UK has been one of ebb and flow. Rather than a singular case of how the UK has progressed the ESE imperative, at an educational policy level in particular there has been a dis-unity and diversity of approach. Our review argues that at the ESE policy level, England in recent years has lagged behind the other home nation countries. But an ESE movement of partnership and co-creation has consistently flourished in the UK through the Third Sector. Charities, voluntary networks and cooperatives within and across the home nations have significantly helped to catalyse, sustain and champion the UK response to both the GAP and the follow-up programme 'ESD for 2030' across the education system in all its formal and informal configurations. An important observation from this review of the Third Sector's contribution has been the need for ESE to seek an interconnected systems-based approach. Just as the SDGs themselves point towards positive moves for environmental and societal change that are interconnected, increasingly movements leading educational change are recognising how ESE needs to be connected across the different education sectors as well as to the closely-related fields of outdoor learning, development education and global learning. Finally, this chapter, in its review of some of the key aspects of this devolved array of grassroots movements for ESE, tentatively offers an emerging integrated model of deep and transformative learning for ESE in the UK. This embraces dimensions of learning nested within an ecological systems approach that include: solutions focused approaches to the SDGs; social learning through dialogue; experiential learning through place-based activation and reflexive learning through contemplative engagement.

References

Bynner, C. 2016. Rationales for Place-Based Approaches in Scotland. Working Paper. Glasgow: What Works Scotland.

Carter, N. and P. Lowe. 2014. Britain: Coming to terms with sustainable development? pp. 17–39. *In*: Governance and Environment in Western Europe. Routledge.

CCEA. 2007. The Statutory Curriculum at Key Stage 3: Rationale and Detail. Belfast: CCEA Belfast.

DCSF and QCA. 2007. The National Curriculum: Statutory Requirements for Key Stages 3 and 4. London: DCSF, QCA.

DCSF. 2008. Brighter Futures – Greener Lives. Sustainable Development Action Plan 2008–2010. The Children's Plan: Building brighter futures. London, Department for Children, Schools and Families.

DCSF. 2009. s3: Sustainable School Self-evaluation. London, Department for Children, Schools and Families.

DfE. 2022. Sustainability and Climate Change: A strategy for the education and children's services systems. Department for Education. Retrieved from https://www.gov.uk/government/publications/sustainability-and-climate-change-strategy/sustainability-and-climate-change-a-strategy-for-the-education-and-childrens-services-systems (Date Accessed 29/06/22).

DfEE and QCA. 1999. The National Curriculum. Norwich: HMSO.

DfES. 2006a. Sustainable Schools for Pupils, Communities and the Environment: An Action Plan for the DfES. London: DfES.

DfES. 2006b. Learning Outside the Classroom Manifesto. Nottingham: DfES Publications.

Donaldson, G. 2015. Successful futures: Independent review of curriculum and assessment arrangements in Wales. Cardiff: Welsh Assembly Government.

EAUC. 2021a. Green Gowns Awards Activity Background. Retrieved from https://www.eauc.org.uk/green_gown_awards1 (Date Accessed 20/03/22).

EAUC. 2021b. The Future Leaders Programme. Retrieved from https://www.eauc.org.uk/shop/mms_single_event.php?event_id=6685 (Date Accessed 20/03/22).

Education Wales. 2020. Curriculum for Wales guidance. Cardiff: Welsh Assembly Government.

Gatley, J. 2020. Can the New Welsh Curriculum achieve its purposes? The Curriculum Journal 31(2): 202–214.

Gruenewald, D.A. 2003. The best of both worlds: a critical pedagogy of place. Educational Researcher 32(4): 3–12. Retrieved from http://ucce.ucdavis.edu/files/filelibrary/5618/12532.pdf.

Gruenewald, D.A. and G.A. Smith (eds.). 2008. Place-Based Education in the Global Age: Local Diversity. Abingdon, Oxon: Lawrence Erlbaum Associates.

Leather, M. 2018. A critique of Forest School: Something lost in translation. Journal of Outdoor and Environmental Education 21: 5–18.

LTS. 2010. Curriculum for Excellence through Outdoor Learning. Glasgow: Learning and Teaching Scotland.

McBride, M. 2018. Place-based approaches to support children and young people. *In*: Glasgow: Children's Neighbourhoods Scotland.

Mills, S. 2014. 'A Powerful Educational Instrument': The Woodcraft Folk and Indoor/Outdoor 'Nature', 1925–75. pp. 65–78. *In*: Informal Education, Childhood and Youth. Springer.

Morgan, A. 2009. Learning communities, cities and regions for sustainable development and global citizenship. *In*: Local Environment: The International Journal of Justice and Sustainability 14(5): 443–459.

Morgan, A. 2012. Inclusive place-based education for just sustainability. International Journal of Inclusive Education 16(5-6): 627–642.

Morgan, A. 2018. Culturing the fruits of the forest: realising the multifunctional potential of space and place in the context of woodland and/or forest schools. Journal of Outdoor and Environmental Education 21(1): 117–130. doi:10.1007/s42322-017-0008-z.

NAEE. 2015. The Environmental Curriculum: Opportunities for Environmental Education across the National Curriculum for England—Early Years Foundation Stage & Primary. Walsall: NAEE.

NAEE. 2018. The Environmental Curriculum: Opportunities for Environmental Education across the Secondary Curriculum for England—Key Stage 3 and 4. Bath: NAEE.

National Curriculum Council. 1990. Curriculum Guidance 7: Environmental Education. York: National Curriculum Council.

NAW and ACCAC. 2000. The National Curriculum in Wales. NAW/ACCAC.

NAW. 2001. Plan for Wales. NAW.

Niens, U. and J. Reilly. 2012. Education for global citizenship in a divided society? Young people's views and experiences. Comparative Education 48(1): 103–118.

OSDE. 2006 Critical Literacy in Global Citizenship Education: Professional Development Resource Pack. Derby: Global Education Derby.

Our Shared World. 2022. The World We Share is Changing. Retrieved from https://oursharedworld.net/ Date Accessed 5/3/22.

Oxfam. 1997. A Curriculum for Global Citizenship: Oxfam's Development Education Programme. Oxford: Oxfam.

QCA. 1998 Education for Citizenship and the Teaching of Democracy in Schools. Citizenship Advisory Group. London: Qualification and Curriculum Authority.

QCA. 2004. Play your part – Post 16 Citizenship. London: Qualification and Curriculum Authority.

Reilly, J. and U. Niens. 2014. Global citizenship as education for peace-building in a divided society: Structural and contextual constraints on the development of critical dialogic discourse in schools. Compare: A Journal of Comparative and International Education 44(1): 53–76.

Rochester, C. 2013. Rediscovering Voluntary Action: The Beat of a Different Drum. Basingstoke: Springer.

Scott, W. and P. Vare. 2020. Learning, Environment and Sustainable Development: A History of Ideas. Milton: Milton: Taylor and Francis.

Scottish Executive. 2004. A Curriculum for Excellence: The Curriculum Review Group. Edinburgh: Scottish Executive.

Scottish Government. 2012. Learning for Sustainability – Report of the One Planet Schools Ministerial Advisory Group. Edinburgh: Scottish Government.

Scottish Government. 2013. Learning for Sustainability. The Scottish Government's response to the Report of the One Planet Schools Working Group. In. Edinburgh: Scottish Government.

Scottish Government. 2016. Vision 2030+: Concluding report of the Learning for Sustainability National Implementation Group. Edinburgh: Scottish Government.

SDGAccord. 2022. SDG Accord The University and College Sector's Collective Response to the Global Goals. Retrieved from https://www.sdgaccord.org/ Date Accessed 20/03/22.

SEEd. 2016. Whole Institution Approach to Sustainability—A How To Guide. Retrieved from https://se-ed.co.uk/whole-institution-school-approach-sustainability/ Date Accessed 20/03/22.

Smith, G.A. and D. Sobel. 2010. Place-and Community-Based Education in Schools. Abingdon, Oxon: Routledge.

TEESNet. 2020. Education as a pedagogy of hope and possibility: The role of Teacher Education in Leading Narratives of Change. Retrieved from http://teesnet.liverpoolworldcentre.org/home/teesnet-2020/ Date Accessed 17/11/21.

UNESCO. 2020. Education for Sustainable Development A Roadmap. ESD for 2030. Paris, United Nations Educational, Scientific and Cultural Organization.

WAG. 2008. National Curriculum for Wales. Wales: Welsh Assembly Government.

Warwick, P. 2023. Towards a Pedagogy of Love. Reflecting upon Thirty Years of Sustainable Education. University of Plymouth (in press).

Welsh Assembly Government. 2008a. Education for Sustainable Development and Global Citizenship: A Common Understanding for Schools. Cardiff: Welsh Assembly Government.

Welsh Assembly Government. 2008b. Education for Sustainable Development and Global Citizenship In the Further Education Sector in Wales. Cardiff: Welsh Assembly Government.

Welsh Assembly Government. 2012. Education for Sustainable Development and Global Citizenship: A Common Understanding for the Adult and Community Learning Sector. Cardiff: Welsh Assembly Government.

Latin America

Chapter 11

Brazilian Environmental and Environmental Education Policies in the Context of the Sustainable Development Goals

Luiz Marcelo de Carvalho[1],* and *Marcelo Gules Borges*[2]

Introduction

Both the environmental movement and Environmental Education (EE) in Brazil emerged in the midst of the military dictatorship (1964–1984). It was an experience marked by intense political resistance to the government's violent and anti-democratic abuses and actions, involving different sectors of society in processes seeking social transformation and the reconstruction of a free and democratic society. These are characteristics of the environmental movement and their emergence marked the birth of EE in Brazil as well as in a number of other countries in Latin America and the Caribbean, all of which experienced similar political changes. In contrast with the prevailing trends in North America, EE practices in these regions were 'driven mainly by non-governmental organizations with strong ties to emancipatory and social transformation' (Briggs et al. 2018, 1631).

---

[1] Instituto de Biociências. Departamento de Educação. Unesp. Rio Claro. Avenida 24A, 1515. Cx. Postal 199.Bela Vista. Rio Claro. CEP: 13506-900.
[2] School of Education, Federal University of Santa Catarina, Campus Universitário Trindade, Florianópolis, SC. Brazil.
Email: marcelo.borges@ufsc.br
* Corresponding author: lm.carvalho@unesp.br

The military government was conservative, technical, and pragmatic, and the hegemonic nature of the Brazilian state had a detrimental impact on this process, with regard both to civil society and any culture of social participation (Lima 2009). Since ecology teaching involved strong naturalistic, behavioral and technical biases, EE practices were seen as a means of fostering desirable ecological behavior at the individual level (Loureiro 2008, Carvalho 2008). However, different ideologies that were directly linked to competing social models and models of the relationship between society and nature undoubtedly had an impact on the discourse around EE as the environmental debate was getting under way in Brazil (Lima 2009).

According to Lima, EE discourse was based on an 'axis polarized by conservatism and emancipation' (152). Although education policy in general was dominated by conservative and technical perspectives, emancipatory practices on the part of citizens became a significant trend in EE. At that time, it was possible to imagine EE becoming a complex, plural and diverse discursive field, guided by different political, ideological and pedagogical perspectives, involving a diversity of stakeholders and social sectors (Lima 2009, Loureiro 2008).

Some sectors of the environmental movement took a critical view of capitalism: this criticism reached the public sphere, and promoted emancipatory ideals (e.g., social movements and NGOs advocating public access to environmental goods). From this time onward, environmental issues gained political momentum (Carvalho 2008).

As a result of the opening and democratization of Brazilian society and politics, the EE movement grew and significant efforts were made to institutionalize EE. Although ideological disputes remain, researchers and environmental educators have worked hard to maintain this critical and emancipatory perspective as a fundamental characteristic of EE in Brazil. At the same time, multiple factors have influenced and contributed to the development of what has become a complex and diversified field in the country.

Some of these provide EE with 'ethical, aesthetic, political-ideological and theoretical perspectives' (Lima 2009, p.16), such as counter-cultural movements (1960s), anarchism and socialism, critical theory, popular education, environmentalism (preservationist and conservationist movements) from North America, and romanticism.

In this chapter, we draw a brief picture of the political and social context that has influenced Brazil's environmental and EE policies. After a very fertile period, during which environmental policies developed and experiences of EE broadened, the country went through a series of political crises, which led to the impeachment of the president who had been legitimately elected by popular vote in 2015, opening the way for groups committed to ultra-liberal economic policies and antagonistic to environmental policies, to assume power in 2016. This trend became even more radical in 2018, when the current president[1], who is on the extreme right of the

---

[1] This chapter was completed in 2021 and analyzed environmental policies, environmental education and the SDGs until Jair Bolsonaro's presidency. In 2022, Luiza Inácio Lula da Silva was elected the new president. Since 2023, significant changes in environmental and education policies have occurred.

political spectrum, took power. This has led to the reversal of many advances in social, educational, and environmental policies in Brazil.

We also outline the socio-political context in which the country committed to the UN Global Action Plan, known as the 2030 Agenda, which, along with its seventeen Sustainable Development Goals (SDGs), reinjected momentum into the notion of sustainability. We argue that the SDGs, on the one hand, have found Brazil to be fertile ground as regards EE state policies since 2010. On the other hand, the EE movement has encountered a high level of resistance, rather than being valued and supported.

Finally, we raise questions and point out the paths forged by Brazilian environmental educators, who are a source of resistance to the current socio-political context. At the same time, we argue that such resistance could open the way to appropriate cracks and opportunities for dialogue with the international community about what international policies and propositions can offer.

The historical context of environmental and environmental education policies in Brazil

During the period known as the Years of Lead in Brazil (1964–1985), there was intense social and political engagement on the part of civil society, which attempted to resist the repressive and coercive atmosphere, and put pressure on the military government to open up political space for civil society. In 1985, the electoral college voted, indirectly, for the country's first civilian president following 21 years of dictatorship, starting a new cycle for the Brazilian Republic—known as the New Republic—and the reconstruction of democracy began. Internal policies influenced this new, politically open, context, but this 'third wave of democratization' was seen as a worldwide movement (Arturi 2001, Nascimento 2015, Santos 2010). At the same time, environmental awareness increased, and EE expanded in Brazil, becoming the object of a significant set of public policies and part of the agenda of social movements (Carvalho 2008).

Brazil's environmental policies were initially implemented in the 1970s during the era of the military dictatorship, when the Special Secretariat for the Environment was created (1973). The country passed a very significant milestone for environmental public policy in 1981, with the approval of the National Environmental Policy Law, which explicitly referenced EE as a way to empower the population through educational practices in school and non-school contexts, 'aiming for active participation in defence of the environment' (Brasil 1981). This political decision was taken after the publication of the documents and agreements signed at the Stockholm Conference in 1972 and, most likely, in response to the need for Brazil to integrate itself into 'an international environmental sphere of governance' (Frizzo and Carvalho 2018, p.117). It was not only an attempt to align the country with an international environmental agenda but also to try to divert the attention of the public from the fight against the dictatorship, focusing on an apparently new conflict, relating to the environment [...]. The dictatorship understood, then, that environmental struggles had nothing to do with political, democratic and class struggles (Acserald 2020, p.1).

Following the period of military government, the promulgation, in 1988, of the Federal Constitution recognized an ecologically-balanced environment as a 'public good for the people's use and essential for a healthy life' (Brasil 1988). The Brazilian Constitution reaffirmed the principles of the National Environment Policy and the State's role in promoting EE at all levels of education (Andrade and Piccinini 2017, Frizzo and Carvalho 2018, Uchoa 2018, Braun 2020).

In the early years after the resumption of democracy in June 1992, one of the most significant events in Brazil's history of environmentalism and environmental policies took place. The country hosted the United Nations Conference on Environment and Development (UNCED – Rio-92 or ECO 92), initiating the cycle of international post-Cold War conferences coordinated by the UN (Lafer 2009).

Beyond the evaluation and outcomes of UNCED, the consequences of the conference for environmental and EE policies in Brazil are clear. The initiative had a positive impact on the country's attitude towards the international community, which had been characterized by defensiveness during the dictatorship, with national sovereignty always to the fore. By hosting and actively participating in the organization of the meeting, the country embraced open diplomacy, switching to abstention on international environmental policies in order to play a collaborative role in developing international law (Lafer 2009).

At the end of the year UNCED (Rio-92) was held, the Special Secretariat for the Environment was granted the status of Ministry, undoubtedly one of the most important public policy initiatives after Rio 92. In a landmark for the institutionalization of EE policy, the EE Division was taken over by the Brazilian Institute of the Environment (IBAMA), now incorporated into the Ministry of the Environment.

In 1993, the Environmental Education Coordination committee was created, with the objective of 'coordinating, supporting, monitoring, evaluating and guiding actions, goals and strategies for implementing EE in education systems at all levels and in all contexts' (Czapski 1998, p.58). According to Czapski, the diversity of these objectives led the federal government in 1996 to put forward the first version of the National Environmental Education Program (PRONEA), whose development would be coordinated by the Ministry of the Environment and the Ministry of Education, working closely with the Ministry of Culture and the Ministry of Science and Technology. Subsequent to this, a series of EE policies were launched, which brought EE up the agenda and led to a number of proposals, programs, and practices relating to EE, as illustrated in Figure 1.

From 1985 to 2016, the country experienced disputes, losses, and the establishment of counter-authoritarian spaces: yet it is possible to identify significant advancements in terms of environmental and EE policies. It is important to remember the severe criticisms of eco-politics and EE policies during the New Republic, mainly after 2007, for espousing an 'environmentalism of results' (Acserald 2004) or 'anti-ecologism' (Layrargues 2018a, b). But it is also important to acknowledge the years of intense debates, the creation of spaces for meetings, the exchange experiences and the implementation of environmental and EE policies and legislation that consolidated legal instruments and spaces such as treaties, laws, resolutions, and academic and educational practices, in school and non-school contexts. Strong

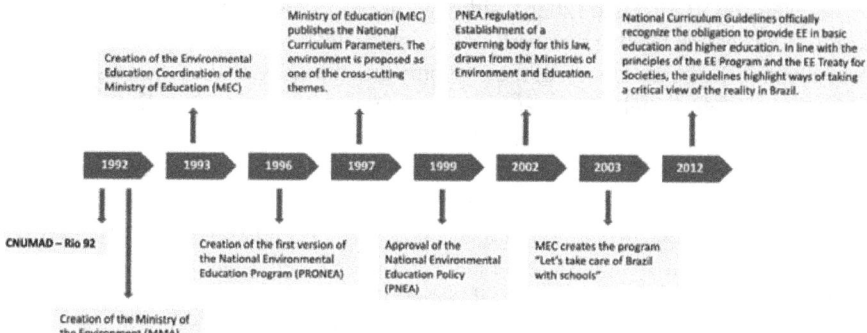

Figure 1. EE policies in Brazil between 1992 to 2012.

social participation meant the whole experience helped to consolidate EE in Brazil. Nevertheless, in the next section of this chapter, we explain the great threat to these legal spaces and practices, and to EE policy achievements.

The current social, environmental and educational context in Brazil

In 2015, Brazilian democracy began to show signs of fragility resulting from a re-emergence of 'patriarchal forces combined with misogyny and economic ultraliberalism' (Wermuth and Nielson 2018, 457). Once again the country faced a political crisis that, inflated by the media during an economic crisis, created the necessary conditions for the impeachment of the president, Dilma Roussef, in 2016.

The Brazilian experience, and the tangible threat to the democratic regime and the Republican movement in the region, can be explained and understood beyond the national context. In a number of countries in different regions around the globe, there is evidence of the advance of neoconservative policies, the 'return of protectionist policies and the expansion of military spending.' Allied to this evidence, Pochmann (2017, 322) cites the UK withdrawal from the European Union (Brexit), and the rise of nationalist forces and ultraliberal economic policies around the world, including the election of Donald Trump in the United States.

This climate of social instability, in conjunction with social conservativism and a wave of economic neo-liberalism, enabled Jair Bolsonaro to be elected president in January 2019. From then on, we experienced 'a kind of liberal-conservative regime supported by an oligarchic alliance of a civil-military character,' with Brazil becoming 'a stronghold of the great Latin American disillusionment and the exhaustion of the democracies that dawned on the horizon with the third wave, about four decades ago' (Oliveira 2018b, 40). The Bolsonaro government gained a hold with broad support from the Pentecostal Evangelical churches, and sectors like agribusiness and arms (Oliveira 2018b, Wemuth and Nielsen 2017). With Bolsonaro's election, the dismantling of the environmental and EE policies implemented in the New Republic began. After first threatening to abolish the Ministry of the Environment, Bolsonaro appointed a minister who is committed to neo-liberal economic interests, with serious consequences for environmental policy.

According to Acserald (2020), although in the 1980s the country had been able to build 'an entire legal framework in the environmental field' (4), the increasing pressures for the liberalization of the economy and the relaxation of the existing rules made this system largely inoperative: 'the process of 'environmentalization' of the Brazilian State has been truncated, the work has been interrupted, left incomplete or has been prevented' (4). In other words, this is the ideology of a 'results-oriented environmental movement' (Acselrad 2010, 106).

This trend, described by Layrargues as a sign of 'anti-ecologism' (Layrargues 2018a, 6) in Latin American political ecology and environmental thought, has influenced and guided environmental policy not only in Brazil, but also in Latin America as a whole. After the dominance of developmental ideology in the 1960s, where economic growth was central to any policy, we experienced four decades considered by Layragues (2018a, 30) as the 'golden age of ecologism,' a phase centered on discursive practices highlighting the notion of natural limits and the finitude of natural resources. However, the golden age came to an end in 2006, with environmental control policies being characterized as obstacles to development and the 'economy abandon[ing] the sustainability pact and resum[ing] the ancient hegemony over ecology' (p.32).

After the impeachment of President Dilma Roussef in 2016, with Vice-President Michel Temer serving as president, there was an onslaught of deregulation with regard to environmental policy. On the one hand, the policies implemented worked against the 'decision to maximize the environmental protection of forests and ecosystem services,' and on the other they maximized the 'profitability of the predatory extraction of natural resources' (Layrargues 2018a, p.33).

As we have pointed out earlier, this political, social and economic climate provided the ground for the election of current President, Jair Bolsonaro (2018). Environmental and EE policies have worsened dramatically under this government. Bolsonaro's government poses a serious threat to social achievements and human rights (Gresse and Engels 2020) and represents an environmental setback, deepening the trend toward anti-ecologism and associated with climate change denial (Layrargues 2020) and suspicion of scientific knowledge in general. Bolsonaro's extreme right-wing government espouses an 'anti-environmentalism of results' (Acserald 2020, p.1), working towards deconstructing any belief in the value of the natural environment while also taking a racialized approach, expropriating the lands, resources, and very lives of indigenous and Quilombola peoples. The movement has been supported by agribusiness practices (making it easy to use pesticides and clear lands) and mining businesses (with few, if any, environmental regulations). The state allows businesses to use and pollute common spaces such as water, the air, forests, soil, and the most dispossessed (Acserald 2020, 1). Additionally, Scantimburgo (2018), Viola and Goncalves (2019), and Fearnside (2020) warn of the current perilous state of the environment (e.g., deforestation and burning, mining, loss of biodiversity and territories by different cultures) and dangers to health (Covid-19 pandemic) in the country.

This is the historical context we face: a government that institutionally deactivates policies and public institutions aimed at promoting environmental protection and EE. We see a government openly questioning the role of international institutions and Brazil's participation in agreements involving international environmental governance. The reality and modus operandi of this government highlights a political agenda that is, explicitly and radically, opposed to the guidelines outlined in the 2030 Agenda (Gresse and Engels 2020, Fernside 2020). The conclusion has to be that, at this historic moment, for reasons related not only to the current government's political options, including ultra-liberal economic policies, the country looks unlikely to be able to achieve the SDGs (Frey et al. 2020).

Current Brazilian Policy on the SDGs: links with the National Education Plan (PNE) and National Common Curricular Base (BNCC)

Although the Brazilian government created the National Commission for the SDGs in 2016 and presented a report to the UN highlighting the country's actions to implement the 2030 Agenda and the SDGs, it has in fact ended up adopting a political agenda that contributes in no way to the achievement of the SDGs (Gresse and Engels 2020). With the support of the Applied Economic Research Institute (Ipea), the National Commission has published the *Cadernos ODS,* a set of documents sharing studies and research, in an attempt to contribute to the national effort to meet the challenges set out in the 2030 Agenda.

The document on SDG4, *Ensuring inclusive and equitable and quality education and promoting lifelong learning opportunities for all* (Brazil 2019), presents a very optimistic prognosis for education in Brazil. The introduction claims that practically all the planned targets for the SDG4 have already been addressed by government action, at the federal, state or municipal level. The document says that the National Education Plan (PNE)—which covers the period from 2014 to 2024—is the policy that has been the most effective in implementing SDG4. According to the team that analyzed the SDG, of the 10 goals in the original document, seven were preempted by the PNE. The optimism expressed in the Ipea document is striking: if the PNE targets were met, this would ensure that 70% of the SDG4 targets on education were met by 2024—that is, six years early.

Another proposition in this document is of great interest to the primary questions that are the concern of this article: the explicit understanding in the Ipea document that Goal 4.7 of SDG4 can be 'accomplished by incorporating it into the National Common Core Curriculum (BNCC) for Brazilian students' (Brasil 2018, p.6). Since the work undertaken by Ipea aims to foster reflection and offer subsidies for the 'formulation and strengthening of national sustainable development policies' (Brasil 2018, p.9), it is important to understand the impact of what the document proposes. It points to the PNE and the BNCC as the instruments for achieving eight of the ten goals proposed by 2030 Agenda, and in particular SDG4.

The absence of state investment in social policies for 20 years, as approved in 2016 by the Temer government, has had a direct impact on the PNE, making its implementation and the fulfillment of its goals highly unlikely (Corsetti 2019, Oliveira et al. 2019). Besides that, given that the current PNE only has only four more years to run, it is not realistic to believe that 70% of the SDG4 goals will be reached within this period.

A further issue is relevant to the analysis of the potential of the PNE and the BNCC to achieve the SDGs. Many authors (Andrade and Piccinini 2017, Behrend et al. 2018, Frizzo and Carvalho 2018, Oliveira and Roger 2019, Silva and Loureiro 2019) have drawn attention to the absence of references to EE in the latest educational policies documents in Brazil. This trend is evident in the PNE and the BNCC, and environmental educators consider it a significant weakness of both policies.

Since the creation of the BNCC has resumed, the process and the validity of the undertaking have been the subject of heated debate in the country. The very notion of a core curriculum and discussion about whether or not to propose one for the whole country has fueled intense discussions between conflicting political and ideological approaches and differing ideas about the curriculum (Lopes 2015, Ribeiro and Craveiro 2017, Lopes 2018, Macedo 2018, Oliveira 2018a). When considering the role of a national curriculum in the implementation of EE and the SDGs, it does not seem appropriate to limit discussion to issues of subject content or even less significant issues, such as the list of skills and abilities to be acquired by elementary school students set out in the BNCC (Aguiar 2018, Macedo 2018, Ribeiro and Craveiro 2017).

These documents have been systematically replacing the term 'Environmental Education' with references to the 'global environmental discourse' and the concept of sustainable development (Frizzo and Carvalho 2018). The question posed by Frizzo and Carvalho deserves our attention: what are the vested interests that are driving the replacement of environmental education with references to sustainability and development?

Beyond centralized power: subversion as a policy for building sustainable societies

In the political context faced by Brazilian society at this historic moment, with the threat of neo-liberalism and the growth of extreme right-wing movements, it is essential to develop strategies to resist the dismantling of environmental and EE policies. It is also important to reflect on and discuss options without falling into the traps of universalization and decontextualization, to appropriate critical dialogues and policies that will assist with the construction of a democratic and sustainable society based on the principles of environmental justice.

Such strategies necessarily involve mobilization of other sources of authority, going beyond national agencies and top down approaches (Frey et al. 2020, p.12). One productive channel is to involve agencies at different levels in the country, such as states and municipalities, but focusing on diverse institutions and sectors of civil society. Social movements and various population groups, especially those seen as marginalized and who are suffering the most from the consequences of environmental degradation, are fundamental.

Frey et al. (2020) consider that local discussion and the participation of civil society are essential to the achievement of the SDGs: they are 'an indispensable engine in the monitoring of government actions, especially in dark times involving attacks on institutions, on science, and, ultimately, on democracy itself' (p.290).

These authors view the lack of action by the federal government to implement the 2030 Agenda as a limiting factor for initiatives and contributions at the sub-national level. They suggest possible ways of ensuring collaborative action by government and the co-production of knowledge, and of strengthening social control over different bodies of power. The 'anti-environmentalism of results' (Acserald 2020) and 'anti-ecologism' (Layragues 2018, 2020), currently prevailing in Brazil will not only completely negate the SDGs, but possibly also lead to the deconstruction and deconfiguration of many initiatives within the country and civil society.

Another issue is especially significant in the Brazilian context: the 2030 Agenda (2015) and its 17 SDGs are aligned with the official stance of the UN since the Brundtland report (1987) in their promotion of sustainability. This was one of the key issues arising out of Rio-92 and a clear driver of the political and ideological positions with which we are aligned.

UNESCO's proposal for the Decade of Education for Sustainable Development received an awkward reception, and was the object of radical criticism on the part of some Latin American environmental educators (Carvalho 2002, Lima 2003, González-Gaudiano 2004, 2005, Sato 2005, Loureiro 2014, Sorrentino and Portugal 2016). In Europe, the proposal was seen as a means to politicize the environmental debate and EE, whereas in Latin America, the proposal was understood as an attempt to mischaracterize those who took a critical stance towards sustainable development: this highlighted the need for radical change in production and social relationships, including with the natural world.

Since the end of the last century, EE educators and researchers in Brazil have been increasingly critical of the idea of sustainable development (Carvalho 2002, Lima 2003, Sato 2005, Loureiro 2014, Sorrentino and Portugal 2016), often avoiding theoretical or political discussion of the broad issues involved with 'environmental sustainability.' Their main critique is based on the idea that, as a discourse, ESD (and all interpretations of it) represents a threat to EE identities in the global south, which are sustained and created mainly by social movements. The concern is that it would trigger the depoliticization of EE.

For some Brazilian environmental educators, the admission of discourses on sustainability opens up work on the Treaty on Environmental Education for Sustainable Societies and Global Responsibility (1992). This document represents a further ideological perspective on the issue, but is guided by principles that are very different from those of sustainable development: the idea of the "sustainable society". For Oliveira (2011), the Global Forum allowed the development of a utopian alternative: the constitution of the 'sustainable society,' which meant the refusal of a merely economic reading of sustainable development and the proposal of a corporate model that takes into account the requirements of the peoples of the Earth.

As Layrargues (2020) proposed, such a perspective implies a return to the origins of ecologism and environmental education, and the associated subversion, which was muted by the sustainable development agenda. For the author, the 'construction of sustainability' must be accompanied by 'the deconstruction of unsustainability.'

When we try in Brazil to connect with the discourses in the Global Plan and the 2030/SDG Agenda, it is very clear that they are as diverse as the ideological positions adopted toward the UN proposal and the ideology of sustainability.

When the proposals and strategies are systematized as suggested by Ipea (see above), what emerges is the task of matching the adaptations recommended by the UN to the reality of each country. There is no debate about the political and ideological perspectives that guide the UN document or about the original concept of sustainable development: this is accepted without questioning the ambiguity of the term or the multiplicity of perspectives it encompasses.

Frey et al. (2020) analyzed the process of 'localization' and reported on specific policies and experiences in the implementation of the SDGs by the Metropolis of São Paulo, the most populous city in the country. Although the authors emphasize the importance of the dialogue with global governance policies, and in particular the Agenda 21 proposals, they clearly take a critical view of the dialogue itself.

In their initial consideration of the main proposal in their collected work, Frey et al. (2020) admit that 'it is difficult to believe it will be possible to achieve the sustainable development goals, especially if the prevailing conservative and liberal approaches to development continue' (p.17). There is thus a need for more profound change and further study to advance the implementation of the 2030 Agenda.

With regard to the focus of the present chapter, and by way of example, we looked in detail at the chapter in the volume dealing with the implementation of SDG4—Quality Education—in the Metropolis of São Paulo. The researchers (Silva and Grandisoli 2020) initially highlight the criticisms of Brazilian environmental educators of the proposal for Education for Sustainable Development.

In addition to understanding that the SDG4 proposal cannot be seen as a model for the diverse contexts that make up the reality of the situation in Brazil and at the same time 'considering the plurality of Brazilian environmental education,' the authors attempt to 'find points of convergence' as well as divergence with the field of environmental education, considering mainly its critical and emancipatory tendencies (p.95).

Finally, it is worth noting the contributions of many environmental educators and researchers, who, for reasons already explained, avoid reinforcing and spreading the ideology of sustainable development. They choose instead to continually update the proposal in the *Environmental Education Treaty for Sustainable Societies and Global Responsibility* (1992), putting the emphasis on the idea of sustainable societies.

Recent Brazilian experience with EE has been productive, creative and constructive in developing counter-arguments for hegemonic discourses. Some of these draw on the dialectical materialism of Marxist theory and/or the perspectives of Critical Theory (Loureiro 2007, Tozoni-Reis 2008, Trein 2012). Others' critiques and resistance are based on more interpretative perspectives, including phenomenology (Passos 2014, Sato 2016, Sato et al. 2017), hermeneutics (Carvalho and Grun 2005, Carvalho et al. 2009) and post-critical/structuralist perspectives (see, for example, Reigota 2009, 2010, Guimarães and Sampaio 2014, Henning and Guimarães 2015, Sampaio 2019a, b). They have often been labeled revolutionary (Sorrentino 2020), undisciplined (Layrargues 2020), insurgent and *desde el sur* (coming from the South) (Plácido et al. 2018, Sànchez et al. 2020, Campos et al. 2020) EE.

Brazilian EE practices and research have taken into account themes such as environmental justice (Herculano 2008, Cosenza 2014, Angeli and Carvalho 2020), climate justice (Manfrinate et al. 2019), social and environmental conflicts (Cozenza and Martins 2012, Cosenza 2014, Sato et al. 2014, Silva et al. 2017, Silva et al. 2018, Santos and Carvalho 2020), and environmental racism (Herculano 2008, Silva et al. 2016). Further striking examples include the indigenous education movement (Gomes 2004, Gomes 2006, Nascimento and Medeiros 2018, Guimarães and Medeiros 2016), rural education (Andreoli and Borges 2020) and *quilombola* education (Melo and Barzano 2020, Santos et al. 2019, Soares et al. 2021), which are based on guidelines guaranteeing rights and access to education in these places, and which are examples of the process of 'localization' of some or all of the SDGs. Beyond that, other movements and initiatives in the country are engaging with issues such as inter-culturality, blackness, and feminism. The theory and practice being deepened and reflected on by many of them, mainly in the context of Latin American thought, leads them towards decolonial perspectives. All of them also explore and explain the limits and potential of the 2030 Agenda proposals.

When we understand that *critical environmental education* was and has been 'built in and for the struggle of popular movements for the rights to land, water, a balanced environment, health and education' and that it takes account of the reflections and practices of such movements on 'the inequality in the use of natural resources and the disproportionate impact of environmental damage on marginalized populations' (Sánchez et al. 2020, p.1), it becomes possible to make critical connections with the SDGs. The fundamental question is to decide on our theoretical and methodological references, and on the political and ideological approaches that could assist with the alignment (or otherwise) of Brazilian experiences of EE and the 2030 Agenda.

Final considerations

Regions around world are currently facing the most severe health crisis of recent times. The policies adopted by federal governments to control or to neglect the pandemic, fighting against or yielding to the Covid-19 epidemic (Brazil has the second-highest number of Covid-19 deaths per million people in the world, https://ourworldindata.org/coronavirus) have revealed what was already known to be the case: the practice of necro-politics, as conceived by Ribeiro (2020) and suicidal politics as proposed by Safatle (2020a, b).

We expect that, going beyond the current government's political initiatives and denialist position, Brazilian civil society will continue to resist and criticize universalized discourses. Decontextualised discourses such as we are seeing at present contribute little to the transformation of production models or the relationships between society and the natural world.

Our hopes rest in the complex of NGO and EE networks and their dynamic development of EE projects, communication practices, and dissemination of the work being undertaken in different regions of the country. The networks in question include:

- The Brazilian Environmental Education Network (Rebea), created in 1992, linked to EE state networks, which has an innovative structure and organizational strategy, and is effective at getting its messages across (Czapski 1998)
- Academic networks, such as the EA Pesquisa, which publicizes the actions of the National Association of Post-Graduate Studies and Research in Education's Environmental Education Working Group (Anped, Brazil)
- The Epea Network, which has played a central role in planning, organizing and implementing Environmental Education Research Meetings in the country (Epea) since 2001, and continues to do so; and
- Observare (https://observatorioea.blogspot.com/p/observare.html), a social movement launched by Brazilian environmental educators to be a militant observatory of public policy.

All these initiatives represent the exercise of resistance to social control by the state (Layrargues 2020).

Allied to this culture of internal resistance, we hope that the SDGs and the resumption of the dialogue on Education for Sustainable Development will help build creative, autonomous, and critical spaces that promote counter hegemonic discourse in order to open pathways for the construction of sustainable and democratic societies founded on the ideals of social and environmental justice. We want to work closely with other countries to pursue these and to counter the resistance to international policies that is only one of the many anti-ecological and anti-environmental policies instituted by the government.

We are confident that the history of EE in Brazil, which has its roots in creative resistance to authoritarian governments, has the potential to trigger critical dialogues with regard to the 2030 Agenda. Our expectations are reaffirmed by the strength and experiences of the social and environmental movements and of environmental educators and researchers.

Acknowledgments

We are grateful for the support and contributions in this chapter by Ralph Levinson, Isabel Carvalho, Jorge Amaro Borges and Janet McVittie.

References

Acserald, H. 2004. A Re-volta da Ecologia Política. Relume-Dumará, Rio de Janeiro.
Acserald, H. 2010. Ambientalização das lutas sociais – o caso do movimento por justiça ambiental. Estudos Avançados 24(68): 103–119.
Acserald, H. 2020. O antiambientalismo de resultados. https://aterraeredonda.com.br/o-antiambientalismo-de-resultados/ Access: January 30, 2021.
Aguiar, M.A. 2018. Relato da resistência à instituição da BNCC pelo Conselho Nacional de Educação mediante pedido de vista e declarações de voto. pp. 08–22. *In*: Aguiar, M.A. and L.F. Dourado (eds.). A BNCC na contramão do PNE 2014–2024: avaliação e perspectivas. [Livro Eletrônico]. ANPAE, Recife, Brazil.

Andrade, M.C. and C.L. Piccinini. 2017. Educação Ambiental na Base Nacional Comum Curricular: retrocessos e contradições e o apagamento do debate socioambiental. Proc. Encontro de Pesquisa em Educação Ambiental. BRAZIL IX: 1–13.

Andreoli, V. and M.G. Borges. 2020. Pesquisas e Práticas em Educação Ambiental e Educação do Campo. Ambiente & Educação 25(2): 3–18. https://doi.org/10.14295/ambeduc.v25i2.11874. Access: January 24, 2022.

Angeli, T. and L.M. Carvalho. 2020. Significados e sentidos de justiça ambiental nas teses e dissertações brasileiras em Educação Ambiental. Actio: Docência em Ciências 5: 1–21.

Arturi, C.S. 2001. O debate teórico sobre mudança de regime político: o caso brasileiro. Revista de Sociologia e Política. 17: 11–31.

Associação Nacional dos Servidores do Meio Ambiente (ASCEMA). 2018/2021. Cronologia de um desastre anunciado: ações do governo Bolsonaro para desmontar as políticas de meio ambiente no Brasil. www.ascemanacional.org.br. Access: January 26, 2021.

Behrend, D.M., C.S. Cousin and M.C. Galiazzi. 2018. Base Nacional Comum Curricular: o que se mostra de referência à educação? Ambiente & Educação 23(2): 74–89.

Berryman, T. and L. Sauvé. 2016. Ruling relationships in sustainable development and education for sustainable development. The Journal of Environmental Education 47(2): 104–117.

BRASIL. 1981. Dispõe sobre a Política Nacional do Meio Ambiente, seus fins e mecanismos de formulação e aplicação, e dá outras providências. Lei N. 6.938, de 31 de agosto. http://www.planalto.gov.br/ccivil_03/leis/l6938.htm. Access February 15, 2021.

BRASIL. 1998. Constituição da República Federativa do Brasil de 1988. Brasília, 05 out. 1988. ttp://www.planalto.gov.br/ccivil_03/constituicao/constituicaocompilado.htm. Access: February 15, 2021.

Brasil. 2018. Base Nacional Comum Curricular. Ministério da Educação, BRAZIL. http://basenacionalcomum.mec.gov.br/images/BNCC_EI_EF_110518_versaofinal_site.pdf. Access: February 15, 2021.

Brasil. 2019. ODS 4 – Assegurar a Educação Inclusiva e Equitativa e de Qualidade, e Promover Oportunidades de Aprendizagem ao Longo da Vida para Todas e Todos: o que mostra o retrato do Brasil? Ministério da Economia, Brasília. BRAZIL. https://www.ipea.gov.br/portal/images/stories/PDFs/livros/livros/190711_cadernos_ODS_objetivo_4.pdf. Access: January 24, 2022.

Braun, J.C.S. 2020. A educação ambiental na última versão da base nacional comum curricular para o ensino médio. Bachelor. Monography. Instituto Federal do Espírito Santo, Vitória, BRAZIL.

Briggs, L., N.M. Trautmann and C. Fournier. 2018. Environmental education in Latin American and the Caribbean: the challenges and limitations of conducting a systematic review of evaluation and research. Environmental Education Research 24(12): 1631–1654.

Campos, B.F., P.M. Bevilaqua and C. Sánchez. 2020. Aprender com as resistências, insistir com as esperanças: de uma herança colonial à construção da utopia. Ensino, Saúde e Ambiente. Special Issue: 412–433.

Carvalho, I.C.M. 2002. O 'ambiental' como valor substantivo: uma reflexão sobre a indentidade da educação ambiental. pp. 85–90. In: Sauvé, L., I. Orellana and M. Sato (eds.). Textos escolhidos em Educação Ambiental: de uma América à outra. Publications ERE-UQAM, Montreal, CANADÁ.

Carvalho, I.C.M. and Grün, Mauro. 2005. Hermenêutica e educação ambiental: o educador como intérprete. pp. 175–188. In: Luiz Antonio Feraro Junior (ed.). Encontros e Caminhos: Formação de Educadoras(es). Diretoria de Educação Ambiental/MMA, Brasilia, BRAZIL.

Carvalho, I.C.M. 2008. A educação ambiental no Brasil. Salto para o Futuro. XVIII(1): 13–20. http://forumeja.org.br/sites/forumeja.org.br/files/Educa%C3%A7%C3%A3o%20Ambiental%20no%20Brasil%20(texto%20basico).pdf. Access: January 24, 2022.

Carvalho, I., M. Grün and M.R. Avanzi. 2009. Paisagens da compreensão: contribuições da hermenêutica e da fenomenologia para uma epistemologia da educação ambiental. Cadernos CEDES 29: 99–116.

Corsetti, B. 2019. Neoconservadorismo e políticas educacionais no Brasil. Educação Unisinos 23(4): 774–784.

Cosenza, A. and I. Martins. 2012. Os sentidos de "conflito ambiental" na educação ambiental: uma análise dos periódicos de educação ambiental. Ensino, Saúde e Ambiente 5(2): 234–245.

Cosenza, A. 2014. Justiça ambiental e conflito socioambientalna prática escolar docente: significando possibilidades e limites. PhD. Thesis. Núcleo de Tecnologia Educacional para a Saúde, Universidade Federal do Rio de Janeiro, Rio de Janeiro, BRAZIL.

Czapski, S. 1998. A Implantação da Educação Ambiental no Brasil, Coordenação de Educação Ambiental. Ministério da Educação e do Desporto, Brasília. BRAZIL. http://www.dominiopublico.gov.br/download/texto/me001647.pdf. Access: January 24, 2022.

Fearnside, P.M. 2019. Retrocessos sob o Presidente Bolsonaro: Um Desafio à Sustentabilidade na Amazônia. Sustentabilidade International Science Journal 1(1): 38–52.

Frey, K., P.H.C. Torres, P.R. Jacobi and R.F. Ramos. 2020. Os Objetivos do Desenvolvimento Sustentável (ODS) no contexto da Macrometrópole Paulista – desafios e perspectivas. pp. 12–19. *In*: Frey, K., P.H.C. Torres, P.R. Jacobi and R.F. Ramos (eds.). Objetivos do desenvolvimento sustentável: desafios para o planejamento e a governança ambiental na Macrometrópole Paulista. EdUFABC, Santo André, BRAZIL.

Frizzo, T.C.E. and I.C.M. Carvalho. 2018. Políticas públicas atuais no Brasil: o silêncio da educação ambiental. Rev. Eletrônica Mestr. Educ. Ambient. Special Issue 1: 115–127.

Gomes, A.M.R. 2004. El proceso de escolarización de los Xakriabá: historia local y rumbos de la propuesta de educación escolar diferenciada. Cuadernos de Antropología Social 19: 29–48.

Gomes, A.M.R. 2006. O processo de escolarização entre os Xakriabá: explorando alternativas de analise na antropologia da educação. Revista Brasileira de Educação 11(11): 316–327.

Gonzáles-Galdiano, E. 2004 ¿Réquiem por un sueño? La educación ambiental en riesgo. Agua y Desarrollo Sustentable. 11: 22–24.

Gonzáles-Galdiano, E. 2005. Education for Sustainable Development: configuration and meaning. Policy Futures in Education 3(3): 243–250.

Grandisoli, E., D.T.P. Souza, P.R. Jacobi and R.A.A. Monteiro (eds.). 2020. Educar para a sustentabilidade: visões de presente e futuros. IEE-USP/Reconectta : Editora na raiz, São Paulo. BRAZIL.

Gresse, E. and A. Engels. 2020. ODS e sociedade civil. pp. 40–54. *In*: Frey, K., P.H.C. Torres, P.R. Jacobi and R.F. Ramos (eds.). Objetivos do desenvolvimento sustentável: desafios para o planejamento e a governança ambiental na Macrometrópole Paulista. EdUFABC, Santo André, BRAZIL.

Guimarães, R.P. and Y.S.R. Fontoura. 2012. Rio + 20 ou Rio – 20? Crônica de um fracasso anunciado. Ambiente & Sociedade XV(3): 19–39.

Guimarães, L.B. and S.M.V. Sampaio. 2014. Educação Ambiental e as pedagogias do presente. Em Aberto. 27(91): 123–134.

Guimarães, M. and H.Q. Medeiros. 2016. Outras epistemologias em educação ambiental: o que aprender com os saberes tradicionais dos povos indígenas. Revista Eletrônica do Mestrado em Educação Ambiental. Special Issue: 50–67.

Henning, P. and L.B. Guimarães. 2015. Educação ambiental: travessias pelo contemporâneo (Apresentação). Revista Eletrônica do Mestrado em Educação Ambiental. Special Issue: 1–3.

Herculano, S. 2008. O clamor por justiça ambiental e contra o racismo ambiental. InterfacEHS 3(1): 1–20.

Jickling, B. and A.E.J. Wals. 2008. Globalization and environmental education: looking beyond sustainable development. Journal of Curriculum Studies 40(1): 1–21.

Lafer, C. 2009. Conferência do Rio. Verbete. Fundação Getúlio Vargas. Centro de Pesquisa e documentação História Contemporânea do Brasil. http://www.fgv.br/cpdoc/acervo/dicionarios/verbete-tematico/conferencia-do-rio. Access: January 21, 2022.

Layrargues, P.P. 2018a. Subserviência ao capital: educação ambiental sob o signo do antiecologismo. Pesquisa em Educação Ambiental 13(1): 28–47.

Layrargues, P.P. 2018b. Quando os ecologistas incomodam: a desregulação ambiental pública no Brasil sob o signo do Antiecologismo. Revista de Pesquisa em Políticas Públicas. Special Issue Human Rights. 1–30.

Layrargues, P.P. 2020. Manifesto por uma Educação Ambiental indisciplinada. Ensino, Saúde e Ambiente. Special Issue: 44–87.

Lima, G.C. 2003. O discurso da sustentabilidade e suas implicações para a educação. Ambiente & Sociedade. VI(2): 99–119.

Lima, G. 2009. Educação Ambiental Crítica: do socioambientalismo às sociedades sustentáveis. Educação e Pesquisa. 35(1): 45–163.

Lopes, A.C. 2018. Apostando na produção contextual do currículo. pp. 23–33. *In*: Aguiar, M.A. and L.F. Dourado (eds.). A BNCC na contramão do PNE 2014–2024: avaliação e perspectivas. ANPAE, Recife, BRAZIL.

Lopes, A.L. 2015. Por um currículo sem fundamentos. Linhas Críticas 21(45): 445–466.

Loureiro, C,F.B. 2007. A questão Ambiental no pensamento crítico: natureza, trabalho e educação. Quartet, Rio de Janeiro.

Loureiro, C.F.B. 2008. Proposta pedagógica. *In*: Brasil, Ministério da Educação (ed.). Educação Ambiental no Brasil. Salto para o Futuro. TV Escola. Ano XVIII boletim 01.

Loureiro, C.F.B. 2014. Sustentabilidade e educação ambiental: controvérsias e caminhos do caso brasileiro. Sinais Sociais. 9(26): 39–71.

Macedo, E. 2018. "A Base é a base". E o currículo o que é? pp. 28–33. *In*: Aguiar, M.A. and L.F. Dourado (eds.). A BNCC na contramão do PNE 2014–2024: avaliação e perspectives. ANPAE, Recife, BRAZIL.

Manfrinate, R., M. Sato and A. Serantes. 2019. Entrelaçamentos entre justiça climática e educação ambiental: diálogos com mulheres de comunidades tradicionais do Mato Grosso e Galícia. Pesquisa em Educação Ambiental. 14: 171–191.

Mattar, H. 2021. O árduo caminho rumo aos Objetivos de Desenvolvimento Sustentável. Folha de São Paulo. BRAZIL.

Melo, A.C. and M.A.L. Barzano. 2020. Re-existências e Esperanças: Perspectivas decoloniais para se pensar uma Educação Ambiental Quilombola. Special Issue: 147–162.

Nascimento, I.O.P. 2015. Transição Democrática e Justiça de Transição: o caso da Argentina. Bachelor. Dissertation, Universidade de Brasília, Brasília, Brasil.

Nascimento, E.C.M. and H.Q. Medeiros. 2018. As contribuições dos conhecimentos tradicionais indígenas para a Educação Ambiental brasileira. Revista Espaço do Currículo 11: 340–355.

Oliveira, A.M.M.C., P.M. Dusek and K.E.S. Avelar. 2019. A trajetória da educação brasileira no contexto econômico. Revista Brasileira de Política e Administração da Educação 35(2): 369–380.

Oliveira, E.T. and M.R. Royer. 2019. A Educação Ambiental no contexto da Base Nacional Comum Curricular para o Ensino Médio. Interfaces da Educação 10(30): 57–78.

Oliveira, I.B. 2018a. Políticas curriculares no contexto do golpe de 2016: debates atuais, embates e resistências. pp. 55–59. *In*: Aguiar, M.A. and L.F. Dourado (eds.). A BNCC na contramão do PNE 2014–2024: avaliação e perspectivas. ANPAE, Recife, BRAZIL.

Oliveira, L.D. 2011. A Geopolítica do Desenvolvimento Sustentável: um estudo sobre a Conferência do Rio de Janeiro (Rio-92). PhD. Thesis. Universidade Estadual de Campinas, Campinas, BRAZIL.

Oliveira, R.P. 2018. A debacle da Nova República brasileira: da desilusão ao encerramento de um ciclo democrático. Espirales. 2(3): 25–42.

Passos, L.A. 2014. Metodologia da pesquisa ambiental a partir da fenomenologia de Maurice Merlau-Ponty. Pesquisa em Educação Ambiental. 9(1): 38–52.

Plácido, P.O., E.M.N.V. Castro and M. Guimarães. 2018. Travessias para educação ambiental 'desde el sur': uma agenda política crítica comum em 'zonas de sacrifício' como o Brasil e América Latina. Ambiente & Educação 23(1): 8–30.

Pochmann, M. 2017. Estado e capitalismo no Brasil: a inflexão atual no padrão das políticas públicas do ciclo político da nova república. Educação & Sociedade 38(139): 309–330.

Prado Júnior, T., M. Cardoso, F. Iacomini Júnior and A.A. Vaz. 2019. O imaginário da polarização política no pré e pós-impeachment de Dilma Roussef: vestígios de cotidianidades inscritos em uma curta-metragem brasileiro. Revista Fronteira – Estudos Midiáticos 21(2): 68–78.

Reigota, M. 2009. Educação Ambiental brasileira: a contribuição da nova geração de pesquisadores e pesquisadoras. Interacções 11: 1–7.

Reigota, M. 2010. A Educação Ambiental frente aos desafios apresentados pelos discursos contemporâneos sobre a natureza. Educação e Pesquisa. 36(2): 539–553.

Ribeiro, G. 2020. Entre armas e púlpitos: a necropolítica do Bolsonarismo. Revista Continentes. 9(16): 463–485.

Ribeiro, W.G. and C.B. Craveiro. 2017. Precisamos de uma Base Nacional Comum Curricular? Linhas Críticas. 23(50): 51–69.

Safatle, V. 2020a. Bem-vindo ao Estado suicidário. N-1 Edições. https://www.n-1edicoes.org/textos/23. Access: February 15, 2021.

Safatle, V. 2020b. Para além da necropolítica. Considerações sobre a gênese e os efeitos do Estado suicidário. https://racismoambiental.net.br/2020/10/24/para-alem-da-necropolitica-por-vladimir-safatle/ Access: February 15, 2021.

Sampaio, S.M.V. 2019a. Como criar uma paisagem em ruínas? Deslocamentos, desconstruções e a insistência de pensar a Educação Ambiental no Antropoceno. Quaestion. 21(1): 19–38.

Sampaio, S.M.V. 2019b. Educação Ambiental e Estudos Culturais: entre rasuras e novos radicalismos. Educação e Realidade. 44: 1–19.

Sánchez, C., B. Pelacani and I. Accioly. 2020. Educação ambiental: insurgências, re-existências e esperanças. Ensino, Saúde e Ambiente. Special Issue: 1–20.

Santos, D.L.M.S., M. Jaber and M. Sato 2019. Resistência quilombola: conflitos socioambientais, injustiça ambiental e luta por direitos. INSURGÊNCIA: Revista de Direitos e Movimentos Sociais. 5: 1–22.

Santos, M.H.C. 2010. O processo de democratização da Terceira Onda de Democracia: quanto pesam as variáveis externas? Meridiano. 47(115): 15–18

Santos, R. and L.M. Carvalho. 2020. Processo educativo e os conflitos socioambientais: construção de possíveis significados e sentidos. Praxis & Saber. 12(28): 1–16.

Sato, M. 2005. Identidades da Educação Ambiental como rebeldia contra a hegemonia do desenvolvimento sustentável. Proc. XII Jornadas da Associação Portuguesa de Educação Ambiental: Educação Ambiental nas Políticas do Desenvolvimento Sustentável. Lisboa. 1–5.

Sato, M., R. Silva and M. Jaber. 2014. Between the remnants of colonialism and the insurgence of self-narrative in constructing participatory social maps: towards a land education methodology. Environmental Education Research 20(1): 102–114.

Sato, M. 2016 Ecofenomenologia: uma janela ao mundo. Revista Eletrônica do Mestrado em Educação Ambiental. Edição Especial: 1–18.

Sato, M., B.D. Moreira and T.C. Luiz. 2017. Educação Ambiental e a narrativa transmídia: pedagogia popular e a fenomenologia recriando espaços da educação escolar. Momento: diálogos em educação. 26(2): 282–296.

Sauvé, L. 2005. Educação Ambiental: possibilidades e limitações. Educação e Pesquisa. 31(2): 317–322.

Scantimburgo, A. 2018. O desmonte da agenda ambiental no governo Bolsonaro. Perspectivas. 52: 103–117.

Silva, M.A., A. Cosenza and V.P.S. Pinto. 2017. Justiça, racismo e conflitos ambientais na literatura sobre educação ambiental: o que dizem os Anais dos encontros nacionais de pesquisa em educação ambiental? Proc. Encontro Pesquisa em Educação Ambiental. 9: 20–36.

Silva, R.F.S. and E. Grandisoli. 2020. ODS 4 – Educação de qualidade. pp. 93–104. *In*: Frey, K., P.H.C. Torres, P.R. Jacobi and R.F. Ramos (eds.). Objetivos do desenvolvimento sustentável: desafios para o planejamento e a governança ambiental na Macrometrópole Paulista. EdUFABC, Santo André, BRAZIL.

Silva, S.N. and C.F.B. Loureiro. 2019. O sequestro da Educação Ambiental na BNCC (Educação Infantil - Ensino Fundamental): os temas Sustentabilidade/Sustentável a partir da Agenda 2030. Proc XII Encontro Nacional de Pesquisa em Educação em Ciências. BRAZIL. XII: 1–7.

Silva, R., M. Jaber and M. Sato. 2018. Social mapping and environmental education: dialogues from participatory mapping in the Pantanal, Mato Grosso, Brazil. Environmental Education Research 24(10): 1514–1526.

Soares, C.C.A., R.A. Silva and M. Sato. 2021. A Arte/Educação no ambiente da escola quilombola de Mata Cavalo: cultura de diálogos e resistência. Revista do Grupo de Pesquisa em Educação e Arte REVISARTE. 8: 10–26.

Sorrentino, M. and S. Portugal. 2016. Educação Ambiental e a Base Nacional Comum Curricular. http://ixfbea-ivecea.unifebe.edu.br/wiew/information/downloads-consulta-publica/3.pdf. Access January 25, 2022.

Sorrentino, M. 2020. Educação Ambiental e Revolução. pp. 176–179. *In*: Grandisoli, E., D.T.P. Souza, P.R. Jacobi and R.A.A. Monteiro (eds.). Educar para a sustentabilidade: visões de presente e futuros. IEE-USP/Reconectta: Editora na raiz, São Paulo. BRAZIL.

Tozoni-Reis, M.F.C. 2008. Educação Ambiental: natureza, razão e história. 2. ed. Campinas - SP: Autores Associados.

Trein, E.S. 2012. A Educação Ambiental Crítica: crítica de que? Revista Contemporânea de Educação. 7(14): 304–318.

Uchoa, R.S. 2018. Análise da Década da Educação para o Desenvolvimento Sustentável (Deds) da Unesco a Partir da Leitura da Pedagogia da Autonomia de Paulo Freire. Rev. Bras. Educação Ambiental. 13(2): 340–350.

Viola, E. and V.K. Gonçalves. 2019. Brazil ups and downs in global environmental governance in the 21st century. Editorial. Revista Brasileira de Política Internacional 62(2): 02–10.

Wermuth, M.A. and J.G. Nielsson. 2018. Ultraliberalismo, evangelicalismo político e misoginia: a força triunfante do patriarcalismo na sociedade brasileira pós-impeachment. Rev. Elet. Curso de Direito da UFSM 13(2): 455–488.

Chapter 12

Environmental Education and Education for Sustainable Development in Chile

An Exploratory Study

Karl Böhmer M., Loreto Aceitón* and *Viviana Contreras*

Introduction

Current conceptions of Environmental Education (EE) and Education for Sustainable Development (ESD) date from the 1972 Stockholm Conference and the 1992 Rio Earth Summit respectively. Nowadays, EE and ESD form the pillars on which cultural, social and economic change occurs in the relationship between human beings and the biosphere.

With that in mind, this study provides an exploratory qualitative overview of how the two concepts operate in Chile. It is divided between a general review of the socio-environmental situation of the country, the historical evolution of EE and the current situation. Emphasis is placed on the particularities of implementation, with special consideration given to the development of ESD. The study concludes by reflecting on the need to deploy ESD that incorporates new methodologies at all levels, drawing particularly on the social dimension of the Sustainable Development Goals (SDGs), in order to develop skills and competences that will reinforce the ability of Chile and the Chileans to ensure democratic co-existence and confront the challenges posed by the need to adapt to and mitigate climate change.

UTEM Diecicho 161, Santiago de Chile.
Emails: loreto.aceiton@utem.cl; vcontrerasc@utem.cl
* Corresponding author: karl.bohmer@utem.cl

The Chilean education system: social, political and environmental context

The American continent in general became part of the world economy and a source of natural resources from the late 15th century onwards. The economic structure that was created as a result, together with the consequences of political and social stratification in Latin America, remains to this day. This process was so decisive that the exploitation of energy, minerals and agricultural resources is no longer regarded as a feature of economic 'underdevelopment' and dependency. Indeed, it has even become an economic strategy of countries with a so-called 'progressive' agenda (for example, Ecuador and Bolivia), to such an extent that 'neo-extractivism'—and all the socio-environmental problems that it entails—has become firmly established in those areas (Gudynas 2009). This not only compromises the quality and availability of natural resources for future generations but also further entrenches conflicts and environmental injustice, the appropriation of common goods leading to the accumulation of resources in the hands of a few, social and ethnic inequality, and human rights violations. Chile is no different and its policies 'represent … the deliberate subordination of the demands of sustainability and environmental regulation to the imperatives of an extractivist economic model geared towards the export of raw materials' (Pelfini and Mena 2017: 270). This is reflected in the fact that a large percentage of the general public views the mining industry as one of the main sources of environmental problems (Marconi 2020). A dramatic demonstration of this can be seen in the number of industrial and agricultural areas that have become sacrifice zones in recent years. One example is Puchuncavi, in the northwest of Valparaiso Region, which is in the emissions zones of the state copper company CODELCO's refinery and of coal-driven power plants, whose toxic emissions and liquid waste are leading to widespread degradation of the soil and the sea. In addition, the surrounding Putaendo Valley (Valparaiso Region) is host to industrial-scale avocado plantations, which significantly affect local groundwater and have a negative impact on biodiversity as a consequence of the related clearing of sclerophyll woodlands and forests in the nearby matorral eco-region.

Despite the above, the environment ranked only fourth among Chileans' priority issues in a 2020 survey. In fact, findings from this survey generated a number of contradictory responses that are a good illustration of the prevailing 'environmental culture' in Chile. In spite of expressing significant concern about climate change, for example, respondents indicated that their preferred means of transport was private car. Similarly, while participants deemed 'the habits of the general public' to be one of the most important factors impacting negatively on the environment, they simultaneously indicated that this negative impact was 'caused by others,' thereby denying any personal responsibility. Moreover, a majority of respondents recognized the value of recycling in improving the environment, while indicating that they personally do 'nothing concrete' to engage in recycling (Marconi 2020).

Regarding education, the Chilean education system is governed by constitutional norms that guarantee the right to education and the freedom to teach. Consequently, the national system consists of a public education sector administered by local municipalities, a public-private sector in which schools are run by not-for-profit educational corporations with state subsidies, and a fully private, fee-paying, for-profit sector. In order to gain state recognition and benefit from state subsidies, educational establishments must put together an Institutional Educational Project (Proyecto Educativo Institucional or PEI), an instrument in which they are required to set out their pedagogical framework. Under this system, establishments have the freedom to define their own values but must comply with all statutory teaching requirements.

EE and ESD in Chile

This reality is why environmental issues became increasingly relevant following the return of democracy in the 1990s (Squella 2000). Chile manages to be the region's most advanced country on this issue, despite the fact that environmental practices are not part of Chileans' day-to-day culture (Chile Sustentable 2020). With its re-democratization and greater involvement in the international community, the country soon acquired a range of new environmental rules, regulations and management policies. Although it did not represent a radical change, EE was institutionalized during the early 1990s through a number of new educational programs and environmental management practices. The biggest step, however, was the enactment in 2009 of Law 20.370, the General Education Act (LGE), which establishes that

> The [education] system will promote respect for the environment and the rational use of natural resources as a concrete expression of solidarity with future generations (BCN 2009).

Ever since, EE has been one of the guiding criteria for state action on education, which is administered at the general level by the Ministry of Education (MINEDUC). Since EE is embedded in the national curriculum framework that derives from the LGE, it forms part of all plans and programmes devised for the education system. It is also part of the Transversal Learning Objectives (OATs) within the national curriculum (MINEDUC 2018). These OATs are seen as general goals for school education and relate to the personal, intellectual, moral and social development of students, although they do not have to be linked directly to any particular subject (MINEDUC 2018). This is undoubtedly positive from the point of view of contemporary approaches to education, since it ensures that the goals are not 'schooled'—what we consider as trained only to repeat slogans or seek the 'right' answer in a test. However, from personal experience and observation, we can affirm that educational establishments do not take OATs into account and they are thus absent from schools and curriculum planning.

Formal education, and thus the PEIs of each educational institution, now face a new challenge. Lawmakers have re-emphasized the importance of educating in a concept of active citizenship through the Citizen Training Plan (Plan de Formación Ciudadana) (BCN 2016). This Plan integrates the key topics of 'rights linked to space... in relation to environmental sustainability' (MINEDUC 2018: 37) and therefore includes a land-based dimension. Moreover, it defines this type of citizenship education as expecting participants to

> develop autonomy, ethical principles, critical thinking, taking responsibility for the defense and promotion of human rights, the protection of and care for the environment and the defense and improvement of democracy (MINEDUC 2016: 28).

The program is a joint initiative between MINEDUC and the Ministry of the Environment (MMA) and represents a clear example of inter-ministerial cooperation on public policy with regard to values and, in particular, environmental education.

This creates an additional challenge for formal education: to put values education (which is rooted in cross-cutting learning objectives) into practice in such a way that it changes the relationship of students with their school and their environment and helps to mould them into the desired perfect citizens, i.e., those who exercise all the functions of citizenship, including the improvement of public spaces and the promotion of land-based equality, including rights to space and housing, free movement and the concept of 'environmental sustainability' (MINEDUC 2016, Muñoz-Pedreros 2014, Castillo-Retamal and Cordero-Tapia 2019). The largest shortcoming in this regard lies in the fact that teachers are not provided with a theoretical framework or the necessary objectives or content to implement the program in the classroom.

Despite an intensive literature review, it has not been possible to identify more than very general qualitative research on the implementation of formal and mainstreamed environmental education in primary and secondary schools in Chile (Muñoz-Pedrero 2014, Castillo-Retamal and Cordero-Tapia 2019). There is clearly a lack of research into the desired mainstreaming of EE and ESD and its impact.

This deficiency is also evident in the lack of environmental training for technicians and professionals in the world of work, most of whom do not have the knowledge and skills necessary to ensure compliance with environmental regulations in their working environments. Consequently, in 2016 the United Nations Economic Commission for Latin America and the Caribbean (ECLAC) and the Organization for Economic Co-operation and Development (OECD) recommended that the Chilean State develop capacity-building schemes to bridge the gap between education and training on the one hand, and the demands of the labour market on the other (ECLAC and OECD 2016).

When Chile began the implementation of Agenda 21 in 1993–1994 (Hajek 1998), EE was incorporated into Law 19.300, the General Environmental Framework Act of 1994, as well as its subsequent extension by means of Law 20.417 of 2010, which created the MMA. This body of legislation strengthens the institutional basis of state-led environmental management and acts as one of the key instruments within the MMA's Division of Environmental Education and Citizen Participation (División de Educación Ambiental y Participación Ciudadana). This Division is composed of the Department of Environmental Education and the Environmental Protection Fund (Departamento de Educación Ambiental y Fondo de Protección Ambiental), and establishes an annual budget for EE-related projects and environmental initiatives in civil society (ECLAC and OECD 2016).

As a consequence, EE has become a central component of State environmental policy and public environmental management and is seen as having an instrumental and conservationist character (Hajek 1998). Accordingly, the Chilean State notes that:

> The educational process, at all levels and through the dissemination of knowledge and teaching of modern concepts of environmental protection, whose aim is to enhance the understanding and awareness of environmental problems, should inculcate the values and develop the habits and behaviors necessary to prevent and resolve such problems (BCN 1994: Title II Article 6).

In 2005, Chile published a National Policy of Education for Sustainable Development (Política Nacional para la Educación para el Desarrollo Sustentable or PNEDS) in order to ensure it was aligned with the UNESCO Decade of ESD (DESD) initiative. The country also devised a PNEDS Roundtable, which was approved in April 2009 by the Council of Ministers of the National Environmental Commission (CONAMA) and brought together different ministries with objectives arising from the Chilean endorsement of the DESD. It should be noted that the DESD scheme ran from 2004 to 2015.

Currently, the Department of Environmental Education, part of the MMA's Division of Environmental Education and Citizen Participation, has openly highlighted the need to re-examine Environmental Education as an area of knowledge with its own concepts, values and methodologies (MMA 2017), thus moving towards compliance with existing state guidelines. In turn, the coverage of its flagship programs has been deepened and expanded to include the National System for the Certification of Educational Establishments (Sistema Nacional de Certificación de Establecimientos Educacionales or SNCAE), the Formal Education Program (Programa de Educación Formal), the Adriana Hoffmann Environmental Training Academy (Academia de Formación Ambiental Adriana Hoffmann), the Network of Environmental Education Centers (Red de Centros de Educación Ambiental) and the Environmental Trainers (Forjadores Ambientales).

All of this is supported and guided by the PNEDS, the main objective of which is:

> To train people and citizens capable of individually and collectively assuming the responsibility of creating and enjoying a sustainable society and contributing to the strengthening of educational processes that allow the inculcation and development of values, concepts, skills, competencies and attitudes in the citizenry as a whole (MMA 2012: 17).

The MMA's approach is framed within the scope of the agreements and social objectives that Chile has pledged to meet in relation to the SDGs, which seek to guide political activities towards improving the well-being of people and caring for the environment (United Nations 2015).

In order to strengthen the implementation of the OATs on environmental training and awareness raising outlined in the national curriculum, MINEDUC, and specifically the Unit for Secondary and Curricular Education (Citizenship Education)

(Unidad de Educación Media (Formación Ciudadana) y Curricular) is developing an Environmental Education Plan. This work is being undertaken jointly with the National Board of Kindergartens (Junta Nacional de Jardines Infantiles) and other bodies responsible for devising and implementing public education policy.

> Its main emphasis is on ensuring that the provision of EE at all formal levels is coherent, that it includes learning objectives, that it ensures effective teaching and learning processes and outcomes (assessment). The MMA stresses that teaching and learning processes are expected to produce significant outcomes over time and thus result in cultural change that supports all areas of State environmental policy.

The MMA's Division of Environmental Education and Citizen Participation and the Department of Environmental Education directly administer a number of EE programmes, which are outlined in the following section.

EE programmes administered by the ministry of the environment

Adriana Hoffmann Environmental Training Academy

The Adriana Hoffmann Environmental Training Academy was founded in 2015 and delivers environmental training to three key groups at the national level: the general public, teachers and civil servants.

The aims of the academy are as follows:

- To meet the growing public demand for EE and ESD through the dissemination of knowledge on environmental protection and modern teaching methods in order to broaden awareness of environmental challenges.
- To guarantee a continuous process of teacher training and continuous development at all levels by empowering educators to disseminate environmental learning content in their educational establishments.
- To update environmental learning content and regulatory and legal processes for public sector employees, particularly members of the Council of Ministers for Sustainability (Consejo de Ministros para la Sustentabilidad).
- To train public and private sector employees on environmental matters related to their work.
- To support and collaborate with national and international partners by strengthening network building with institutions providing training in related subjects (MMA 2020).

At present, the academy provides training courses via an e-learning format and has a large number of enrolments (see Table 1) (MMA 2020).

Table 1. Enrolments in the Adriana Hoffmann Environmental Training Academy 2020.

Training courses available	Total enrolments
14	**1900**

Source: In-house research based on MMA data (2020).

National system of environmental certification of educational establishments

The National System of Environmental Certification of Educational Establishments (SNCAE) is an initiative inspired by the principles of Agenda 21. As such, the Chilean State enacted Decree 90 on 6 May 1998 to formally adopt the Agenda as its guiding instrument for environmental management. Although the creation of the SNCAE was inspired by several international examples, including the Danish Foundation for Environmental Education's Eco-Schools program, and Agenda 21 for Schools, it incorporates key features of national educational management.

As stipulated by Agenda 21, the program is a voluntary scheme that enables all participating schools to integrate environmental management into their educational activities. It operates on an inter-sectoral basis and includes the participation of the MMA, the National Forestry Corporation (CONAF), UNESCO, and the General Directorate of Water (DGA), the Ministry of Energy and MINEDUC.

The SNCAE promotes the design, approval and implementation of environmental education, promotion and dissemination programmes which aim to raise national awareness of environmental protection, development, sustainability and the preservation and conservation of the local environment and its natural resources. Table 2 sets out the number of establishments certified since 2011, the year in which the system came in to being.

Therefore, according to the latest statistics of the MMA approximately 2,065 primary and secondary education establishments in Chile had been certified by the SNCAE, showing the interest of Chilean schools to participate in traditional EE programs (waste management, recycling, etc.) (MINEDUC 2020: 40).

The SNCAET is coordinated by the National Committee for the Environmental Certification of Educational Establishments (CNCA), and includes active participation from the MMA founding institutions of CONAF, UNESCO and MINEDUC. The DGA resumed its participation in the Committee in 2017, the same year in which the Ministry of Energy and the Chilean Energy Efficiency Agency (AChEE) were also invited to participate.

It should be noted that the SNCAE is primarily an exercise in environmentally sound management in each of the participating educational establishments. As a result, EE is one of many instruments that make up the system and is not its

Table 2. Educational establishments certified by SNCAE 2020.

Year	2011	2012	2013	2014	2015	2016	2017	2020
Total certified establishments	546	903	909	935	1,115	1,249	1,476	2,065

Source: In-house research based on MMA findings, 2020, Santiago, Chile.

over-riding objective. This may be one reason for the very limited research to date into the way in which the system is integrated into the pedagogical work of participating establishments.

Taking the aforementioned into consideration, it can be stated that, in accordance with the System, there is no educational planning in Chile that incorporates

> [...] the concepts, content and methodologies of environmental education relevant to the local context and environment; as stipulated by senior-level bodies (ministries, international institutions, corporations, municipalities, etc.), [they] are not efficiently implemented [...] (Quinceno 2011: 69).

There is also a lack of appropriate teaching methods, teacher training and institutional support in general (Quiceno 2011, Molina 2017), and the involvement of local stakeholders in education is weak. For example, no in-depth work is undertaken with local communities and there is no adequate evaluation of educational establishment and their programs, or the impact of the SNCAE on the local environment, or its impact on schools as institutions (Molina 2017). Moreover,

> the application of programs of this nature [environmental education] fails to provide teachers with the extra time required to adequately plan coherent activities, and such activities are not subject to appropriate remuneration to incentivize those implementing them to ensure they are correctly executed (Burgos et al. 2010: 18).

In other words, school administrators appoint teachers to coordinate and deliver these schemes as part of their existing workload and such teachers do not receive adequate remuneration for these activities.

We agree with Burgos et al. (2010) in that the weakness of the system lies in its excessive focus on school administrators, which weakens its own continuity. In addition, many educational communities are largely indifferent to the program. Indeed, there is frequently a shortage of competent professionals to coordinate environmental certification within schools (Molina 2017). The most recent research into the SNCAE was undertaken by Barazarte et al. (2014) in Valparaiso and compared the environmental knowledge and pro-environment attitudes of secondary school students from different socio-economic groups, types of schools and in both certified and non-certified establishments. It should be noted that the authors of this research focused solely on waste management, energy efficiency and similar aspects, thus restricting environmental awareness to specific procedures and environmental management support. Notwithstanding this, the authors found that

> [...] students from certified schools do not demonstrate greater environmental knowledge or pro-environment attitudes than students from non-certified schools (Barazarte et al. 2014: 81).

The research concludes that the variable that has the greatest influence on students' level of environmental knowledge is the type of educational establishment they attend, i.e., whether it is a public, private or semi-private institution, and the factor that influences pro-environment attitudes is their socio-economic group (Barazarte et al. 2014: 81).

Despite the ongoing challenges, the SNCAE has undoubtedly embedded the idea of environmental management within participating educational establishments. It has also helped to disseminate the environmental management policies promoted by the state and proactively supported by the MMA, and has generated imitation in other educational establishments that do not necessarily belong to the network but that are delivering environmental public policy objectives.

Nevertheless, our literature review reveals that, although the SNCAE has established the idea of environmental management in participating educational establishments within the network, it has not been able to overcome the more general challenges that face Chilean educational institutions as a whole: a lack of teacher training in contemporary methods appropriate to cross-cutting themes; communication difficulties, in this case between the establishments that belong to the network (i.e., rather than behaving as part of a coherent network, they each tend to behave like an isolated entity); and the challenges of transforming students' cognitive structures and value frameworks.

Network of Environmental Education Centers

The Network of Environmental Education Centers was created to increase reflection by the general public on environmental problems and to bring the relevant stakeholders together to identify common solutions.

The Network was officially inaugurated in March 2017 and its aim is to

> [...] help, to the broadest extent possible, the Environmental Education Centers (hereinafter "the Centers") to improve their environmental education programs, promote the work they do and support them in all their activities, in order to ensure their full and efficient operation. This support will take the form of training, knowledge sharing and other tasks to strengthen links between the Centers and increase the number of visitors (MMA 2016).

According to the MMA, the Network's objective is to enhance the visibility of the Centers across the country by promoting their activities and/or environmental education programs. The aim of this is to facilitate public access to open-air environmental education. The main objectives of the Network are as follows:

1. To build public recognition of the work of each Center.
2. To act as a meeting place for network members to communicate, share experiences and develop joint initiatives and projects.
3. To support and promote Network programs at the regional and national levels.
4. To support innovation in the new programs designed and implemented by the Centers (MMA 2017b).

In addition, an overriding priority of the network is to 'promote coordination, define agreed objectives and enhance the visibility of the different forms of environmental education provided around the country' (MMA 2017b). In other words, during this stage the Network is seeking to identify stakeholders with distinct educational approaches from across Chile, who will be able to work together to promote sustainability nationwide.

The Centers are characterized on the one hand as learning-friendly spaces for school-children, researchers (preferably on weekdays) and families (at weekends) and, on the other, as places that promote environmental education. The goal is for the Centers to provide:

> public and private educational spaces with facilities and infrastructure that enable innovative practices in environmental education, through training sessions, workshops, educational outings, field activities and other initiatives targeting different audiences (MMA 2017b)

At present, there are 61 Centers across the country working in a variety of fields (Figure 1).

As Figure 1 shows, 50% of the Centers are dedicated to EE in the form of conservation of natural areas, which in turn implies addressing the ownership of large-scale tracts of land with a significant presence of flora and fauna. A total of 20% of Centers promote sustainable development, which is understood as relating to production; 17% work to promote Education for Sustainable Development, but do not indicate their understanding of the subject or the nature of their specialism in pedagogical strategies; 7% provide social support to the lowest socio-economic groups using natural contexts as a physical backdrop for recreation and the delivery of community projects; 3% claim to engage in sustainable tourism; and the final 3% work to foster the value of social and/or natural environments, specifically by actively engaging with schools and institutions that need natural spaces in which to

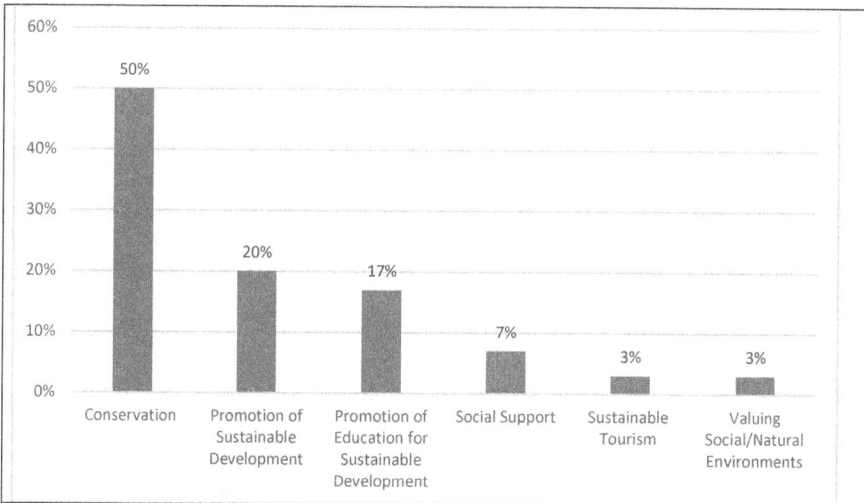

Figure 1. Breakdown of activities of Environmental Education Centers. Source: In-house research based on MMA and UAHC figures (2019).

do activities. This reaffirms once again the importance that EE in Chile places on the conservation of local flora and fauna (MMA and UAHC 2019). However, it has not been possible to identify research into the Centers' plans and programmes related to the promotion of sustainable development or the promotion of ESD (MMA and UAHC 2019).

According to MMA and UAHC (2019), the Centers focus on:

- Experiencing the environment through interpretive trails and guided tours.
- Recognizing local culture, flora, fauna and geology.
- Promoting renewable energy.
- Promoting bioconstruction.
- Disseminating the concept of 'reduce, reuse, recycle.'
- Training in organic farming, vermiculture, composting and nurseries.
- Holding workshops in bee-keeping, pottery and handicrafts, among others (MMA and UAHC 2019).

Evaluations of the Centers highlight the following needs and challenges with regard to EE:

- *Materials*: poor infrastructure and teaching materials (no magnifying glasses, maps, etc.).
- *Teacher training*: minimal training of education workers, poor group management or appealing learning environments for young learners.
- *Inclusion of specific issues in educational activities*: lack of knowledge on how to balance the use of scientific and general language on topics including climate change, environmental management, etc.
- Lack of knowledge on potential issues such as climate change, etc.
- Lack of coherent teacher training.
- Frequent absence of a clear definition of educational models (MMA and UAHC 2019).

In general, the overall evaluations are somewhat inconclusive with regard to the Centers' EE programmes and do not provide a clear picture of overall strengths and weaknesses.

Environmental Trainers

In 1999, CONAMA launched the campaign 'Chile, I'll take care of you (Chile, yo te cuido)' which led to the creation of Environmental Trainers Clubs (Clubes de Forjadores Ambientales). Nowadays, Environmental Trainers is a network of activists run by the MMA as a participative environmental education and action programme. Club members include school students, scouts, neighborhood groups and university groups, and there are even clubs specifically for adults and older persons. The MMA states:

> The Environmental Trainers program promotes team work and collective responsibility for solving environmental problems, strengthening the social cohesion of communities, the quality of the environment and the creation and recognition of local leaders (MMA 2020a).

It should be noted that these groups are self-organizing and self-managing entities, formed collectively and which carry out local and targeted campaigns. According to the MMA, their membership is predominantly made up of people who wish to 'contribute directly to improving the quality of their environment' (MMA 2020b).

Interestingly, the State has named 18 October the National Day of Environmental Trainers and the MMA website claims that there are currently over 2,300 environmental trainers clubs in the country.

Joint international initiatives to promote EE

Finally, it is worth noting the work undertaken by the MMA in recent years with regard to international support and collaboration in the field of EE.

Evaluation of the international course on education for sustainability - land as an educational space

Together with the Japanese International Cooperation Agency (JICA) and the Chilean International Cooperation Agency (AGCID), the MMA defined the terms of reference for the evaluation of the cycle of three international courses known as Land as an Educational Space (El Territorio como espacio educativo). The courses were held in 2014, 2015 and 2016, with 15 scholarship-holders from across Latin America and the Caribbean participating in each cycle.

United Nations environment programme environmental training network

The MMA's Division of Environmental Education and Citizen Participation is a member of the Environmental Training Network of Latin America and the Caribbean, coordinated by the United Nations Environment Programme (UNEP). The platform aims to strengthen regional exchange of knowledge and experience on environmental education. Its membership of the Network means that the MMA plays an active role in shaping a shared vision and environmental education strategies for the region.

EE in higher education

The leading universities in Chile have all joined the Sustainable Campus Network. According to its mission statement, the Network is

> an association of higher education institutions and professionals dedicated to the promotion of sustainability in higher education which aims to generate, through this channel, the transformation required if society is to become sustainable (RCS 2020).

However, the 'environmental' aspect of higher education is generally only present in technical degrees linked to management and, to a minimal extent, in teacher training courses. Only the Universidad Tecnológica Metropolitana (UTEM), which is renowned for its focus on sustainability, offers a broad-based undergraduate workshop that examines the related challenges of the SDGs, gender and sustainability (UTEM 2020).

Postgraduate learners have only two options for acquiring training in environmental education. The first is through the Master's in Education with a specialization in Environmental Education, Educational Evaluation or Educational and Vocational Training offered by the Universidad de la Frontera (Temuco). The second is the four-month postgraduate Diploma in Environmental Education offered by the Universidad Alberto Hurtado (Santiago), which emphasizes sociological reflection and ESD.

Chile–specific definition of EE, ESD and the Global Action Program on ESD for 2030

Our literature review and field research have not uncovered a Chile-specific definition of either EE or ESD. However, we did identify two official documents that refer to EE and ESD. The first, which refers specifically to ESD, is from 2009 and therefore written prior to the creation of the MMA in January 2010. It is titled National Policy on Education for Sustainable Development (Política Nacional de Educación para el Desarrollo Sustentable), and uses the definition of ESD in the UNESCO Global Action Program on ESD, outlining its objectives as follows:

> To train people and citizens to be able to take individual and collective responsibility for creating and enjoying a sustainable society and to contribute to the strengthening of educational processes that enable the instilling and development of values, concepts, skills, competences and attitudes across society as a whole (MMA 2020: 9).

The second document is a manual titled Environmental Education. Perspectives from the Chilean institutional framework (Educación Ambiental. Una mirada desde la institucionalidad ambiental chilena) and critiques ESD, noting that it aims to 're-work the productivity-centered model of development' (MMA 2020b). This is thus in direct contradiction with the 2009 ESD policy and MINEDUC's adherence to it in its curricular guidelines. Indeed, the MINEDUC guidelines characterize ESD as a:

> (a) training process that facilitates the development of the attitudes, skills and knowledge that will enable children and young people to make informed decisions, take responsibility for their own and collective actions, to reflect on and to act to improve the conditions of their surroundings and address the social, economic, cultural and environmental problems of their local community, country and society in general (MINEDUC 2020: 38).

This blatant contradiction is a clear demonstration of the lack of senior-level discussion and research on ESD and its emancipatory character. One reason for this from an educational point of view is the prevailing lack of interest on the part of teacher training centres in conducting research and training into EE and ESD; this, crucially, is linked to the absence of knowledge of non-Spanish language literature in the field.

EE and ESD in Chile: Emerging issues, current and future trends and requirements

As the MINEDUC can play no role in the formation or implementation of PEIs, it cannot impose any obligation on educational establishments to incorporate cross-cutting learning objectives into their teaching. Consequently, there is no authoritative and technical point of reference on the basis of which to support, advise on and help integrate EE within classrooms at a national level. The Chilean State has thus yet to implement the ECLAC/OECD suggestion that the MINEDUC should partner with the Council of Ministers for Sustainability in order to play a more active role and bring its expertise to bear in the design of a national environmental education plan (ECLAC and OECD 2016: 133). What the MMA and other institutions involved in EE processes have achieved is to instill in certain sections of the population an awareness of environmental management (understood by the state as recycling, energy saving, the efficient use of drinking water, among others) and green issues, albeit excluding the social aspects.

Regarding the CNCA, critical reflection is needed on its weaknesses: how it is implemented by the MMA and the process it uses to certify participating establishments. Only by taking on board constructive criticism will the MMA be able to fulfil its objectives in relation to the SNCAET; objectives that the Ministry itself proposed.

Our literature review found no research into the content or achievements of the laudable initiative of the Adriana Hoffmann Environmental Training Academy, which provides the general public and civil servants with free training courses, and so are unable to we pass judgement on it.

Generally, we have observed that the flagship EE programmes have a series of weaknesses that stem largely from existing deficiencies in the Chilean educational system. These include inequality in the quality of teaching and learning depending on the type of establishment involved (private or semi-private, municipal or state-run), and a lack of teacher development that leads to the application of traditional 'teacher talk' methods which, in turn, impedes the incorporation of value-based objectives.

Furthermore, as the OECD report Environmental Performance Reviews: Chile 2016 states, it is not only important to improve the quality of teachers' environmental awareness, update the curriculum in line with new knowledge and strengthen non-governmental stakeholders through non-formal education to enable them to become pro-active voices in the development of environmental policies. Rather, the most urgent task is to train workers and business owners on environmental issues (ECLAC and OECD 2016: 57, 133).

The efforts of individual stakeholders on EE and ESD to date, and the small-scale but significant progress made by the general public on environmental issues, should be acknowledged. Of course, there remain a number of challenges to achieving EE more broadly in Chile, particularly those related to improving teaching methodologies, increasing opportunities to learn outside the classroom (outdoor education), and embedding cross-cutting strategies and issues within the national curriculum. Moreover, there remains a profound lack of knowledge in the country about the potential of ESD to act as the basis for fundamental reforms to teaching and learning processes and to raise levels of public awareness on sustainable development.

One final area that needs to be addressed is the lack of opportunities for research into, reflection on and training in EE and ESD, particularly given that there is no body of research or conceptual work in relation to the latter.

Final reflections

Humanity is currently experiencing multiple simultaneous crises: social, environmental, health and economic, both nationally and globally. These crises pose significant challenges and require the collective creation of a conscious, proactive society that collaborates to shape a more sustainable world for future generations. Chile is no different. In order to bring about a more sustainable world and achieve SDG 4.7, it is crucial that the country trains a critical mass of educators specializing in EE and ESD. Such a critical mass of transformative change agents, in conjunction with a general public that adopts sustainable practices and lifestyles, will help to mitigate any socio-environmental impact generated by their actions.

It is also essential to bring together the diverse and interdisciplinary ministerial agents within collaborative strategies setting out concrete steps to achieve greater sustainability. Alongside this, the integration of EE into the training of education professionals is vital, as is the opening up of academic spaces to encourage new research into, reflection on and the development of a definition of ESD that combines the criteria proposed by UNESCO on the one hand, with pressing national and regional challenges on the other. Such spaces require investment and funding from the state and/or higher education institutions themselves.

The researchers interviewed for this paper also noted the need for greater emphasis to be placed on adaptation to and mitigation of climate change in all economic, social and cultural contexts. To this end, it is important that all related knowledge, skills and competences are dealt with in a cross-cutting manner within formal training processes, and in continuing and non-formal education.

References

Barazarte, R., A. Neaman, F. Vallejo and P. García. 2014. El conocimiento ambiental y el comportamiento proambiental de los estudiantes de la Enseñanza Media, en la Región de Valparaíso (Chile) (Environmental knowledge and pro-environmental behavior of secondary school students in the Valparaiso Region (Chile)). Revista de Educación. 364: 12–34.

Burgos P., O. Perales, F. Palacios and J. Gutiérrez. 2010. Evaluación de la calidad de los establecimientos educativos incorporados al Sistema Nacional de Certificación Ambiental de la provincia de Bio Bio (Chile) (Evaluation of the quality of educational establishments incorporated into the National System of Environmental Certification in the Province of Bio Bio (Chile)). Profesorado. Revista de Currículum y Formación de Profesorado. 14(2): 213–241.

BCN. 2006. Law 18.962 Organic Constitutional Teaching Act. Accessed 2 November 2019, from BCN website: https://www.bcn.cl/leychile/navegar?idNorma=30330.

BCN. 2009. Law 20.370 General Education Act. Accessed 2 November 2019, from BCN website: https://www.bcn.cl/leychile/navegar?idNorma=1006043.

BCN. 2010. Law 20.417 Creation of the Ministry of the Environment, the Environmental Assessment Service and the Superintendence of the Environment. Accessed 2 November 2019, from BCN website: https://www.bcn.cl/leychile/navegar?idNorma=1010459.

BCN. 2016. Citizenship Education Plan in educational Establishments. Accessed 12 January 2018, from BCN website: https://www.bcn.cl/leyfacil/recurso/plan-de-formacion-ciudadana-en-establecimientos-educacionales.

Castillo-Retamal, F. and F. Cordero-Tapia. 2019. La educación ambiental en la formación de profesores en Chile (The environmental education in the training of teacher's [sic] in Chile)). UC Maule. 56: 9–28.

Chile Sustentable. 2020. Accessed 30 October 2020: http://www.chilesustentable.net.

Chilean Library of Congress (BCN). 1994. Law 19.300 General Bases of the Environment Act. Accessed 2 November 2019, from BCN website: https://www.bcn.cl/leychile/navegar?idNorma=30667.

Economic Commission for Latin America and the Caribbean (ECLAC) and Organisation for Economic Co-operation and Development (OECD). 2016. Environmental Performance Reviews: Chile 2016. United Nations, Santiago, Chile.

Gudynas, E. 2009. Diez tesis urgentes sobre el nuevo extractivismo. Contextos y demandas bajo el progresismo sudamericano actual (Ten urgent discourses on the new extractivism. Contexts and demands in line with current South American progressivism). pp. 187–225. *In*: Centro Andino de Acción Popular and Centro Latino Americano de Ecología Social Extractivismo (eds.). Extractivismo, política y sociedad (Extractivism, politics and society). Albazul Offset, Quito, Ecuador.

Hajek, E. 1998. La agenda 21 en Chile: algunas consideraciones (Agenda 21 in Chile: Observations). Accessed 30 November 2020, from Ecolyma website: https://www.ecolyma.cl/documentos/la-agenda-21-en-chile-1998.pdf.

Marconi, M. 2020. IX Encuesta de Actitudes hacia el Medio Ambiente Universidad Andrés Bello: chilenos apoyan incluir el medio ambiente y la ciencia como ejes en la nueva Constitución (IX Survey on Attitudes towards the Environment by Universidad Andrés Bello: Chileans support the inclusion of the environment and sciences as core components of the new constitution). Accessed 30 November 2020, from Universidad Andrés Bello news website: https://noticias.unab.cl/ix-encuesta-de-actitudes-hacia-el-medio-ambiente-unab-chilenos-apoyan-incluir-el-medio-ambiente-y-la-ciencia-como-ejes-en-la-nueva-constitucion.

Ministry of Education (MINEDUC). 2016. Orientaciones curriculares para el desarrollo del plan de formación ciudadana (Curriculum guidelines for the implementation of the Citizenship Education Plan), Santiago, Chile. Accessed 20 January 2018, from MINEDUC website: https://bibliotecadigital.mineduc.cl/handle/20.500.12365/467.

MINEDUC. 2018. Bases Curriculares 1o a 6o Básico (Curriculum objectives for 1st to 6th grade students). Santiago de Chile Accessed 20 January 2018, from MINEDUC website: https://www.curriculumnacional.cl/portal/Documentos-Curriculares/Bases-curriculares/.

MINEDUC. 2018b Indicadores de la educación en Chile 2010-2016 (Education indicators in Chile 2010–2016). Santiago de Chile Accessed 20 January 2018, from MINEDUC website: https://centroestudios.mineduc.cl/wp-content/uploads/sites/100/2018/03/INDICADORES_baja.pdf.

MINEDUC. 2020. Formación integral y convivencia escolar (Holistic education and school coexistence). Santiago de Chile Accessed 20 November 2020, from MINEDUC website: http://convivenciaescolar.mineduc.cl/formacion-para-la-vida/desarrollo-sustentable/.

Ministry of the Environment (MMA). 2009. Política Nacional de Educación para el Desarrollo Sustentable (National Policy of Education for Sustainable Development). Santiago de Chile Accessed 15 January 2020, from Biblioteca Digital del Gobierno de Chile (Chilean Government Digital Library) website: https://biblioteca.digital.gob.cl/handle/123456789/1406.

MMA. 2012 Política Nacional de Educación para el Desarrollo Sustentable (National Policy of Education for Sustainable Development), Internal Ministerial Draft Report, Santiago, Chile. Accessed 10 November 2020, from MMA website: https://.mma.gob.cl%2Fwp-content%2Fuploads%2F2017%2F10%2FPNEDS-PDF.

MMA. 2016a Encargada de la Red de Formación Ambiental del PNUMA visitó Ministerio del Medio Ambiente Santiago de Chile Accessed 17 October 2020. Website: https://educacion.mma.gob.cl/encargada-de-la-red-de-formacion-ambiental-del-pnuma-visito-ministerio-del-medio-ambiente/.

MMA. 2016. Bases del Concurso Público "Creación de una Red de Centros de Educación Ambiental a nivel nacional" (Rules for the public tender "Establishment of a national network of Environmental Education Centers)." Santiago de Chile Accessed 10 April 2017, from MMA website web: https://mma.gob.cl/wp-content/uploads/2016/11/Res-Ex1286-aprueba-bases-y-anexos-concurso-publico-creacion-de-una-red-de-educacion-ambiental.pdf.

MMA. 2017. Planificación Departamento de Educación Ambiental Ministerio del Medio Ambiente (Planning of the Department of Environmental Education in the Ministry of the Environment), Internal Ministerial Report, Santiago, Chile.

MMA and Universidad Academia de Humanismo Cristiano (UAHC). 2019. Diagnóstico de los Centros de Educación Ambiental (Assessment of the Environmental Education Centers), Internal Ministerial Report into Draft Version of MMA 2017a, Santiago, Chile.

MMA. 2020. Cuenta Publica Participativa Santiago de Chile Accessed 20 October 2020, from MMA website: https://cuentaspublicas.mma.gob.cl/.

MMA. 2020a. Forjadores Ambientales (Environmental Advocates). Santiago de Chile Accessed 20 October 2020, from MMA website: https://forjadoresambientales.mma.gob.cl/.

MMA. 2020b. Educación Ambiental. Una mirada desde la institucionalidad chilena (Environmental Education. Perspectives from the Chilean institutional framework). Santiago de Chile, Accessed, 30 October 2020, from MMA website: https://bibliotecadigital.mineduc.cl/handle/20.500.12365/17610.

MMA. Date unknown. Red Nacional de Centros de Educación Ambiental (National Network of Environmental Education Centers), Internal Ministerial Report, Santiago, Chile.

Molina, B. 2017. Evaluación del proceso de certificación ambiental de los establecimientos educacionales de la comuna de Concepción (Evaluation of the environmental certification process of educational establishments in the municipality of Concepcion.). Power Point. Accessed 20 June 2018, from Prezi website: https://prezi.com/cti82o96mhju/certificacion-ambiental-exitosa-de-los-establecimientos-educ/.

Muñoz-Pedreros, A. 2014. La educación ambiental en Chile, una tarea aún pendiente (Environmental education in Chile, a task pending). Ambiente & Sociedad. XVII(3): 177–198.

Pelfini, A. and R. Mena. 2017. Oligarquización y extractivismo. Cerrojos a la democratización de la política ambiental en Chile (Oligarchization and extractivism. Obstacles to the democratization of environmental policy in Chile). Perfiles Latinoamericanos. 25(49): 251–276.

Quiceno, E. 2011. Pertinencia y conceptualización de educación ambiental en escuelas municipales certificadas de la Región Metropolitana (Importance and conceptualisation of environmental education in certified municipal schools in the Metropolitan Region). Accessed 18 October 2018, from Universidad de Chile Faculty of Agricultural Sciences website: http://repositorio.uchile.cl/handle/2250/112357.

Red Campus Sustentable. RCS 2020. Santiago de Chile, Accessed 30 November 2020: https://redcampussustentable.cl/quienes-somos.

Squella, M. 2000. La educación ambiental en Chile: Un estudio exploratorio (Environmental education in Chile: An exploratory study). Lit VERLAG, Münster, Germany.

United Nations. 2015. Transforming our world: Agenda 2030 for Sustainable Development. Accessed 15 December 2020, from United Nations website: https://www.un.org/ga/search/view_doc.asp?symbol=A/RES/70/1&Lang=.

Universidad Tecnológica Metropolitana UTEM. 2020. Programa de Sustentabilidad Santiago de Chile Accessed 4 October 2020. Website https://sustentabilidad.utem.cl/compromiso-institucional/programa-de-sustentabilidad/.

Chapter 13

Environmental and Sustainability Education in Ecuador in the Context of the Sustainable Development Goals

Patricia M. Aguirre,[1] *Freddy Villota González*[2,]* and *Silvia Mera Pincay*[3]

Introduction

Environmental Education (EE) is a dynamic and participatory social initiative that seeks to respond to concerns about humanity's relationship with nature in the context of economically and socially sustainable development (Bustos 2011). It aims to transform ways of thinking and feeling, both individually and collectively, and inculcate the values necessary to achieving a sustainable balance between the environment and society (Andrade 2015: 7). In Ecuador, EE is advancing slowly, but it has had an impact on daily habits in recent decades, transforming the behaviour of the population to the benefit of the environment in the context of a holistic and transdisciplinary vision of society, where learners interact harmoniously with their environment, understanding that their actions both leave a mark in and transcend the present.

Proposals for EE have been put forward by international summits on development seeking consensus on the need for action by the world community in the face of the undeniable environmental crisis facing the planet (United Nation [UN] 2022).

[1] Instituto de Posgrado, Universidad Técnica del Norte, Ibarra, Ecuador.
[2] Centro Universitario de Tonalá, Universidad de Guadalajara, Jalisco, México.
[3] Universidad de Vechta, Vechta, Alemania.
Emails: pmaguirre@utn.edu.ec; silviamera1@gmail.com
* Corresponding author: freddyvillota@gmail.com

The global education policy known as the Decade of Education for Sustainable Development (ESD) (2005–2014) included environmental issues and emphasizes the promotion of responsible consumption (United Nations Educational, Scientific and Cultural Organization [UNESCO] 2005). In 2015, the UN approved the 17 global Sustainable Development Goals (SDGs). The fourth of these is Quality Education, a strategy for the dissemination of information, awareness raising and training, aimed to support and strengthen the concept of sustainable development, taking a long-term perspective and moving the focus from the exploitation of natural resources to adequate management of them, and consideration of the needs of future generations.

In the context of Ecuador, EE, ESD and the SDGs are the reference framework, especially in the field of education, as they provide the basis for management, planning tools, the development of educational programmes, and proposals for curricula to promote the knowledge, values and practical skills necessary to address the environmental crisis and propose appropriate solutions (UNESCO 2020).

Ecuador's political, economic, cultural and social circumstances, including its education system

The Republic of Ecuador is a constitutional State based on rights and justice, a social, democratic, sovereign, independent, unitary, intercultural, pluri-national and secular country with a decentralized structure of government (Constitution of the Republic of Ecuador 2008: 1). It has an area of 283,561 km² and is located in the northwest of South America, bordering Colombia to the North, Peru to the East and South, and the Pacific Ocean to the West. Its territory also includes the Galapagos Islands. It consists of 24 provinces and its capital is Quito. Ecuador has a population of 17.23 million inhabitants (National Institute of Statistics and Censuses 2019), and the official language is Spanish. Quichua, Shuar and other ancestral languages spoken by indigenous peoples also have official status. The official legal currency has been the United States dollar since 2000.

In the political context, Ecuador has seen transformative changes over the last four years, amongst which is the implementation of policies aimed at promoting economic renewal through entrepreneurship, production and exports, and other activities (see Plan de Creación de Oportunidades 2021–2025, National Planning Secretary 2021). These strategies have been lent greater urgency by the health emergency (Covid-19), which has encouraged several countries, including Ecuador, to reconsider their policy guidelines in order to stabilize and improve their economies (Maldonado et al. 2021). As state education budgets have been severely impacted, educational management and quality improvement measures have been directly affected, causing extensive weakening of the Ecuadorian education system (Campos and Ortíz 2021).

Ecuador's main income comes from the export of non-renewable resources such as oil and agricultural products including bananas (Banco Central del Ecuador 2019). On the basis of 2018 figures of the National Secretariat for Planning and Development (SENPLADES), part of government expenditure has a social purpose,

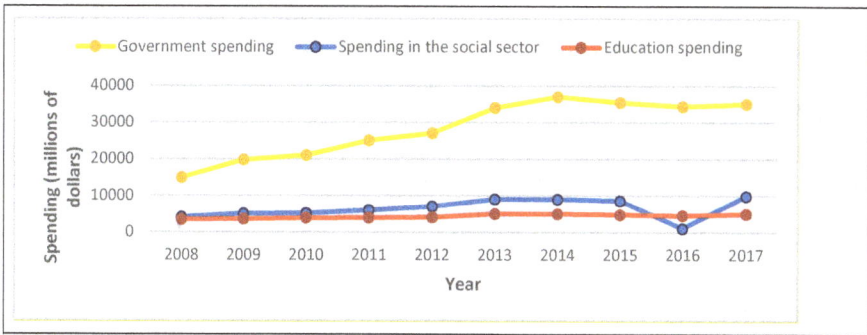

Figure 1. Evolution of government spending, social sector spending, and education spending 2008–2017. Source: Ministry of Economy and Finance (2018).

and spending on education increased by 151% between 2008 and 2017, rising from 1,911 to 4,792 million dollars (Figure 1). The years 2015 and 2016 saw decreased investment, but in 2017 the government ratified its commitment to education and, as part of the implementation of the 'A Whole Life' ('Toda una Vida') plan, budget allocations to the education sector by the Ministry of Education of Ecuador (2018) and Instituto Nacional de Evaluación Educativa (2018) were increased.

In Ecuador, government mechanisms such as the Constitution of the Republic of Ecuador (2008), the 'Toda una Vida' National Development Plan 2017–2021 (SENPLADES 2017), the Organic Law on Higher Education – LOES in Spanish – (Asamblea Nacional 2010) and the Organic Law on Intercultural Education – LOEI in Spanish – (Asamblea Nacional 2011) guarantee the inter-culturality of teaching processes at all levels of education, safeguarding the rights of students to be instructed in their native language. Ecuador is a multi-ethnic country and the education system reflects that.

LOES also guarantees intercultural, inclusive and diverse higher education for students (Rieckmann et al. 2021), promoting the strengthening and development of the languages, cultures and ancestral wisdoms of the peoples and nationalities in Ecuador within an intercultural framework (Alonso et al. 2019).

With regard to the social context, the country is currently facing a crisis, especially in unemployment and education. This is due to several factors, such as socio-economic inequality and the ineffective coverage of public services, but the issue making the greatest impact today is the health emergency caused by Covid-19.

In the educational context, the LOEI enacted by the Constitution of the Republic of Ecuador (2008) lays down constitutional guidelines for education. It sets out the responsibilities of the National Education System—SNE in Spanish—and its employees for developing individual and collective capacities and the potential of the population, requiring them to facilitate learning, and promote the generation and use of knowledge, techniques, arts and culture (Instituto Nacional de Evaluación Educativa 2018, Rieckmann et al. 2021).

'The SNE comprises all institutions, programmes, policies, resources and individuals involved with the education process, including Pre-school Education (PE), General Basic Education (GBE) and the Unified General Baccalaureate

(UGB)' (Instituto Nacional de Evaluación Educativa 2018: 11). The LOEI also governs the education provided by the state to indigenous peoples and nationalities through the higher education system and the bilingual intercultural education system (SITEAL 2019).

The types of education that are delivered are: formal education (pre-school, basic and high school), which is cumulative, progressive, leads to a degree or certificate, and is governed by specific standards and curricula; and non-formal education, which is culturally and linguistically appropriate and provides citizens with lifelong training and development, and it is not related to the curricula. Dividing educational institutions by sources of financing, there are a number of different types (Table 1).

Table 1. Types of educational institutions in Ecuador.

Institution type	Description
Public or municipal, armed forces or police	Institutions managed by the state offering free, secular education
Public/religious	Institutions run by congregations, orders or any other religious or secular denomination. Totally or partially financed by the state (costs regulated by the Ministry of Education)
Private	Institutions managed by natural or legal persons under private law. Education may be religious or secular. Authorized in accordance with the regulations established by the Ministry of Education to charge for the educational services they provide.

Source: SITEAL (2019).

In response to the coronavirus pandemic, the Ministry of Education has implemented a Covid-19 Education Plan, which includes the following strategies:

1. Access to: https://recursos2.educacion.gob.ec, with around 840 Covid-19 Contingency Plan digital resources.
2. A self-directed course for teachers provided by the Ministry of Education and Universidad Central del Ecuador called 'My Online Classroom.'
3. Microsoft Teams to facilitate interaction between members of the education community.
4. Training course for teachers via the 'mecapacito.gob.ec' platform.

Virtual education would be ideal if all students had internet access. However, data from the National Multipurpose Household Survey conducted in December 2018 by the National Institute of Statistics and Censuses (INEC) revealed that, while 37.23% of households nationwide have a computer, in rural areas the percentage is lower (23.27%). In addition to this, it was made clear that not all students had access to the necessary technology to enable them to take part in this new mechanism proposed by the Ministry of Education.

The student population in rural areas is the most vulnerable and it has been severely affected by the pandemic. Teachers have been also struggling to handle the now indispensable online education platforms. For these reasons, the structure of education has gone through significant change in Ecuador, and depending on the situation, it will be necessary to develop new curricula.

Historical background and development of environmental and sustainability education in Ecuador

The inclusion of EE and sustainability in Ecuador's education system has been underway since the early 1980s, following international action to promote it (UNESCO 2009, Bucaram and Intriago 2017, Arroyo 2013, Santillán 2012, Bustos 2011, Crespo 2004):

- The Environmental Education Conference in Tbilisi, 1977.
- The International Congress on Environmental Education and Training in Moscow, 1987.
- The Earth Summit in Rio de Janeiro - Brazil, 1992.
- The Ibero-American Congress on EE in Guadalajara - Mexico 1992.
- The World Summit on Sustainable Development, Johannesburg, 2002.
- The Latin American and Caribbean Program for EE (PLACEA in Spanish), in Panama - 2003.
- The Andean Amazonian Communication and EE Action Plan (PANACEA in Spanish) in Lima - Peru, 2005.
- The regional strategy, Building Education for Sustainable Development in Latin America and the Caribbean, in San José - Costa Rica, 2006.
- The UN Decade of Education for Sustainable Development, 2005–2014.
- The Andean Environmental Agenda (Andean Community [CAN] in 2006–2010).
- SDGs - Agenda 2030.

These actions reflect an anthropocentric understanding of the relationship between society and nature, according a strong role to EE and the promotion of sustainability. International organizations guide member countries in the implementation of action plans within their education policies.

In the course of incorporating EE and sustainability into its education system, Ecuador has gone through several implementation stages, reviewing study plans and programmes at the school and non-school level, and training educators (Falconi and Hidalgo 2019, Santillán 2012, Bustos 2011) (Table 2).

Drawing on the list in Table 2, the most outstanding events with respect to the uptake of EE and sustainability in Ecuador are highlighted below.

Between 1983 and 1993, the EDUNAT Programmew put measures in place to promote EE, incorporating it into study plans and programmes, teacher training and teaching guides and other materials focused on the subject (Arroyo 2013, Santillán 2012). By 1994, the CAAM had produced The Basic Principles for Environmental Management in Ecuador, the Ecuadorian Environmental Plan and the Basic Environmental Policies of Ecuador, which established EE as a high priority, and

Table 2. Activities that included and transformed EE in Ecuador.

Year	Activities related to EE	Promoting institution
1983	Education for Nature Programme (EDUNAT)	Fundación Natura, Ministry of Education and Culture (MEC) and the United States Agency for International Development (USAID)
	Forestation programme for high school students	MEC, Ministry of Agriculture and Livestock (MAG)
1991	Strategies for the Development of Environmental Education in Ecuador Workshop	MEC and UNESCO
1992	Creation of the Department of Environmental Education	National Directorate of Regular and Special Education
1994	Ecuadorian Agenda for Environmental Education	MEC, Ecociencia and UNESCO
	Basic principles of environmental management, Ecuadorian environmental plan and basic environmental policies for Ecuador	Environmental Advisory Commission (CAAM)
1995	Regulation of environmental education, training and communication policies	MEC
1996	Unified curricular reform of Ecuadorian basic education (elementary, primary, and middle school)	MEC
	National Division of Environmental and Road Safety Education, dealing with ecology and biosphere, natural resources, ecotourism, environmental quality and quality of life, and road safety education Ecuadorian Basic Education: Consensual Curriculum Reform	MEC MEC
1998	Development of a Bilingual-Intercultural Education Model within EE	National Directorate of Indigenous Intercultural Bilingual Education (DINEIIB)
1999	National Environmental Education Plan (PNEA in Spanish) and teaching guide	MAE (Ecuadorian Ministry of Environment)
2000	Implementation of education, training, and environmental awareness and communication programmes	Cooperation agreement between MEC and MAE
2005	Basic Education and Baccalaureate National Environmental Education Plan (2005–2016)	MEC
2006	National Environmental Education Plan (2006–2016): key component of the education policies under the Ten-Year Education Plan (2006–2015)	MEC
2008	National Education Programme for Democracy and Good Living: EE is integrated into all national education programmes and reforestation projects are introduced as a part of extracurricular activities	Ecuador Ministry of Education

Table 2 contd. ...

...Table 2 contd.

Year	Activities related to EE	Promoting institution
2010	Curriculum reform: update and strengthening	Ecuador Ministry of Education, MAE, Ministry of Health of Ecuador and the Navy
	Creation of green schools, certification of educational centres promoting environmental awareness.	
	Proposal to include coastal and marine environmental education in the curriculum for fourth, fifth, sixth and seventh grades of elementary school	
	Toda una Vida National Plan	SENPLADES
	EE promotes awareness, learning and teaching of knowledge, skills, values, duties, rights, and behaviours in the population	Organic Environment Code
2017	National Environmental Education Strategy for Sustainable Development 2017–2030 Tierra de todos EE programme	Ecuador Ministry of Education y MAE
2020	3rd and 6th grades of primary school curriculum content assessment	UNESCO Regional Office

Source: Falconi and Hidalgo (2019), Santillán (2012), Bustos (2011).

integrated it into all phases, forms and subjects of formal and informal education, and training in general (Bucaran and Intriago 2017, Arroyo 2013, Santillán 2012, Bustos 2011, PNEA 2006, Crespo 2004).

In 2005 one of the most important changes to the curriculum was the National Environmental Education Plan for Basic Education and High Schools (2005–2016). This promoted the institutionalization of EE to support sustainable development, developing curricula to incorporate and/or strengthen its environmental content, and promoting critical and committed involvement by the educational community in environmental management and sustainability (Bustos 2011, Santillán 2012). Further to this, in 2010 the curricula was reformed, updated and strengthened to include environmental protection as a guiding principle. El Buen Vivir is a key aspect of the Ecuadorian education system, and one of the cross-cutting themes that develop learners' values. The subject of maritime awareness, integrated coastal management, and coastal marine environmental education is integrated into the curriculum of the fourth, fifth, sixth, and seventh years of basic education (Arroyo 2013).

Finally, in 2017 Ecuador launched the National Strategy for Environmental Education for Sustainable Development 2017–2030. Among its objectives are the promotion of EE in initial, general and high school education, strengthening the environmental content of teacher training and linkages with higher education institutions and communities (MAE 2018: 29). To support this strategy, the Organic Code of Environment was created, stating in Article 16 that 'environmental education will promote awareness, teaching knowledge, skills, values, duties, rights and behaviours to ensure the protection and conservation of the environment and to promote sustainable development. It will be a cross-cutting theme of strategies, programmes and plans at all levels and in all forms of formal and non-formal education.' Complementing this strategy, the Tierra de todos programme strengthens the environmental approach in the National Education System, with the aim of mainstreaming environmental education.

As can be seen in this chronology of the integration of EE and sustainability in the Ecuadorian education system, there have been crucial moments where EE has been to the fore, but it has also been built into the concepts of Buen Vivir and into quality initiatives aiming to ensure that education promotes sustainable development.

Understanding environmental education and sustainability in Ecuador

The understanding of EE and sustainability in Ecuador 'has its own peculiarities because the Constitution (2008) establishes 'Good Living' as a development model, implying a kind of nationalization of these concepts' (Zambrano and Isch 2015: 109).

In Ecuador, teachers at the basic and high school levels appear to be indifferent to EE. This is due to the fact that 'traditionally, EE has been considered a secondary school subject, since environmental matters were included in subjects such as natural sciences or environmental sciences, and in general with content that is unrelated to daily lives' (Félix 2011: 50).

The understanding and implementation of EE and sustainability teaching is more complex in higher education. The four relevant dimensions here are (Zambrano 2015):

- **Concepts:** Learners are taught that Ecuador has been taking into account the concept of sustainable development since 2008, which leads to confusion about the meanings of sustainable and long-lasting.
- **Institutional issues:** Teachers do not have the autonomy to modify curricula, and the integration of EE therefore becomes complex.
- **Curriculum issues:** Confusion in the application of the parameters and limited teacher training.
- **Practices:** Teachers use experiential methods, looking at case studies and performing experiments in order to connect students with the reality of environmental issues.

In 2019, a research project interviewed 20 professors from a range of Ecuadorian universities (Mera 2022), asking how they understood the concepts of sustainable development and sustainability and what they associated them with. The highest percentage (30%) of those interviewed related the concepts to balance, equity, progress, and processes (Figure 2). Respondents considered that these concepts should promote balanced development processes, to ensure an equitable relationship between society and the natural world. Of those interviewed, 10% associated the concepts with a utopian ideal, stating that they were multi-faceted and that no specific objectives had yet been provided for the higher education system, with the result that every teacher adjusted the concepts depending on their needs and circumstances.

Despite the complexity of promoting a parallel understanding of EE and sustainability in the educational system, the government, international organizations

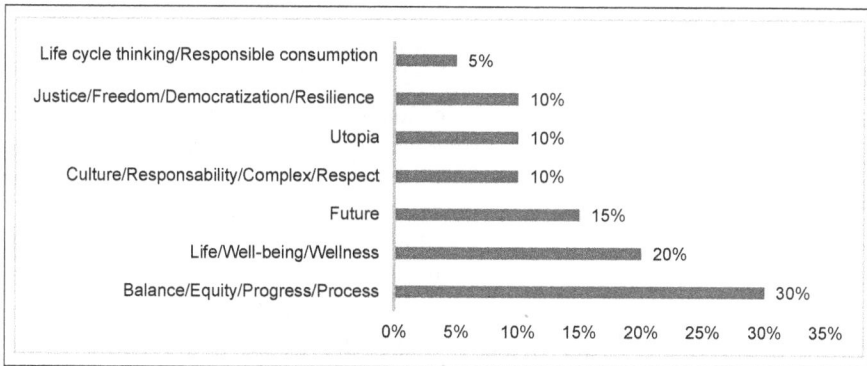

Figure 2. Understanding of sustainable development concepts by university professors. Source: Mera (2022).

and private companies have supported projects seeking to foster involvement with—and application of—the two concepts. The initiatives include:

- Green Point (2010) implemented by MAE, which grants incentives to all national development activities whose processes optimize natural resources and promote good practices with regard to the environment and the well-being of citizens.
- The TiNi Project (2017), implemented by ME, which aims to mainstream environmental issues and institutionalize the environmental approach in pre-school, basic and high school education through the application of active teaching methods.
- Sustainable Ecuador (2019): Ecuadorian companies and multinationals operating in the country have committed to boosting and promoting the SDGs. According to Deloitte Ecuador (2019), more than 70% of companies are prioritizing sustainability in their planning proposals.
- The MAE (2020) project, The Itinerant Virtual Classroom for Environmental Education, runs a number of training courses on EE and sustainability issues.

These initiatives promote active participation and encourages all stakeholders and sectors to take responsibility in their respective spheres of competence, to ensure that national and international EE and sustainability objectives are achieved.

Environmental and sustainability education and the SDGs in Ecuador

In 1998, UNESCO declared that EE should be directed towards sustainable development, in order to educate people to become proactive and aware of the environment, and ensure that they have the capacity to get involved with social issues and identify appropriate solutions. It also declared that teaching must be inclusive, conscious, and cooperative, so that sustainable development ceased to be an ideal, and became a tangible goal (Guarga 2004).

In Ecuador, initiatives to incorporate environmental education and sustainability into the educational curriculum derive from the National Environmental Education Plan for Basic Education and High Schools; this is the 'navigation chart' for the institutionalization of environmental education to support sustainable development (González and Del Pozo 2020). Ecuador adopted the 2030 Agenda and the Sustainable Development Goals (SDGs), formalizing them as public policy in Executive Decree No. 371 of April 19, 2018 (Art. 1), presenting its first Voluntary National Exam to the United Nations High Level Political Forum on Sustainable Development in 2018, and committing to providing an annual compliance report with contributions from the public and private sectors, civil society, academia, local government and agencies, among others (Technical Secretariat Planifica Ecuador 2019: 9).

The MAE's National Strategy for Environmental Education (2017–2030) is based on the National Development Plan 'Toda Una Vida' 2017–2021, and a number of international agreements, including:

- The International Convention on Biological Health
- The United Nations Framework Convention on Climate Change
- The RAMSAR Convention on Wetlands
- The Andean Environmental Agenda 2006–2010 (Comunidad Andina 2006)
- Andean Environmental Management 2012–2016 (Comunidad Andina 2012), and in particular
- Agenda 2030 and the Declaration from the Buenos Aires Regional Meeting of Ministers of Education of Latin America and the Caribbean 'E2030: Education and Skills for the 21st century' (UNESCO 2017).

The National Strategy highlights the importance of these strategies for the achievement of Ecuador's international commitments and their influence on the public policies introduced to implement SDGs 4, 12, 13 and 15.

EE in Ecuador is derived from environmental regulations, policies and the SDGs, and reflects the need for the country's people to develop an environmental identity and raise their awareness, encouraging them to take a responsible approach to their relationships with society and the environment (Torres and Calderón 2017). One important conduit for EE is higher education, which Torres and Calderón (2017) analyze as follows:

- HEIs have insufficient funding to promote programmes and projects focused on the environmentally-sound management of natural resources.
- 94.12% of HEIs offer undergraduate courses focusing on environmental issues.
- 70% of HEIs offer master's degrees in environmental issues. Two institutions have a doctoral programme in water resources; and only one offers specializations in environmental issues.
- 53% of HEIs have an approved document setting out the institution's environmental policy. Of this total, 55.56% do not have a formally-approved plan or programme that is aligned with the HEI policy and can be adapted and implemented.

- With regard to research (innovation, technology transfer, research groups, strategies to promote, research dissemination and use, among others) the performance of HEIs was classified as average, scoring 4.29 out of 7 points.
- Community engagement (community programmes addressing environmental problems, participation in environmental networks, communication programmes or projects, etc.) was on average rated 3 out of a possible 6.
- 59% of HEIs confirmed that they had a department, unit, office, or service of a technical and/or administrative nature that was dedicated exclusively to environmental issues; 90% of these have an assigned budget.

In response to the current situation with regard to EE in HEIs, a range of alternatives have been proposed, including:

- Professional training with an environmental focus, based on the policies promoted by the governing bodies
- Strengthening the environmental content of initial teacher training (pedagogical institutes, faculty of educational sciences, national university of education)
- Promotion of inter-disciplinary educational entrepreneurship on environmental issues of local and national interest
- Creation and implementation of plans to incorporate environmental and sustainability into the administrative management, professional training and external representation of universities
- Promotion of research into EE, in accordance with the LENIA No. 5 national strategic guidelines for environmental research (Torres and Calderón 2015)
- Integration of university development plans and business improvement programmes developing sustainable production and consumption
- Promotion of active participation of the academic sector in meeting the challenges of the 2030 Agenda
- Strengthening of environment and sustainability networks in national universities and their incorporation to international networks, and
- Promotion of environmental management as an element to be taken into account in the accreditation of HEIs by the Council for Evaluation, Accreditation and Quality Assurance in Higher Education (CEAACES in Spanish).

The Ecuadorian higher education system is unlikely to see action at a national level to adjust professional training curricula to include EE and sustainability. However, studies by Torres and Calderón (2016), Cadme et al. (2017), and Falconí and Hidalgo (2019) have identified gaps and highlighted some progress, such as the commitment of universities to implement and mainstream EE concepts and sustainability. HEIs are not likely to introduce changes to the design of their curricula, since they do not have the necessary budget or access to the relevant advice.

Falconí and Hidalgo (2019: 8) emphasize that the limited training of HEIs on environmental issues is based on a 'lack of research, little feedback from educational programmes and the absence of evaluation (impact, cost-profit, etc.).' Despite the operational limitations in HEIs' strategic development plans (PEDI in Spanish), LOES 2018 seeks to promote the integration of sustainable development, based on research, international cooperation and teaching, as well as other activities.

This chronology of the integration of EE and sustainability shows that its incorporation into the concept of 'Buen Vivir' fosters educational quality initiatives aimed at promoting and achieving the SDGs. Implementation is underway, and involves not only students but also faculty staff, parents, and the community. The three cases below illustrate the implementation of EE and ESD.

Environmental education for 3 to 5-year-olds in Alfredo Pérez Guerrero educational unit

Castro (2018) designed methodological strategies to introduce EE for students at initial and preparatory sub-levels in the Alfredo Pérez Guerrero Educational Unit. The Educational Unit's curriculum was evaluated, identifying EE-related issues. Methodological strategies were designed and then communicated, and the results obtained through the research were evaluated.

This work revealed that children have a clear idea of what the environment is, but did not understand what they needed to do to protect the natural world. This was evident in the students' disrespectful attitudes towards the institution's physical spaces, and especially in the inappropriate disposal of garbage. To address this problem, environmental education strategies were designed, introducing 20 activities related to water care, cultivation, green areas, plant care, waste classification, and empathy with animals and the environment. The strategies were based on ludic games, allowing children to develop their knowledge, and learn how to behave and act.

In this way, it was determined that curriculum content relating to environmental education at the initial and preparatory sublevels is scarce, and that this results in a lack of environmental awareness and poor habits among students as regards the use of natural resources.

Universidad San Francisco de Quito Office of Innovation and Sustainability (USFQ-OIS)

USFQ created an Office of Innovation and Sustainability in 2004, with the aim of developing strategies that would allow the university to grow and foster sustainable development. The carbon footprint of the university campus has been calculated since 2012, and sustainability reports for the institution have been regularly generated. Some of the projects implemented by the Office of Innovation and Sustainability are detailed in Table 3.

Table 3. Projects implemented by the USFQ Office of Innovation and Sustainability.

Project	Description
Water use efficiency	Environmental engineering students were involved in a study whose results served as the basis for the creation of policies and strategies to promote responsible and sustainable consumption of water.
PanchoBus	A mobility survey among university groups led to the creation of new routes with the aim of reducing the carbon footprint of students, faculty staff, and administrative staff.
Energy efficiency	Strategies were developed to reduce the consumption of electrical energy in the USFQ Cumbayá campus through policies, technology, and the generation of alternative energies.
e-waste recycling project	This project separates and recycles electric devices and objects that may contain heavy and precious metals, seeking to provide employment for people who need it, as well as to prevent these dangerous compounds from mixing with general waste, which could eventually lead to soil and water pollution.

Sustainability report 'Towards a sustainable university' by Universidad Técnica del Norte

Between 2009 and 2011, the Universidad Técnica del Norte (UTN) cooperated with Leuphana University of Lüneburg in Germany on the Teaching and Learning Sustainability programme (ENSU), which was funded by the German Academic Exchange Service (DAAD). This initiative trained teachers and professors to integrate the concept of sustainable development into all university programmes (Barth and Rieckmann 2012). The project also organized an exchange programme where faculty staff, university authorities and students from both universities shared their various experiences. In addition, a number of seminars were held on sustainable development and the role of higher education in this context.

The ENSU project also introduced an eight-module specialization in Education for Sustainable Development (ESD) (32 credits in total), involving professionals from all departments in the university. Professors from UTN and invited speakers such as Prof. Dr. Walter Tenfelde, from the University of Hamburg conducted short workshops for students from different disciplines.

Academic events promoting sustainability and its principles, with the involvement of a number of universities from Germany, Ecuador and Latin America, were an important part of the ENSU project, with 10 scientific publications resulting from this programme.

Other notable UTN projects promoting sustainability are:

- *The Sustainable University project*: The objective here was to change the university structure to make the institution more sustainable. Teaching practices, research activity, financial management, and community engagement projects were analyzed to determine whether their processes met sustainability criteria. Proposals for improvement were formulated and implemented and the results obtained evaluated.

- *The Sustainable University Club project*: This was a student initiative, derived from the Sustainable University project and aimed to contribute to sustainable development in university life by continuously promoting the responsible management of natural resources. The club put together a sustainability guide for the university community, and ran campaigns to reduce water consumption, save energy, and encourage sustainable waste management and the proper disposal of batteries.

It is important to note that a number of the sustainable development research projects at UTN have arisen out of its framework agreement with the Deutsche Gesellschaft für Internationale Zusammenarbeit (GIZ). As well as working toward the achievement of the SDGs, undergraduate degree projects must comply with Ecuadorian government regulations (GIZ 2013).

These activities represent the implementation of Education for Sustainable Development to date. ESD is a fundamental pillar of the work of Universidad Técnica del Norte and has been integrated across management, research and teaching.

Environmental and sustainability education in Ecuador: Emerging issues and trends, and current and future needs

EE and ESD are key mechanisms for raising national awareness of the responsible use of renewable and nonrenewable natural resources. The emerging issues and trends with regard to current and future needs for responsible advocacy and enforcement are therefore addressed in the context of different themes.

Contextualized educational reform

The absence of formal proposals on EE and ESD in Ecuador is a weakness despite the country's good intentions. Programmes are often not adapted to the reality of educational institutions and therefore tend not to be given responsible or serious enough consideration. Educational reforms focusing on the environmental realities and the need for sustainability are essential in order to ensure changes to programmes and curricula that have long-lasting relevance.

Although a multitude of curricula and educational projects reference ESD and EE, 'there is an insufficient relationship and not enough coordination between the stakeholders involved in those projects or programmes, which means there is only limited potential to generate synergies' (Isch and Zambrano 2008: 51).

Ways have been found over the years to integrate EE and ESD into educational reforms, but poor follow-up and support has led them to be forgotten or simply ignored due to the lack of the appropriate conditions and advice. This requires a theoretical framework, a pedagogy and a curriculum that defines and frames how EE and ESD will be introduced.

Need for updated skills for teachers and technical support

Changes to education cannot be guaranteed by a well-structured and well-founded educational model and curriculum organization alone; 'it is also necessary to have the human capacity to implement them; in other words, education administrators and teachers' competencies need to be built across all areas, that is, to cover the cognitive, procedural, attitudinal, habitual and evaluative issues' (Merino et al. 2017: 91).

Educators have to engage in three types of interactions: with content, with their students, and with reality itself. Their profile plays a key role in enabling them to deliver EE and ESD, so appropriate training and professionalization for educators is essential. According to Merino et al. (2017), the first step is to analyze the current profile of educators, and their theoretical, conceptual and operational awareness.

Several Ecuadorian authors including Zambrano (2015), Merino et al. (2017), and Falconí and Hidalgo (2019) highlight the importance of support from experts in EE and sustainability issues for the implementation of projects and teacher training. Technical support is necessary to ensure teachers' skills are updated because educational administrators are responsible for decision-making in education management, and it is the role and responsibility of teachers to bring about fundamental changes in current and future generations.

Citizen training through spaces for exchange and awareness-raising

The Ministry of Education, in cooperation with MAE, has introduced a national strategy take forward and improve sustainability over time. It has put forward several programmes setting out all EE initiatives and those involved with them, and explaining national objectives and formal and informal agreements (Velasteguí 2020). These initiatives use EE as the tool to manage three key areas of sustainable development: respect for—and appreciation of—nature (environmental), community integration and participation (social) and environmental entrepreneurship (economic) (Ecuador Technical Secretariat Plans 2020).

Despite these efforts, it is evident that people are not connecting or engaging with the environmental problem (Hill 2020), demonstrating a lack of responsibility and commitment to the country's environmental difficulties. It is thus necessary to introduce national awareness programmes to educate citizens, since the efforts of educational institutions will have no effect if they are not supported by society in general.

In the same way, it is necessary to provide space to facilitate debates and promote exchanges of knowledge on EE and sustainability issues. It is also important to establish a clear vision and objectives through an open and participatory national process on the development and implementation of clear and coherent actions with respect to EE.

Research and studies on EE and sustainability issues

Unfortunately, there are not enough studies and investigations on EE and ESD in Ecuador, which means that knowledge of the EE situation is poor. A multi-disciplinary approach is essential, starting with studies and research that address the issues from the point of view of different environmental sciences (Zambrano 2015). Consideration must also be given to the dissemination of research findings.

State policies

It is not sufficient to be party to international treaties on EE and sustainability; it is also vital to implement national policies that fit the country's current situation (Arroyo 2013). From this point of view, it is the responsibility of the state to provide guidelines and leadership on the country's development and well-being in its various contexts. It is important to develop national policies that promote the introduction of EE and ESD. Change at the national level requires the government to provide support for a range of entities, in coordination with the ministries that will implement state strategies.

The history of EE in Ecuador is the story of the increased invisibility of the concept: various 'regulations [have] removed and restored [it], and also imposed [it]' (Bustos 2011: 99) with no methodological basis or order, and above all without any state policy to guide the process and lead to its productive and successful application.

Conclusion

This chapter highlights three aspects of environmental and sustainability education in Ecuador: first, the influence of international policy on the curricula of the Ecuadorian education system; second, the relationship between the agreements and disagreements that have been made visible in order to institutionalise EE in the education sector; and finally, the implications of the 2030 Agenda and the SDGs for the reconceptualization of EE and sustainable development and for the implementation of ESD.

There are interesting initiatives on EE, ESD and the SDGs in Ecuador. At a normative and legislative level, several measures have been implemented. However, teacher training, EE and ESD teaching and learning methodologies and budgets remain challenges that must be addressed if Ecuador is to achieve real progress in terms of changed attitudes and measurable results on sustainable development.

State policies and discourses are opportunities which are presented, taken away, and imposed. This is the history of EE in the education sector: at times it has been invisible and downplayed (Bustos 2011), but it must also be recognised that these opportunities are temporary and that EE, like the recognition of certain social interests, is forged from the bottom up, from social need. It is in these spaces at the base of society that young people, teachers, environmental educators, educational institutions, communities, universities and civil society organisations, have to recover EE as a social project with axiological and political dimensions (Gonzalez and Pozo 2020, Murriagui et al. 2018, Andrade 2015, Arroyo 2013, Bustos 2011). It is therefore important to raise awareness and motivate all those involved with education to actively participate in a debate that will help reinstate environmental and sustainability education at the heart of Ecuadorian education.

References

Alonso García, S., Y. Roque Herrera and V. Juárez Ramos. 2019. La educación intercultural en el contexto ecuatoriano de educación superior: un caso de innovación curricular. *Tendencias Pedagógicas* 33: 47–58.

Arroyo Cisneros, L. 2013. Information Flows on Environmental Education at the Inter-institutional Level. Analysis of the National Environmental Education Plan for Basic Education and Bachelor's Degree 2006–2016. M.Sc. Thesis. Facultad Latinoamericana de Ciencias Sociales Sede Ecuador. Quito, Ecuador.

Asamblea Nacional. 2010. Organic Law of Higher Education (LOES). Official Registration Supplement 298 of 12-oct.-2010. Quito, Ecuador.

Asamblea Nacional. 2011. Organic Law on Intercultural Education (LOEI). Official Register 417 of 31-mar-2011. Quito, Ecuador.

Banco central del Ecuador. 2020. Mining Report. Analytical Management and Data Intelligence. Quito, Ecuador.

Barth, M. and M. Rieckmann. 2012. Academic staff development as a catalyst for curriculum change towards education for sustainable development: an output perspective. Journal of Cleaner Production 26: 28–36.

Bucaran Intriago, C.T. and M. del Rocío Intriago. 2017. La educación y la gestión ambiental contemporánea en Ecuador. Una mirada a la universidad. Revista científica especializada en Ciencias de la Cultura Física y del Deporte 14(32): 126–139.

Bustos Lozano, H. 2011. Environmental Education and National and Global Educational Policies for the New Baccalaureate (2000–2011). M.Sc. Thesis. Simón Bolívar Andean University, Ecuador Seat.

Cadme, M., R. Herrera Anangonó, R. García, B. Cerezo Segovia, M. Sandoval Cuji, L. Saltos Velásquez, F. Contreras Estelvina, L. Simba Ochoa, B. González Osorio, J. García and M. Carrillo 2017. Environmental education in university formation before the climatic change in Quevedo, Ecuador. International Journal of Humanities and Social Science Invention 16: 2319–7714.

Campos, M. and M. Ortíz. 2021. Pandemic Times in Ecuador: The Right to Education in Intensive Care? In Times of Pandemic in Ecuador, is the Right to Education in Intensive Care? (pp. 23–46). Editorial Abya-Yala. Quito, Ecuador.

CEPAL. 2021. COVID-19 Observatory in Latin America and the Caribbean Economic and social impact. United Nations. https://www.cepal.org/es/temas/covid-19.

Comunidad Andina. 2006. Andean Environmental Agenda 2006–2010. Secretary General of the Andean Community. Lima, Peru.

Comunidad Andina. 2012. Gestión Ambiental Andina (2012–2016). Secretary General of the Andean Community. Lima, Peru.

Constitution of the Republic of Ecuador. 2008. Official Registry 449 of October 20th, 2008. Quito, Ecuador.

Crespo Coello, P. 2004. Final report: Proposal from Ecuador for the environmental education and awareness program in the Andean paramo project. Paramo Andino Ecoscience Project.

Deloitte. 2019. Sustainability Trends Study 2019. https://www2.deloitte.com/ec/es/pages/about-deloitte/articles/estudio-de-tendencias-de-sostenibilidad-2019.html.

Deutsche Gesellschaft für Internationale Zusammenarbeit (GIZ). 2013. Environmental Education in GIZ Ecuador Systematization of approaches of the GESOREN Program Series of systematizations. Fascicle 9. Quito, Ecuador.

Executive Decree No. 371. 2018. https://observatorioplanificacion.cepal.org/sites/default/files/instrument/files/decreto_371_71305.pdf.

Falconí, F. and E. Hidalgo. 2019. Environmental Education and Teacher Training in Ecuador. Educational Policy Notebook No. 7. UNAE Observatory.

Faro Group. 2018. Achievements and Challenges in the Implementation of the SDGs in Ecuador. Annual Sustainable Overview No. 1. Quito: SDG Territory Ecuador.

Félix Ríos, V. 2011. Genesis of Environmental Education and Advances in Environmental Education in The Educational System of Ecuador. Bachelor's Degree. Thesis, Pontificia Universidad Católica del Ecuador. Quito, Ecuador.

González, S. and F. Del Pozo. 2020. Education and environmental sustainability challenge and university worldview. Case of the Central University of Ecuador. Educere 24(77): 95–101.

Guarga, R. 2004. París+ 5 Follow-up or revision of the UNESCO World Conference on Higher Education, Paris, 1998? Universities, 27: 3–13.

Instituto Nacional de Estadística Y Censos (INEC). 2018. National Multipurpose Household Survey – Information and Communication Technologies (Follow-up to the National Development Plan).

Isch, E. and A. Zambrano. 2008. Education for sustainable development in the Andean region: Some significant experiences in Ecuador and Venezuela. UNESCO Regional Bureau for Education for Latin America and the Caribbean.

Maldonado, D., J. Vinueza, J. Oviedo and A. Ramirez. 2021. Strategies for the economic reactivation of Ecuador. FIPCAEC (Edición 23) 6(1): 685–695.

Mera, S. 2022. Analysis of the Supportive Factors and Obstacles Experienced by University Teachers in Integrating Sustainable Development Concepts into Teaching and Learning Processes at Universities in Ecuador and Germany. Ph.D. Thesis. University of Vechta, Faculty of Education and Social Sciences] Mensch und Buch Verlag.

Merino Alberca, V., E. Loaiza Carrión and D. Vilela Mora. 2017. The problem of environmental education in Ecuador and the need for a new teaching profile to face it. Olympia: Scientific publication of the Faculty of Physical Culture of the University of Granma 14(44): 83–94.

Ministry of Education and Culture and the Ministry of Environment. 2006. National Environmental Education Plan for Basic Education and Bachelor's degree (2006–2016). Quito, Ecuador. https://1library.co/document/oz1okkdq-plan-nacional-de-educacion-ambiental-ecuador.html.

Ministry of Education of Ecuador and the National Council of Education (2006–2015) second year of its execution. 2017. Ten-Year Education Plan of ECUADOR 2006–2015. Quito, Ecuador. https://www.siteal.iiep.unesco.org/bdnp/966/plan-decenal-educacion-ecuador-2006-2015.

Ministry of Education. 2017. Organic Law on Intercultural Education. Supplement to the Official Register No. 572 of August 25, 2015. Ecuador.

Ministry of Education of Ecuador. 2018. Land of all environmental education programme. https://educacion.gob.ec/educacion-ambiental/.

Ministry of Education. 2018. Education budget for 2018 grows by 16%. Ministry of Education of Ecuador. https://educacion.gob.ec/presupuesto-de-educacion-para-2018-crece-en-un-16/.

Ministry of Education. 2020. COVID-19 Educational Plan. https://educacion.gob.ec/plan-educativo-covid-19/.

Ministry of Energy and Non-Renewable Natural Resources. 2019. Strategic Plan 2019–2021. https://www.recursosyenergia.gob.ec/wp-content/uploads/2019/02/Plan-Estrategico-Institucional-2019-2021-MERNNR.pdf.

Ministry of Environment of Ecuador. 2017. National Environmental Education Strategy for Sustainable Development 2017–2030. First edition. Quito, Ecuador.

Ministry of the Environment. 2017. National Environmental Education Strategy. Dirección de información, seguimiento y evaluación - DISE. Quito, Ecuador.

Murriagui Lombardi, S., M. Viteri Gordillo and S. Barreno Freire. 2018. Education for sustainable development in the University curriculum. Challenges of Science 2(3): 47–54.

National Institute of Educational Evaluation. 2018. Education in Ecuador: Achievements and New Challenges. Educational Results 2017–2018. First Edition. Quito, Ecuador.

National Institute of Educational Evaluation. 2020. UNESCO: Ecuador must advance concepts such as gender equity and knowledge of the world in the curricula of third and sixth grade.

National Planning Secretary. 2021. Opportunity Creation Plan 2021-2025. Quito, Ecuador. https://www.planificacion.gob.ec/wp-content/uploads/2021/09/Plan-de-Creación-de-Oportunidades-2021-2025-Aprobado.pdf.

National Secretariat of Planning and Development. 2013. National Development Plan / National Plan for Good Living 2013-2017. Quito: SENPLADES.

Particular Technique of Loja University. 2020. UNESCO Chair of Sustainable Development. https://www.utpl.edu.ec/es/catedras.

Plan Ecuador Technical Secretary. 2020. Expertos analizan los desafíos de la educación para el desarrollo. https://www.planificacion.gob.ec/expertos-analizan-los-desafios-de-la-educacion-para-el-desarrollo/.

Rieckmann, M., S. Flores, M. Pabón Ponce, E. Vélez and S. Mera Pincay. 2021. The education system in Ecuador: Moving towards quality education. *In*: Jornitz, S. and M. Parreira do Amaral (eds.). The Education Systems of the Americas. Cham: Springer, First Online: 3 March 2021.

Santillan Egas, F. 2012. Environmental Education Management for Sustainable Development in Ecuador. M.Sc. Thesis. International University of Andalusia. Ecuador.

SENPLADES. 2018. Report of the Physical Execution of the General State Budget. Quito, Ecuador.

Sistema de Información de Tendencias Educativas en América Latina (SITEAL). 2019. Ecuador: perfil del país https://www.siteal.iiep.unesco.org/sites/default/files/sit_informe_pdfs/dpe_ecuador-_25_09_19.pdf.

Technical Secretariat Plan Ecuador. 2019. Progress report on the implementation of the 2030 Agenda for Sustainable Development. First edition. Quito, Ecuador.

Torres, R. and E. Calderón. 2015. Diagnosis on the inclusion of environmental and sustainability considerations in the universities of Ecuador – first phase. AMBIENS 1(2): 101–119.

UNESCO. 2005. United Nations Decade of Education for Sustainable Development (2005–2014): international implementation scheme. https://unesdoc.unesco.org/ark:/48223/pf0000148654?posIn Set=13&queryId=329f6549-de93-40ad-b09f-57946bc1e56c.

UNESCO. 2009. Regional, subregional and national policies, strategies and plans in education for sustainable development and environmental education in Latin America and the Caribbean. https://unesdoc.unesco.org/ark:/48223/pf0000181906_spa.

UNESCO. 2017. Declaration of Buenos Aires. Regional Meeting of Ministers of Education of Latin America and the Caribbean "E2030: Education and Skills for the 21st century". https://cerlalc.org/wp-content/uploads/2019/10/20-Declaración-de-Buenos-Aires_2017.pdf.

UNESCO. 2018. Conferencia Mundial Sobre Educación Superior. Sección especial. 9(2).

UNESCO. 2020. Curricular analysis. Regional Comparative and Explanatory Study (ERCE 2019) Ecuador: National Results Document. https://unesdoc.unesco.org/ark:/48223/pf0000373963.

United Nations. 2022. Conferences: Environment and sustainable development. https://www.un.org/en/conferences/environment.

Zambrano C. 2015. Enfoque de la educación para el desarrollo sostenible en la formación docente en el Ecuador. *In*: Ortiz, M.E., C. Crespo Burgos, E. Isch and E. Fabara Garzón. (eds.). Reflexiones sobre la formación y el trabajo docente en Ecuador y América Latina. Quito, Ecuador.

Chapter 14

Environmental and Sustainability Education in the Context of the UN Sustainable Development Goals

Trajectory and Challenges in Mexico

Rosalba Thomas Muñoz,[1,*] *Helio García-Campos*[2] and
Teresita Maldonado Salazar[3]

Introduction

The emergence of environmental education (EE) is associated with the emergence of the global environmental crisis. As a field of study, it comprises the contributions of different social stakeholders committed to the environment and theoretical and methodological ideas from a range of disciplines. Its focus has gradually shifted from ecological conservation to sustainability, in order to promote permanent learning that allows citizens to be more critical, responsible and committed to their environment.

Almost twenty years ago, EE stakeholders and academics in Mexico resisted UNESCO's attempt to assimilate the subject and replace the term 'environmental

[1] University of Colima.
[2] Veracruz Intercultural University.
[3] National Pedagogical University, Unit 094.
Emails: gc.helio@gmail.com; petite_thereses_rouge@hotmail.com
* Corresponding author: rosalbathomas@gmail.com

education' with 'education for sustainable development' (ESD). This process triggered a broad discussion both in Mexico and in the international sphere (including some UNESCO consultations) and finally the term 'environmental education for sustainability' (EES) was chosen, a diplomatic convention reflected in the official documents of the Mexican State (SEMARNAT 2006). This is an example of the challenges faced by those who are committed to EE, with contradictory initiatives issuing from the environmental bodies under the umbrella of the United Nations (UN).

This chapter reflects on EES in Mexico, and starts by recognizing the diversity of the country's conceptual approaches and its openness towards this worldwide trend. It also seeks to substantiate both national and Latin American contributions to theory, which are critical of a notion of development that is restricted only to the idea of economic growth (Sauvé 1997 in González Gaudiano 2008).[1]

Drawing on our experience, we have tried to bring together the multiplicity of approaches, proposals and initiatives in the fields of education, culture and communication in response to the mega-crisis or syndemic of 2020 and 2021. We agree that, although education will not achieve change in and of itself, it does contribute to the development of those who 'will be capable of transforming it, making it more humane and supportive, more just, democratic and egalitarian' (Caride and Meira 2020).

Environmental, political, economic, cultural and social conditions and contexts

Mexico is a mega-diverse country in terms of both biology and culture. It is one of the select group of nations with the greatest diversity of animals and plants, which account for almost 70% of the world's biological diversity: amphibians, reptiles, birds and mammals and vascular plants (CONABIO 2020). With an area of 1,972,550 km², Mexico is the 14th largest country in the world, and is located in the tropical zone with the greatest diversity of species in the world. However, the Sectoral Program derived from the National Development Plan 2019–2024 indicates that some of the Republic's states that are particularly rich in biodiversity, including the Gulf of Mexico, have lost up to 80% of their original ecosystems to agricultural land (Semarnat 2016). Between 2010 and 2015, the estimated average rate of deforestation was 92,000 hectares per year (FAO 2015). As a consequence of the loss and degradation of ecosystems, a significant number of species—around half of the mammals, amphibians and reptiles in the country—are today at risk of extinction (*Ibid*).

The Report on the Situation of the Environment in Mexico (Semarnat 2016) highlights climate change, the loss of ecosystems and biodiversity, and the scarcity and contamination of water resources, among other issues, to the list of the most pressing problems. The Mexican territory has suffered sustained degradation and loss of natural heritage due to a range of social and economic phenomena resulting from

---

[1] For a review of the 10 years of intense debate, see: González Gaudiano 2008.

its transformation into agricultural fields, induced grasslands and urban areas. The ecosystems that still persist are already showing clear signs of alteration (Semarnat 2016: 61).

The international community, for its part, continues to try to close the gap between so-called poorly-developed and under-developed countries (we do not proposed to unpack here the large number of euphemisms that have been used to describe socially and ecologically impoverished countries at various points in time). This led in 2015 to the development of the 2030 Agenda, which, in addition to continuing the work of its predecessor campaign—the 8 millennium goals—seeks to establish a development model that facilitates economic growth, reduces poverty and increases the well-being and quality of life of all inhabitants, without mortgaging the natural resource base (Semarnat 2016: 1).

Historical background and development of environmental education for sustainability

In 1946, Beltrán Castillo recommended 'the inclusion of basic conservation concepts in each and every one of the subjects that make up the curriculum at the primary and secondary education levels' (cited in González 2003). Ecology was first taught in Mexican primary education in the state of Veracruz in the mid-1970s, but it was not until 1993 that the necessary analysis was undertaken and strategies developed that enable the field of EE to be founded.

Formally, EE in Mexico dates back to the early 1980s. In the 1990s it gained visibility and its impact on public policies increased; however, since the first decade of the 21st century, 'a weakening has been noted in both national and municipal and community initiatives, more due to the direct impact of national policies, than as a result of their work' (Reyes et al. 2016).

In 1986, the Ministry of Public Education (SEP), the Ministry of Urban Development and Ecology (SEDUE), and the Ministry of Health and Assistance (SSA), jointly published the National Environmental Education Program (PRONEA). However, the program was not renewed by the next federal administration (SEMARNAT 2006). Finally, in 1988, the General Law of Ecological Balance and Environmental Protection (LGEEPA) was approved, recognizing EE as a means of raising the ecological awareness of the population, promoting the incorporation of ecological content into curricula and the training of children and young people at various educational levels.

At the end of 1999, the environmental sector was acknowledged in the structure of government with the creation of the Ministry of the Environment, Natural Resources and Fisheries (SEMARNAP) at the federal level and its counterparts in the federal bodies and municipal governments. Within this framework, the Center for Education and Training for Sustainable Development (CECADESU) was also created to formulate, organize, direct, supervise and evaluate the development of education and training programmes and projects to promote sustainable development (González 2000).

Special attention was given to the linkage of CECADESU with the SEP, with the two bodies signing collaboration agreements in the late 1990s. In basic education, environmental content was included in some subjects and teaching materials were designed. In addition, the National Association of Universities and Institutions of Higher Education (ANUIES), established the ANUIES-SEMARNAP Committee, with a view to implementing an environmental training programme for higher education, and thus linking the country's environmental and educational problems (González 1999). As a result, Institutional Environmental Plans (PAI), Environmental Agendas and other initiatives were developed to facilitate the environmental management of campuses and the greening of the curriculum.

With the support of CECADESU and the participation of many members of the National Academy of Environmental Education (ANEA), state environmental education programmes were promulgated, collectively forming a National Strategy for Environmental Education for Sustainability (CECADESU-SEMARNAT 2006). The National Development Plan 2007–2012 set out guidelines on EES, seeking to promote social awareness and environmental culture based on the protection of the environment and the sustainable use of natural resources. The cross-cutting programme integrates the environmental dimension across all areas and levels, focusing on sustainability in the National Educational System.

Within the framework of the Integral Reform of Basic Education (IRBE), the SEP and SEMARNAT worked together to incorporate EES into study plans and programmes, standards documents, free textbooks, study materials and teacher training at all three levels of Mexican basic education. This work was undertaken by the SEMARNAT-SEP Transversality Agenda Group was formed, led by CECADESU with representatives from the federal agencies' education bodies (Semarnat and Cecadesu 2012).

During this period, SEMARNAT, through CECADESU, designed the Green School Program, to promote environmental management initiatives at basic education schools with the participation of the educational community, helping to reduce schools' environmental impact and promote responsible citizenship (Maldonado 2010). This initiative was coordinated jointly between the SEP and local educational authorities. More than 5000 schools participated, generating wide recognition in educational communities and, despite the lack of follow-up, some entities and municipalities are continuing this work with the help of environmental and education authorities and citizen organizations. Coordination between educational institutions and authorities at the federal, state and municipal levels has also led to more widespread teacher education and modernization.

In the National Development Plan 2013–2018, the basis of environmental policy was what was known as 'green growth,' focusing on low carbon consumption. However, the Plan did not target any specific strategic sectors, so it did not have a great deal of practical impact. During this period, the Strategy for Environmental Education for Sustainability was not updated; however, it continued to be a fundamental reference point despite this.

Public policy gradually relegated EE to second place, and the CECADESU became less of a focus. It was a period of decline in which the government structures that had coordinated EES were reduced, and education policy focused only on labor reforms. Collaboration agreements between SEP and SEMARNAT were not signed, which limited joint working.

The most constant and visible efforts to achieve the objectives of EES were made by the non-formal sector and university institutions, which are today the primary sources of professional training for environmental educators.

Today, both EE and the entirety of the Mexican government's environmental structures are in sharp decline and at risk of having less and less impact. Sadly, not even the recent incorporation of EES into the General Education Law will be sufficient to avoid the anticipated centralization of various areas of educational practice.

Understanding of environmental education for sustainability

Neither the emergence of alternatives to the socio-environmental crisis nor eco-development (Sachs 1974) nor sustainable development have succeeded in reducing social inequality, suffering or the ongoing and potential collapse of human and animal populations and the entire biological matrix. Since the emergence of the notion of sustainable development and its gradual establishment as a benchmark for UN policy, the concept has attracted a broad consensus, but also some criticism. The term was inspired by the idea of the durability and permanence of the natural heritage on which humanity depends for its survival. It was bolstered by concepts derived from the environmental sciences, such as the carrying capacity and resilience of ecosystems. However, the notion of development continues to attract significant criticism, because it is considered to mask the ongoing 'mantra' of unlimited growth that has been the norm over the last forty years.

The term sustainability also refers to the utopia of the harmonization of relationships between human beings and the natural world. For González-Márquez and Toledo (2020), the concept has begun to reveal its weakness and is in crisis, since a fundamental criterion for scientific paradigms (Kuhn 1962) is their potential to allow solutions to problems to be identified; the authors emphasize that the first sign of crisis in the sustainability sciences paradigm is the proliferation of increasingly complex problems that are resistant to solutions. This does not mean that all the effort invested should be considered worthless, but it does highlight the need to address the nature of the crisis through new and rigorous debates and by encouraging the involvement of academics from disciplines that have not yet contributed sufficiently to the science of sustainability.

The international discourse, for its part, continues to try to bridge the gap between developed and underdeveloped countries (not counting the large number of euphemisms with which socially and ecologically impoverished countries have been called, at various historical moments), hence the creation in 2015 of the 2030 Agenda, which, in addition to continuing the work promoted by its predecessor campaign 'the 8 millennium goals,' seeks a development model that allows economic growth, reduce poverty and increase well-being and quality of life for all inhabitants, without mortgaging the natural resource base (Semarnat 2016: 1). There is a glaring disconnect between the educational initiatives and structures of the UN's

Table 1. Educational initiatives and structures of the UN's main environmental bodes and their equivalents within the Government of Mexico.

International entity	Institutionally assigned name	Description/comment	Reference institution in Mexico	Name of official program in Mexico
UNESCO	Education for Sustainable Development	Has distanced itself from environmental education, considering it to have been overtaken by its emphasis on the bio-ecological dimension. Aims to exceed EE by integrating social and economic aspects.	Secretary of the Environment and Natural Resources- CECADESU National Commission of Protected Natural Areas- CONANP	Environmental Education for Sustainability Conservation Education
Regional Office for Latin America and the Caribbean. UNEP-ROLAC	Environmental Education	Claims to be deploying environmental education	SEMARNAT- CECADESU	Environmental Education for Sustainability
United Nations Framework Convention on Climate Change- UNESCO	Action for Climate Empowerment- ACE	Addresses six areas: education, training, public awareness, public participation, public access to information and international cooperation	National Institute of Climate Change	Climate Change Education
United Nations Framework Convention on Biodiversity	Capacity-building	The process by which people, organizations and society as a whole unleash, strengthen, create, adapt and maintain the ability to manage their affairs over time.	National Commission for the Knowledge and Use of Biodiversity. CONABIO	Education, communication and environmental culture
UNDEP	Sustainable Development Goals	Education is integral to SDG 4.7. Mainstreaming with the other SDGs is expected.	Government Programme of the Presidency of the Republic (2018–2024)	It is posited that the SDGs have been an important reference for the preparation of the National Development Plan (2018–2024).
FAO	Farmer Field Schools capacity building	Participatory methods are used to create an environment conducive to learning: participants can share knowledge and experience in a safe environment.	National Forestry Commission	Forest Culture
UN Water Decade Program on Capacity Development.	Capacity development	Capacity development through training, education, and institutional development.	National Water Commission	Water Culture

Source: Authors' own, drawing on data from the institutions mentioned.

main environmental bodies and their equivalents within the Federal Government of Mexico. Table 1 shows the various forms that the same strategy takes.

Political and administrative organization within Mexico suffers from a similar difficulty, because although the difficulties of integrating the various institutions with an environmental focus are recognized, government activity is heavily siloed, leading to a lack of overview that would have helped to resolve problems. It is thus important that initiatives are brought forward to help achieve the objectives of ESD, but there is no aspiration to design a single overarching curriculum in Mexico: the country's cultural and ecological characteristics are too diverse.

The Political Constitution of the United Mexican States, under the principle of concurrence provided for in Article 73, Section XXIX-G, empowers the Congress of the Union to issue laws on environmental protection and preservation and restoration of the ecological balance. The Constitutional Reform of May 2019 added an important paragraph to Article 3, which sets out the national educational policy framework, establishing the plans and study programs of basic education as key aspects of an integrated education:

> Preschool, primary and secondary education are part of basic education; The Federal Government will determine the plans and study programs of said educational levels throughout the Republic, which will have a vision of respect for gender equity and will include knowledge of the sciences and the humanities: teaching will cover mathematics, literacy, history, geography, civics, philosophy, technology, innovation, the indigenous languages of our country, foreign languages, physical education, sports, the arts, especially music, the promotion of healthy lifestyles, sexual and reproductive education and care for the environment, among other things, and the State will guarantee that teaching materials, educational infrastructure and learning environments are suitable, and appropriately maintained and contribute to the goals of education.

Article 30 Section XVI of the General Education Law, as amended in September 2019, states for the first time that EES, the term agreed upon in Mexico, must include:

> Knowledge of the concepts and principles of environmental sciences, sustainable development, prevention and combatting of climate change, and the raising of awareness about the need for management, conservation and use of natural resources that guarantees social participation in environmental protection.

This advance is consistent with responses to public consultations on EE. However, the LGEEPA describes EE as a:

> training process directed at the whole of society, both in schools and in an extracurricular environment, to raise environmental awareness in order to ensure more rational behavior with regard to social development and the environment; Environmental education includes the assimilation of knowledge, the inculcation of values, the development of competencies and behaviors with the purpose of guaranteeing the preservation of life; it is considered of public utility.

The Mexican EE Strategy for Sustainability thus justifies the term it has selected on the basis of its having attracted a greater consensus among the country's environmental educators; retaining the term EE follows the trajectory and the capital built in the field and also makes explicit that its strategic focus is sustainability rather than sustainable development (CECADESU and SEMARNAT 2006: 35). Although, as the strategy points out,

> there is no intention to attempt to impose its adoption in Mexico; the debate must continue and it would be a mistake to try to close it down now when different positions are being in a more defined (*idem*).

We recognize the advances made by Mexican EE; however, they are not enough. The State must genuinely to make EE policy a central focus, led jointly by the environmental and education sectors, academic institutions, and civil society organizations, with environmental educators playing a truly leading role.

Implementation of environmental education for sustainability

Mexico's Specialized Technical Committee for the Sustainable Development Goals (SDGs) (CTEODS), which falls within the framework of the National Statistical and Geographic Information System (SNIEG), was formed in 2015 in order to coordinate the generation, monitoring and updating of data and other indicators of the country's progress towards Agenda 2030. The National Institute of Statistics and Geography (INEGI) mapped global indicators for the 2030 Agenda and determined that Mexico could monitor 169 of the 232 international indicators, so, in 2017 the National Council for the 2030 Agenda for Sustainable Development was established with participation from a range of sectors.

At the sub-national level, INEGI data indicates that the reality is somewhat different. States such as Chiapas, Oaxaca, and Guerrero have high illiteracy rates, which are particularly high for women, indigenous peoples and people living in poverty. The average level of schooling completed in these areas is primary, while in the rest of the country it is secondary. Among the indigenous population, 17.2% of school-age pupils (13–17 years old) do not attend school (INEGI 2015).

As many as 76.6% of girls, boys and adolescents do not have access to the internet and, of these, only 11% of primary school students and 23% of secondary school students attend public institutions with computers, internet access and adequate infrastructure (CONEVAL 2018b, cited in Gobierno de la República 2018, 40). The implications of the pandemic for the education system, including the need for students to learn at home on occasion, make this a pressing issue.

This is the reality of Mexico in the context of the 2030 Agenda. There have been insufficient achievements on equity and quality, and there thus continue to be limited opportunities to enter the labour market. The abandonment of studies by upper secondary school pupils is one of the biggest challenges of the educational system. This phenomenon not only has high economic and social costs, but also perpetuates poverty and the exclusion of vulnerable groups (Federal Government 2018).

The government's proposals under the 2030 Agenda and the SDGs are constructed around seven strategic axes:

1. Leave no one behind, leave no one out. Assume that education is a human right, not merely a means to develop useful capacity.
2. Promote and expand early childhood education. Expand preschool and initial education services to ensure they can deliver on their obligations.
3. Ensure excellence in all types and levels of education and across all sectors of the National Educational System. Develop relevant, regionally differentiated curricula.
4. Incorporate education for sustainable development into schemes of work and programs at all educational levels.
5. Promote comprehensive, long-term teacher professionalization processes. Strengthen normal schools and public teacher training institutions.
6. Analyze and understand the causes of school dropout, taking into account factors such as poverty, lack of resources, distance from schools, adolescent and early pregnancy, domestic violence, among others.
7. Deepen the links between the education sector and the labour market. Encourage alignment between professional training and the needs of the labour market.

By 2030, the Federal Government aims to guarantee that all girls, boys, adolescents and young people will have access to compulsory, free education from preschool to higher education. Teachers are vital agents of change in the educational process and contribute to social transformation; their professionalization and training, and that of their managers and supervisors, will be promoted; schemes of work and programs of study will be kept up-to-date with a view to ensuring that education is based on the relevant science, takes a sustainable approach and develops human potential (Government of Mexico 2019: 40). To date, however, there is no operational strategy for achieving these aims.

One of the most important challenges for the delivery of the 2030 Agenda is ensuring that teachers are given appropriate initial and ongoing training, to enable them to help children and adolescents acquire knowledge and skills that will enhance their well-being and promote responsible behaviour. According to data from the National Institute of Educational Evaluation (INEE 2015), a third of teachers do not create schemes of work and study programs and a quarter do not have enough textbooks for their classes.

The support of the Office of the Presidency and GIZ Mexico (the German Society for International Cooperation) enabled the Sustainable Development Solutions Network to be created and launched nationally in February 2019. The network is coordinated by two higher education institutions, namely The National Autonomous University of Mexico and the Monterrey Institute of Technology and Higher Education, and mobilizes scientific and technological expertise to promote practical solutions that will deliver the SDGs and the Paris Agreement. Its main initiatives include the creation of a database of innovative projects, which shares university initiatives, encouraging replicability and connecting projects with technical and/or financial support to enable them to be implemented. There are currently 23 projects addressing a range of topics relating to education, the environment, social issues and security.[2]

In addition, an alliance has been established between the SEP, Educación para Compartir, the STEAM Movement, Cemex and the UNESCO Office in Mexico, to promote EES through the implementation of SDG 16. This goal seeks to encourage the adoption of inclusive, participatory and representative decisions at local level and in so doing to follow the Federal Government's principle of 'leaving no one behind.' The alliance has put together supplementary material on the 2030 Agenda and official sources indicate that it will soon be available to 2.5 million sixth grade pupils.[3] Materials will be prepared as follows:

1) Supplementary material.
2) A distance learning course aimed at teachers on how to incorporate the 2030 Agenda into lessons.
3) Review of basic education schemes of work and programs to ensure they are aligned with the 2030 Agenda (Government of Mexico 2020).

It is important to acknowledge the ideological and political background to the 2030 Agenda. The underlying assumption of the SDGs is that societies operate on the basis of an urban agro-industrial model. Even the most realistic proposals remain rhetorical and their approach, which is highly technical, does not start from an epistemological and ethical analysis of the natural world and thus fails to address the causes of the socio-environmental crisis. Unfortunately, the National Strategy for the Implementation of the 2030 Agenda does not accord a meaningful role to EES.

The underlying logic of the design and application of the SDGs is circular. Their goals and indicators are technical and productivist and do not take account of the system as a whole; they do not recognize the complexity and crucial importance of the environment. When experts, politicians and administrators engage in such technocratic thinking, they fail to open up spaces to consider what the future will really look like.

We propose that Mexico and the countries of the world should not to subscribe to these initiatives, but should instead focus on proposals that have emerged from grassroots discussions and interactions, constructing alternatives that reflect our realities and based on dialogue and interculturality: defending the world's biocultural heritage, sharing knowledge and developing innovative ethical and aesthetic models in order to protect habitats and life itself. That is the environmental education we propose: hopeful, rebellious, decolonizing.

Environmental and sustainability education: Emerging themes and trends, and current and future needs

In Mexico, EE became relevant to institutions with effect from June 2005, when the Environmental Education Strategy for Sustainability was approved. Even today, it is

---

[2] For more information see https://sdsnmexico.mx/.
[3] Further information: https://conaliteg.sep.gob.mx/cuadernillo_ODS.pdf.

considered 'a general planning tool, which enables the articulation and development of macro-level guidelines in order to direct education policy' (SEMARNAT 2006: 17). It was the product of analysis involving a large number of stakeholders, which highlighted some key changes and trends:

- EE has been gradually moving from a reductionist approach to a system-wide vision, which recognizes that the fight to conserve ecosystems cannot be isolated from the economic, political and social context.

- EE arises out of the concern about the accelerated impact of human activity on ecosystems; it is one of a number of programmes and projects designed to inform society about ecological problems.

- The reductionist vision, in which concern to develop time-limited solutions to individual ecological problems prevails, is being abandoned in favor of the creation or strengthening of social organizations that defend principles such as equity, social justice and democracy.

- There is also a move away from isolationist approaches that are not interested in discussion of concepts towards awareness and recognition of the conceptual differences within the field.

- EE is thus shifting, at the urging of some of the country's most influential thinkers on the topic, from practices and actions that are focused on solving problems, to broader perspectives where problems are analyzed and evaluated from a more complex perspective.

- Over the last three decades, EE has also shifted from voluntary, time-limited and atomized activities and projects to a recognition of the need to place the field at the center of political debate and accord it public visibility.

The universities have made important advances, gradually adopting strategies to promote sustainability on their campuses; they also have the most consistent and consolidated offer in terms of postgraduate training for environmental educators, professionalization more generally, and research on a variety of topics. The same can be said of the significant but diffuse efforts of environmental organizations and movements, and human rights organizations, which have been described as being part of a 'silent transformation' (Toledo 2016). These go beyond partial correction of the development model that is causing many of our problems; the contribution they are making to urgently-needed social transformation deserves study and support.

A public policy agenda is required if EE is to be successfully introduced across all levels and modalities of the National Educational System. It is essential to regionalize content and establish links with schools' environmental management models. It is also vital to strengthen EE in higher education institutions, promoting curricular greening and the incorporation of the environmental paradigm in both the substantive functions and operational work of HE, in line with national policy guidelines and the principles of sustainability. Research must also be strengthened, in dialogue with other fields of knowledge.

One of the most pressing tasks is to promote joint working and inter-institutional exchange, to develop responses to the social, political, economic, environmental and institutional needs arising from the current historical context. This will require the committed participation of all sectors as well as that of all relevant government agencies at all levels; we need to generate new forms of social and institutional co-responsibility, and seek to transform citizens' actions, behaviors, knowledge, values, thoughts and reflections in order to establish a better relationship with our environment.

Institutions' proposals must be built in dialogue with the residents of regions and local communities, to ensure that curriculum content, teaching methods and subject knowledge are aligned with territories' identities, needs, imaginations and ecological particularities. An alternative educational project must start with the protection of life, building a new science linked to the realities of the world we now live in, and generating new forms of knowledge in the context of complexity. Starting from critical environmental theory, we need to rethink many of our established pedagogical approaches, recognize human multidimensionality and emphasize that human happiness does not require us to dominate ecosystems or take an all-or-nothing approach.

EES should promote a politics that encourages broad citizen participation in the construction of new realities on the basis of a comprehensive review and re-articulation of the development model, transforming not only EE but also the world we live in.

Conclusions

EES is a familiar concept in Mexico, mainly to civil society and academia, but it requires the support of the state and the private sector to address the accelerating environmental deterioration in all regions. Beyond the challenges faced by EE as a field and a consolidating movement, it is first important to identify the characteristics of Mexican EE theory. Have conceptual advances been made that delimit, define or deepen the potential solutions to the problems highlighted throughout this chapter? If not, how could they be achieved?

It is necessary to review the investigations undertaken throughout the country, as well as the results of national initiatives. In this regard, it is worth considering the general view of EES as a methodology. What do experiences of non-formal education contribute to school-based EES? A study is currently being undertaken on EE in Mexico, so we hope we may soon have answers to these questions.

The challenges of incorporating EES into basic and intermediate level schemes of work and teacher training have already been discussed. However, it is important to recognize the additional challenges presented by the budgetary shortfalls faced by all educational institutions in this regard. Added to this is the avalanche of digitalized information and the increase in the virtual delivery of education arising out of the pandemic.

Environmental education had greater potential and has faced greater challenges. What processes have been developed successfully? Where has it flourished and at which educational levels? It is worth connecting with and learning from these experiences.

Art and spirituality are also areas of growing interest among Mexican environmental educators. On the one hand, they are an effective means of raising awareness; and, on the other, they act as tools for action and cultural and communicative empowerment. It would be useful to identify who is active in this field and what knowledge has been built up. What qualities, limitations and challenges do they face? What theoretical analysis has been undertaken and what type of initiatives stand out? There remain lines of investigation that it would be beneficial to explore and expand.

Rural and urban community initiatives have also made progress and faced challenges, either in the context of regional or local interventions or through efforts to build an environmental culture with an explicit educational component. Where has this happened, what are the key features of such efforts? Who are the stakeholders promoting the generation of knowledge on the subject and what initiatives have they introduced?

Public EES policies that cut across areas such as social development, territorial organization and sustainability policy require collaborative projects involving government sectors at the local, state and federal level, and collective monitoring and evaluation of the programmes developed. To achieve this, it is necessary to reinforce environmental and educational institutions, and to increase the federal, state and municipal budgets allocated to EES. It is also necessary to undertake critical analysis of the policies, projects and initiatives implemented, and to design strategies, plans, programmes and educational projects that are aligned with the principles of sustainability, in particular the following priorities:

- Eradication of poverty and increased equality of opportunity
- Promotion of a healthy lifestyle and universal access to health care services
- Gender equality and equity
- Sustainable use of biodiversity
- Socially and environmentally sustainable urban development
- Dynamic, inclusive and sustainable rural development, and
- Promotion of the intercultural approach and recovery of the heritage and legacy of indigenous peoples.

If it were not subject to intense pressure and significant budget cuts, and promoting a decentralized approach to the environmental sector as a whole, SEMARNAT would be the appropriate body to direct policy on EES at all levels of government, reactivating the national, thematic and regional advisory councils, implementing the Environmental Education Strategy for Sustainability and preparing guidance on biocultural diversity at the municipal, state and national level. A task that must not be postponed is ensuring social participation in the design, operation, and evaluation of public policies on EES and identifying key stakeholders to collaborate on its development.

Another important aspect is the intercultural dialogue between the societies and cultures that make up humanity. The nature of the discourse needs to be clarified, moving from a global perspective to one more in tune with the heritages of more than 6,000 living cultures, many of them resistant to the most predatory aspects of globalization (even where it is well-intentioned).

More effective community and territorial integration will provide better opportunities for a dialogue around eco-social transition, which has recently been highlighted as an area in need of attention at state and global level and is critical to addressing the planetary crisis.

References

ANEA. 2014. Reflections and outlines on the situation of Environmental Education in Mexico. Seminar on Environmental Training Programs and Processes: Debates (1st ed.). ANEA. Pátzcuaro, Mich. Mexico.

ANEA. 2015. National Academy of Environmental Education. Retrieved November 14, 2020, from http://www.anea.org.mx.

ANEA. 2015. Statutes, first amendment, December, 2015. National Academy of Environmental Education A.C. Retrieved December 08, 2020, from http://www.anea.org.mx.

Carmona, M. 2003. General Law of Ecological Balance and Environmental Protection, comments and concordances. Virtual Legal Library of the Institute of Legal Research of the UNAM. Retrieved December 04, 2020, from https://biblio.juridicas.unam.mx/bjv/detalle-libro/542-ley-general-del-equilibrio-ecologico-y-la-proteccion-al-ambiente.

CECADESU-SEMARNAT (ed.). 2006. National Strategy of Environmental Education for Sustainability. (1st ed.). Mexico.

COMPLEXUS. 2020. Mexican Consortium of Higher Education Institutions for Sustainability. Mexican Consortium of Higher Education Institutions for Sustainability. Retrieved December 04, 2020, from http://complexus.org.mx/.

CONABIO. 2020, 06 02. National Commission for the Knowledge and Use of Biodiversity. Retrieved 10 28, 2020, from https://www.biodiversidad.gob.mx/pais/quees.

Figueroa, E. 2006, January. Cultural policies for development in a globalized context. Politics and Culture 26: 157–183. SciELO Analytics. 0188-7742.

Food and Agriculture Organization of the United Nations. 2015. Forest Resources Assessment (FAO ed.). FAO. Rome.

Gómez, T. 2020, 01 14. Mexico's environmental challenges for 2020. Monagabay Latam. Retrieved 10 28, 2020, from https://es.mongabay.com/2020/01/los-desafios-ambientales-de-mexico-para-el-2020/.

González Gaudiano, E. 1993. Strategic Elements for the Development of Environmental Education in Mexico. University of Guadalajara/WWF. Mexico.

González Gaudiano, E. 2000. Environmental Education in Mexico: Achievements, Perspectives and Challenges Facing the New Millennium. Iberoamerican Congress of Environmental Education. Caracas Venezuela.

González Gaudiano, E. 2003. Glimpsing the conceptual construction of environmental education in Mexico. *In*: Bertely Busquets, María (Coord.). Education, Social Rights and Equity. Educational Research in Mexico 1992–2002. Mexican Council for Educational Research.

González Gaudiano, E. (Coord.). 2008. Education, Environment and Sustainability. Eleven Critical Readings. Siglo XXI. Mexico.

González-Márquez, I. and V.M. Toledo. 2020. Sustainability science: A paradigm in crisis? Sustainability 12(7): 2802. MDPI AG. Retrieved from http://dx.doi.org/10.3390/su12072802.

Government of the Republic. 2018. Voluntary national report for the high-level political forum on sustainable development.

Government of Mexico. 2019. The National Strategy for the Implementation of the 2030 Agenda in Mexico (1st ed.). The Office of the Presidency of the Republic.

Government of Mexico. 2020. Activity Report 2019–2020 (1st ed.). Executive Secretariat of the National Council of the 2030 Agenda.

INEGI. 2018. National Survey of Household Income and Expenses. INEGI. https://www.inegi.org.mx/programas/enigh/nc/2018/.

Official Journal of the Federation. 2020. Sectoral program derived from the national development plan 2019–2024. H. Congress of the Union. https://www.dof.gob.mx/nota_detalle.php?codigo=5596232&fecha=07/07/2020.

Reyes, J., E. Castro and J. Esteva. 2016. Environmental Education Research Guide. Preparation of the Protocol. University of Guadalajara, Mexico.

Sachs, I. 1974. Ecodevelopment: a contribution to the definition of development styles for Latin America. International Studies Year 7, No. 25 (January–March 1974), Institute of International Studies University of Chile.

Semarnat. 2016. Report on the Situation of the Environment in Mexico. SEMARNAT.

Senate of the Republic. 2014, November 25. Gazette: LXII/3PPO-59/51411. Senate Gazette. Retrieved December 03, 2020, from https://www.senado.gob.mx/64/gaceta_del_senado/documento/51411.

World Resources Institute. (s/f). Forest monitoring designed for action. Global Forest Watch. Retrieved on October 28, 2020, from https://www.globalforestwatch.org.

North America

Chapter 15

The State of Environmental and Sustainability Education in Canada

A Review of Past, Current, and Future Directions

Kristen Hargis,[1,*] *Marcia McKenzie*[2,3] and *Nicola Chopin*[3]

Introduction

Manoeuvring the 'many currents [that] stir and animate the waters' of environmental and sustainability education (ESE) across Canada is a formidable task (Russell et al. 2000, 203). These currents manifest within formal, non-formal (e.g., zoos, museums), and informal (e.g., media) education (CEGN 2006) and are the result of international trends as well as national and sub-national contexts (Hart and Hart 2014, Hopkins 2013).

This chapter particularly focuses on the state of formal ESE in Canada during the timeframe of the Global Action Programme (GAP) on Education for Sustainable Development (ESD), the UN 2030 Agenda, and the ESD for 2030 framework. The five-year GAP programme was launched in 2015 as a continuation of the UNESCO Decade of Education for Sustainable Development ('DESD' or 'the Decade')

---

[1] School of Environment and Sustainability, University of Saskatchewan, Canada.
[2] University of Melbourne, Australia.
[3] University of Saskatchewan, Canada.
Emails: marcia.mckenzie@usask.ca; nicola.chopin@usask.ca
* Corresponding author: kristen.hargis@usask.ca

(UNESCO 2014). Its overarching goal was 'to generate and scale up action in all levels and areas of education and learning to accelerate progress towards sustainable development' (UNESCO 2014, 14). Adopters of the 2030 Agenda committed 'to end poverty, protect the planet and ensure that all people enjoy peace and prosperity by 2030' by committing to achieving 17 Sustainable Development Goals (SDGs) (UNDP 2021). The ESD for 2030 framework aimed to scale up ESD action following the Decade and GAP on ESD (United Nations General Assembly resolution 74/223 2020).

This chapter reports results from research in Canada by the Sustainability and Education Policy Network (SEPN), as well as other published reports and studies, to provide a partial overview of the state of formal ESE in Canada. SEPN is an international partnership of researchers and leading policy and educational organizations advancing sustainability in education policy and practice. In 2012, SEPN began the first large-scale research collaboration to collect and analyze comparable data across Canada's formal education system. Between 2012 and 2020, SEPN conducted a series of comparative research projects including research syntheses, document analyses, a national survey, and site analysis case studies. The research examined sustainability uptake in primary (Grades Pre-K-6) and secondary (Grades 7–12) ministries of education, school divisions, schools, and higher education institutions (HEIs) in Canada. Evidence was collected of sustainability uptake in policy and practice, including relationships between the two. This research also examined how sustainability uptake relates to other policy and practice priorities and explored the development and enactment of sustainability in education policy and practice.[1]

The sections that follow provide details on Canada's unique national and sub-national education system, including in relation to historical considerations. The chapter also provides an overview of how ESE is understood and implemented within primary and secondary education as well as higher education across the country, before considering current trends and future directions.

Political, economic, cultural and social contexts and conditions in Canada

Understanding ESE uptake across Canada necessitates understanding its unique socio-political-cultural-economic context. Education in Canada is constitutionally demarcated to provincial jurisdictions, with minimal national level policy directing formal education (CMEC 2012). Ministries or departments of education (hereafter ministries of education) within Canada's ten provinces and three territories exercise autonomy over educational policy-making, including curriculum development, student assessment, and financial overview (CMEC 2012). Ministries of education are led by ministers who collaborate through their participation in the Council of Ministers of Education Canada (CMEC), discussing matters of shared concern, including sustainability (CMEC 2010). Separate ministries of education exist for primary and secondary education as well as higher education within some jurisdictions (CMEC 2010).

Across primary, secondary, and higher education levels, there are key differences in how the country's education systems operate in relation to access and governance. Regarding access at the primary and secondary level, public education is free for those of a certain age and residency status (Hopkins 2013). Alternatively, students can attend private schools, which governments may subsidize, but student guardians typically pay for tuition (Statistics Canada 2015). Governance across the 13 jurisdictions in relation to school administration, operations, financial accountability, and curriculum implementation, is overseen by 374 regional school divisions or boards (hereafter school divisions) (McKenzie and Aikens 2021).

At the higher education level, Canada has some of the highest student participation levels globally (Jones 2014). Canada's 220 HEIs are relatively autonomous (e.g., setting admissions standards and degree requirements as well as overseeing financial management) (CMEC 2012). That said, regarding teacher education, ministries of education at the primary and secondary level can mandate that certain pre-service requirements align with ministry curriculum, which can influence higher education programming (CMEC 2012). HEIs in Canada include public and private institutions, vocational colleges, and pre-university colleges (Jones 2014). Traditionally, degree-granting privileges were exclusive to universities, but since the 1990s, this status has been extended to the college sector (Jones 2014).

Education systems in Canada at all levels operate within particular geographic, historical, and cultural contexts (Bieler and McKenzie 2017). At almost 10 million square miles, Canada is the second largest country in the world but has one of the lowest population densities at 3.9 people per square kilometre (Statistics Canada 2017). This geographic reality makes providing quality education for urban, rural, and Indigenous communities challenging (Bentham et al. 2019, Hopkins 2013). Adding to this complexity, the Canadian population is highly concentrated around the United States border, with two out of three people living within 100 miles of its southern neighbor (Statistics Canada 2017). For example, the province of Ontario, located in the southeast, resides at one extreme (population: 14.6 million; HEI students: 889,000), while Canada's three far northern territories (combined population: 125,000; combined HEI students: 4,000) reside at the other (Statistics Canada 2021a, Statistics Canada 2021b).

Canada's cultural and historical origins as a settler-colonial nation is reflected in its diverse population of Indigenous peoples, as well as early European settlers (Gebhard 2017). Due to this history, there are regular tensions over policies in Canada related to bilingual, multicultural, and religious concerns (Patrick 2017). For example, though the Canadian constitution guarantees that citizens can receive education in the official colonial languages of English and/or French (Canadian Charter of Rights and Freedoms 1982), a federal plan does not exist to ensure access to Indigenous language instruction (McIvor and Ball 2019). This reality exists, despite Canadian commitments to ensure Indigenous language education through its signing of the UN Declaration on the Rights of Indigenous Peoples, and Canada's Truth and Reconciliation Commission calls to action (Bentham et al. 2019). Policies protecting

Indigenous language education are needed due to early education attempts to 'civilize' Indigenous peoples through residential schooling where they endured forced foreign language acquisition and religious conversion as well as abuse from their caretakers (Jones 2014). The residential schools resulted in 'successive generations with deep wounds from familial separation, cultural and community displacement, shame and sadness' (McIvor and Ball 2019, 18). The marginalization of Indigenous education continues, in part, because education is legislated provincially, whereas Indigenous services fall under federal jurisdiction (OECD 2020). Additionally, Indigenous-managed schools within Canada are severely under-funded (Bentham et al. 2019).

Global shifts toward a neoliberal political economy have also affected primary, secondary, and higher education systems in Canada (Fisher et al. 2009, Metcalfe 2010, Schuetze et al. 2011, McKenzie et al. 2015, Virone 2016). At the primary and secondary level, several jurisdictions have enacted neoliberal agendas by 'fostering private schools ('increasing choice'), introducing a number of market mechanisms into the public school system, imposing standardized testing, enhancing competition between schools, and allowing private companies to advertise their products in schools' (Schuetze et al. 2011, 62). Parker (2017) notes neoliberal objectives are achieved through a 'curriculum of accountability', wherein measurable targets are set to meet provincial standards. Similar trends exist at the higher education level (Fisher et al. 2009, Metcalfe 2010). HEIs operate in competitive atmospheres wherein research is valued for its profitability (Jeppesen and Nazar 2012), and education is valued for its marketability and capability to train students for a global economy (Olssen and Peters 2005). To up the ante, HEIs are called upon to increase rank and reputation amid dwindling provincial funding (Harden 2017).[2]

Canada's political economy is also strongly tied to extractive fossil fuel industries (Erickson and Lazarus 2014), which has resulted in varying levels of commitment to reduce overall greenhouse (GHG) gas emissions. For instance, Canada ratified the 1997 Kyoto protocol in 2002 (under the Liberal party) only to withdraw in 2011 (under the Conservative party) (Burch and Harris 2014, Bieler and McKenzie 2017). Shortly after the Liberal party gained power in 2015, Canada adopted the UN 2030 Agenda for Sustainable Development (Government of Canada 2019) and signed the Paris Agreement (Government of Canada 2020).[3]

Canadians have paid increasing attention to climate change in recent years due, in part, to the 2018 IPCC report, which warned that just 12 years remained to address climate change, as well as increasing frequency of extreme weather events, the Fridays for Futures climate strikes, and other initiatives. In 2019, education systems in Canada largely responded positively to the school climate strikes, with several school divisions (such as Toronto District School Board, Vancouver School Board) and HEIs (such as the University of British Columbia, and Dawson College) not penalizing student strike attendance (CBC News 2019, The Canadian Press 2019).

---

[2] In 1982, government funding covered 82.7% of universities' operating revenue. That number dropped to 54.9% in 2012 (Harden 2017).

[3] Signatories to the Paris Agreement committed to limiting global warming below 2°C (UN 2015).

The lasting impact on education policy and practice in Canada of these school climate strikes, as well as documented (e.g., IPCC) and observed (e.g., natural disasters) evidence of climate change remains to be seen.[4]

Historical background/development of ESE in Canada

ESE history in Canada is an expression of the aforementioned national contexts and international movements. ESE programming in Canada has existed since the 1970s (Hopkins 2013), with environmental education (EE) becoming 'a definite part of the school experience' during the mid-1980s (Hart 1996, 57). Provincial specialist associations for EE were also established to serve as resources for schools and educators (e.g., EEPSA 2022). The 1987 publication of the Brundtland report marked the beginning of an international movement towards ESD. Canada participated in this movement, in part, through its involvement in Preparatory Committee meetings in advance of the 1992 Earth Summit (Hopkins 2013). Canada's involvement with ESD continued through participation at the 1992 Earth Summit meeting in Rio de Janeiro and the resulting work programme, Agenda 21 (Russell et al. 2000, Hopkins 2013), as well as through subsequent UN programmes (e.g., the DESD, GAP programme, UN 2030 Agenda). This involvement was partly fueled by increased interest in EE by educators, politicians, and parents in Canada, which began as early as 1990 due to increased awareness of 'resource scarcity, environmental deterioration, and failed economic policies' (Hart 1990, 45).

Implementing the Earth Summit goals was met with several challenges in Canada. The first challenge was related to debates regarding acceptance, adaptation, or rejection of ESD (which continue today[5]) (see Jickling 1992, McKenzie and Aikens 2021, Sauvé et al. 2005). Canadian scholars also considered possibilities and potential implications for the 'interwoven spheres of culture, environment, and education' (McKenzie et al. 2009, 1), relationships between outdoor education and EE (McClaren 2009), and ethical and cultural motivations for EE (Bai 2009, Fawcett 2009). A second challenge was related to the aforementioned geographic complexities: ESE leaders were dispersed across 6,400 kilometres without a national ESE organizational structure to foster cohesion and direction (Hopkins 2013).

Some have argued, however, that a lack of national education standards facilitated new and creative approaches (Russell et al. 2000). Despite an overall hands-off approach by the national Canadian government in relation to ESE, some national direction was provided early on through the publication of a national ESE framework (Government of Canada 2002) and a range of work by the Council of

---

[4] For more on how primary and secondary education systems in Canada can and are responding to climate change (see Hargis and McKenzie 2020).

[5] Critiques include that the ESD term is: (1) vague and thus susceptible to manipulation, (2) a logically inconsistent oxymoron, and (3) ill-defined such that it is impossible to determine what an educator is educating 'for' (Jickling 1992). There are also misgivings about education aiming 'for' sustainable development or any other type of behavior, as opposed to empowering individuals to think for themselves (Jickling 1992).

Ministers of Education, Canada (CMEC) (McKenzie and Aikens 2021). A third challenge to implementing the Earth Summit agenda was its co-occurrence with a global recession (Hopkins 2013). Due to financial cutbacks, formal education leaders tended to focus on core subjects, such as mathematics, which often meant exclusion of ESE (Hopkins 2013), with some exceptions (for example, through integrated environmental studies programs) (Russell et al. 2000, Breunig 2013).

Prior to and during the Decade, ESE work was largely led by passionate champions, non-governmental organizations (NGOs), and school-based programming (Russell et al. 2000, Hart and Hart 2014, Hopkins 2013). For instance, the Canadian Network for Environmental Education and Communication (EECOM) and the *Canadian Journal of Environmental Education* were founded (in 1993 and 1996, respectively) to support formal and non-formal ESE education (EECOM 2019, Russell et al. 2000). National and provincial Green School initiatives also led ESE work through school-based eco-certification programs, such as the national SEEDS Foundation's Learners in Action program, which recognizes school participation in environmental activities, as well as the provincial Destination Conservation program (in Alberta and British Columbia), which emphasizes energy and waste auditing (Russell et al. 2000).

With the launching of the Decade in 2005 came several ESE initiatives at national and sub-national levels, which were not always in alignment. For instance, at the national level, the Government of Canada published its objectives for the UNESCO Decade of ESD in 2005. At the same time, CMEC committed to incorporate 'sustainable development themes into formal, non-formal, and informal education and to report on this implementation' (CMEC 2012, 16). CMEC's commitment to ESD was related to its involvement with the United Nations Economic Commission for Europe (UNECE) (CMEC 2012), as well as the influence of ministerial policy actors (in Manitoba) who held high-level national CMEC positions while active in global UNESCO initiatives (McKenzie and Aikens 2021). Within sub-national contexts, there was also adaptation and rejection of national-level commitments to ESD (Hart and Hart 2014, McKenzie and Aikens 2021). For instance, the Manitoba Eco-Globe programme, an eco-certification programme administered by the Manitoba Ministry of Education, remains largely faithfully to ESD conceptualizations, whereas the Ontario EcoSchool programme has maintained alignment with ministerial and divisional commitments to EE (McKenzie and Aikens 2021). During the decade, Regional Centres of Expertise on ESD also provided some coherence for ESE activity through their focus on 'greening' HEI campuses, teacher education, and community initiatives (Hart and Hart 2014).

Significant changes occurred during and beyond the Decade within primary, secondary, and higher education systems in Canada in relation to how ESE is understood and implemented, some of which are summarized in the next section.

The understanding and implementation of ESE in Canada

This section overviews some of the current aspects of ESE implementation at primary, secondary, and higher education levels in Canada based on SEPN's comparative research of ESE in formal education between 2012–2020.

Primary and secondary education

At the primary and secondary level, approximately half the ministries of education and school divisions had sustainability-specific policies (Beveridge et al. 2019). Sustainability conceptualizations within the ministry of education policies fell into three clusters (i.e., EE, ESD, and Indigenous Education), each of which had implications for ESE pedagogy and action (Aikens and McKenzie 2021). Other major findings included: climate change education responses were insufficient to meet Canada's Paris Agreement and SDG commitments (Bieler et al. 2018), a whole institution approach was effective but under-utilized (Beveridge et al. 2019, McKenzie and Chopin 2023b) and having a sustainability education policy made a focus on sustainability practice more likely (McKenzie and Aikens 2021). Details related to these findings are provided in the sections below in relation to policy documents and practices on the ground, where possible.

Around half the ministries of education and school divisions have a sustainability policy

A national census of sustainability-specific policies in Canada within primary and secondary education found that only 7 of 13 (54%) ministries of education, and 219 of 374 (59%) school divisions had a sustainability-related policy (Beveridge et al. 2019). Beveridge and colleagues (2019) developed Sustainability Initiative scores based on the presence of three policy initiatives at the school division level: (1) sustainability-related policies, (2) eco-certification programmes, and (3) sustainability staff. The provinces of Ontario and Nova Scotia received the highest scores (see Figure 1). Relationships between these policy initiatives were also analyzed, finding that participation in an eco-certification programme and the presence of sustainability staff were weakly correlated with having a sustainability policy (Beveridge et al. 2019).

Primary and secondary ESE policy in Canada in relation to national and international contexts

SEPN's policy analysis and site visit research highlighted the significance of UN initiatives in influencing how ESE was framed and understood within some Canadian provinces at the primary and secondary level in ways that reflect Canada's unique national context (Beveridge et al. 2019, McKenzie and Aikens 2021). A systematic literature review of ESE policy research across 71 countries, found a global shift in the language framing sustainability within education policies away from 'environment' and towards 'sustainable development' and 'sustainability' (Aikens et al. 2016). This move aligns with global policy initiatives, such as the Decade, the GAP programme, and the ESD for 2030 framework. Despite the primacy of ESD in international policy arenas, ESD has not mobilized uniformly across Canada. SEPN's census of school division policies found that 'environment' was the dominant language used in policy titles in Canada, with exceptions in Manitoba and Québec, where 'sustainable development' or 'développement durable' were common (Beveridge et al. 2019). Nationally, the CMEC adopted language related to ESD, and ESD working groups

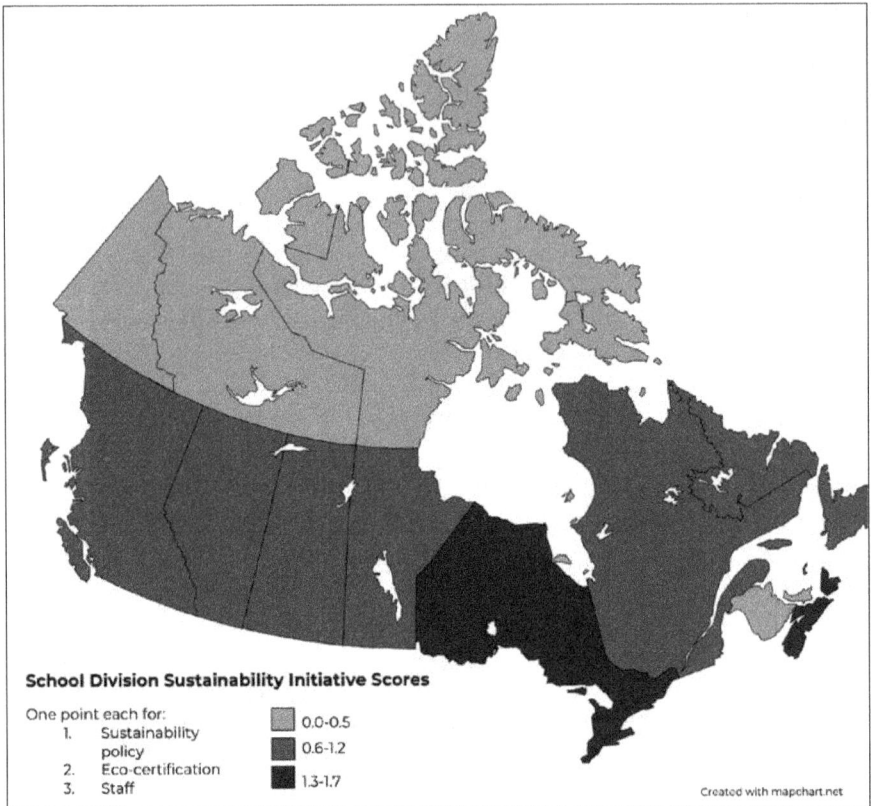

School Division Sustainability Initiative Scores

One point each for:
1. Sustainability policy
2. Eco-certification
3. Staff

■ 0.0-0.5
■ 0.6-1.2
■ 1.3-1.7

Created with mapchart.net

Figure 1. Map of Sustainability Initiative scores across Canadian provinces. Adapted from Beveridge et al. 2019.

were established within the CMEC and across the 13 provinces/territories (CMEC 2012). The language of ESD was particularly taken up within the province of Manitoba (though resistance and adaptation of ESD in some provinces were noted, for example, in British Columbia and Ontario who preferred to use EE) (McKenzie and Aikens 2021).

A content analysis of sustainability-related policy documents from all 13 Canadian provinces and territories found three dominant clusters of ESE understanding in primary and secondary education policy in Canada: Indigenous Education, ESD, and EE (Aikens and McKenzie 2021). Each cluster brought distinct approaches, which shaped the cluster's orientation, action, and pedagogy (Aikens and McKenzie 2021). For instance, within the Northwest Territories and Nunavut, there was strong alignment between Indigenous knowledge systems and Indigenous education policy, which was likely due to 'localized, relevant and appropriate policy development' (Aikens and McKenzie 2021, 69). In Manitoba and Québec, there was strong alignment between global initiatives around ESD, which was likely due to Manitoba's role in leading Canada's work in ESD mentioned above (Aikens and McKenzie 2021). Adaptation and rejection of global ESD within policies from

British Columbia and Ontario was also noted, with a focus on EE instead, which was likely due to a longer history of EE in those areas (Aikens and McKenzie 2021). Across all three clusters, an explicit focus on climate change was missing.

Climate change education responses in primary and secondary are insufficient

Within primary and secondary policy documents in Canada, SEPN's research indicated that current climate change approaches lacked sufficient depth to meet Canada's Paris Agreement commitments (McKenzie and Chopin 2023b). A census of sustainability-related policies from all the ministries of education and school divisions found that only 22 (7%) focused on climate change (Beveridge et al. 2019). SEPN also assessed alignment of climate and education policy across all 13 provinces and territories, finding *climate* policies more likely to address climate change education than *education* policies (Bieler et al. 2018). While all 13 provincial climate policies mentioned that education was essential to address climate change, only six included specific targets, which usually focused on operational energy efficiency instead of focusing on actions within other whole institution domains (i.e., governance, curriculum, community outreach, and research) (Bieler et al. 2018).

On the ground, climate change was rarely mentioned when participants were asked open-ended questions about sustainability initiatives at their setting on SEPN's national survey and during site visits (McKenzie and Chopin 2023b). When climate change was mentioned, it was most commonly done so by students and was referenced as a barrier and a driver for sustainability uptake (McKenzie and Chopin 2023b).

A whole institution approach is effective but under-utilized in primary and secondary education

SEPN's analysis of sustainability uptake found that institutions commonly focused on curriculum and operations, with less attention paid to other whole institution domains (governance, community outreach, and research) (Beveridge et al. 2019). The ministries of education usually addressed sustainability in curriculum documents, whereas school divisions usually addressed sustainability in operations-related documents (Beveridge et al. 2019). While all the ministries of education addressed sustainability in some capacity, only five out of 13 had a sustainability-related policy in more than one domain (Beveridge et al. 2019). Of these five, only Manitoba included policy across all five domains, though British Columbia and Ontario also included engagement within several domains (see Figure 2).

SEPN's analysis of policy flows across international, national, and sub-national levels found that policy presence at 'higher' levels made it more likely that policy would exist at 'lower' levels across whole institution domains (McKenzie and Aikens 2021). Within ministries that adopted a whole institution approach (i.e., Manitoba, British Columbia, and Ontario), mutually reinforcing relationships between ministry and school division policy with respect to whole-institution domains were found (McKenzie and Chopin 2023b). Specifically, if sustainability was included in a

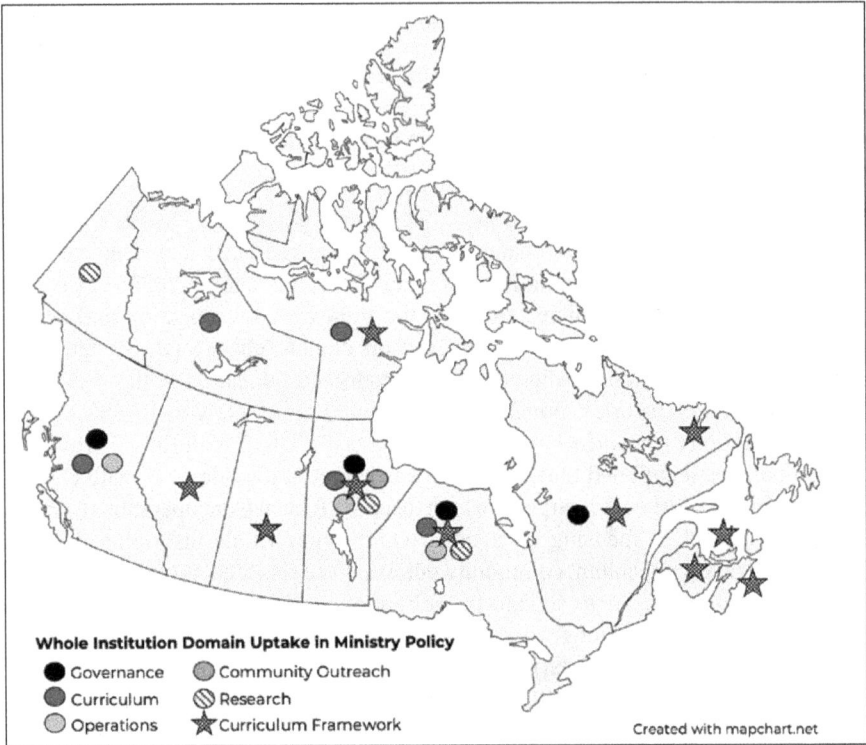

Figure 2. Whole institution domain uptake in ministry of education policy in Canada. Figure created based on data from Beveridge et al. 2019.

particular policy domain at the ministry and school division levels, it was more likely that policy would exist at lower levels (McKenzie and Aikens 2021).

Non-education-related sustainability policy was also found to influence sustainability uptake at the provincial and municipal levels (McKenzie and Aikens 2021). For example, a policy analysis by Beveridge and colleagues (2019) found that provinces with broader government sustainability mandates were more likely to have higher sustainability uptake in education policy (also see McKenzie and Aikens 2021). Site visits also identified cases where broader provincial policy had supportive or reciprocal relationships with policy at the ministry level (McKenzie et al. 2017). Specifically, four of the six provinces visited which had broader environment-specific acts also had higher sustainability policy uptake in their ministry of education and school divisions as indicated by a critical policy analysis (McKenzie et al. 2017).

Whole institution uptake within Indigenous primary and secondary education contexts differed slightly from the findings mentioned above. Within these contexts, there was a strong focus on curriculum and community outreach, with schools viewed by participants as the focal point for community activities (Bentham et al. 2019). Descriptions of sustainability education in these contexts viewed whole institution domains as enmeshed (Bentham et al. 2019). There was a strong focus on language acquisition and cultural revitalization, with participants associating greater fluency in Indigenous language with higher sustainability engagement in Indigenous communities (Bentham et al. 2019).

High ESE policy presence makes practices more likely in primary and secondary education

Site visits indicated that high policy and high practice contexts tended to co-occur at the primary and secondary levels[6] (McKenzie and Chopin 2023b). Sustainability Initiative scores were used to categorize site visit school divisions as Sustainability 'Learners' (low policy uptake) or 'Leaders' (high policy uptake). Participants at Sustainability Leader contexts ranked policy higher than participants at Learner institutions across whole institution domains (see Figure 3). Additionally, participants at Learner (12 of 24) and Leader (12 of 21) institutions felt policy was positively associated with sustainability practice uptake across all domains (McKenzie and Chopin 2023b).

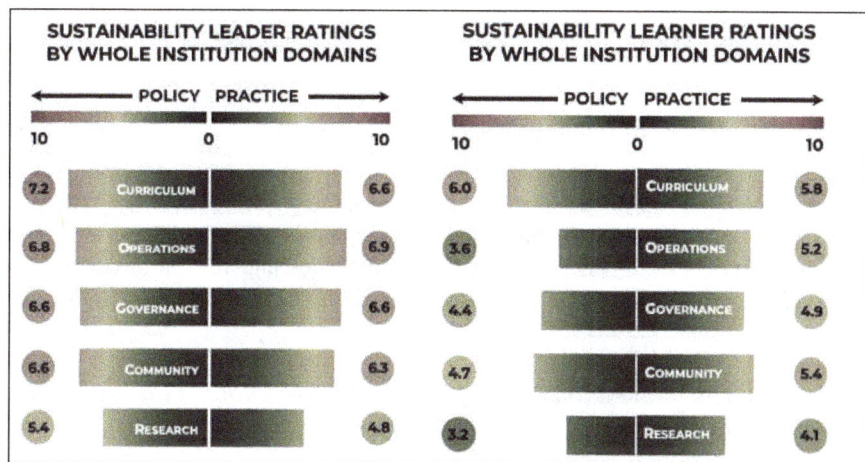

SUSTAINABILITY LEADER RATINGS BY WHOLE INSTITUTION DOMAINS			SUSTAINABILITY LEARNER RATINGS BY WHOLE INSTITUTION DOMAINS		
← POLICY		PRACTICE →	← POLICY		PRACTICE →
10	0	10	10	0	10
7.2	CURRICULUM	6.6	6.0	CURRICULUM	5.8
6.8	OPERATIONS	6.9	3.6	OPERATIONS	5.2
6.6	GOVERNANCE	6.6	4.4	GOVERNANCE	4.9
6.6	COMMUNITY	6.3	4.7	COMMUNITY	5.4
5.4	RESEARCH	4.8	3.2	RESEARCH	4.1

Figure 3. Sustainability Leader and Learner Heat Diagram ratings of sustainability inclusion in policy and practice. Participants rated whole institution domains between 0–10 for policy and practice, higher scores indicate 'hotter' domains. Source: McKenzie and Chopin 2023b.

Higher education

Half of the accredited HEIs in Canada were found to have a sustainability policy (Beveridge et al. 2015). Within strategic plans from a representative sample of HEIs, sustainability was most commonly understood and implemented through accommodative (vs. transformative) measures, wherein sustainability is included as one of several policy priorities as opposed to responses that deeply question

---

[6] We found one outlier wherein participants indicated two domains were sustainability hotspots despite being in a low sustainability policy context (McKenzie and Chopin 2023b).

the status quo of educational paradigms in relation to sustainability (Bieler and McKenzie 2017). Other major findings include: shallow engagement with climate change (Henderson et al. 2017), unbalanced uptake of a whole institution approach (Vaughter et al. 2016), and policy can drive practice (McKenzie and Chopin 2023a). Additional details are provided below regarding each of these findings in relation to policy and practice.

Half the number of HEIs have a sustainability policy

Beveridge and colleagues (2015) found that 110 out of 220 accredited HEIs in Canada (50%) had a sustainability-related policy (Beveridge et al. 2015). Sustainability Initiative scores were used to analyze whether HEIs had a sustainability-related policy, a sustainability assessment, a sustainability office(r), and/or a sustainability declaration (Beveridge et al. 2015). Only 32 (15%) institutions received a score of four, indicating all four initiatives were present (Beveridge et al. 2015). In contrast, 60 (27%) institutions received a score of zero, indicating they had not undertaken any of the high-level sustainability policy initiatives (Beveridge et al. 2015).

Beveridge and colleagues (2015) also analyzed relationships among the four sustainability initiatives, including in relation to geographic and institutional characteristics, and found strong linkages between three sustainability initiatives: assessment, office(r), and policy. This suggests that the uptake of one might encourage the uptake of others. The existence of sustainability-specific policies in higher education was strongly related to province, with the majority of institutions in Québec (85%) and British Columbia (67%) having policies (Beveridge et al. 2015). In contrast, only 14% of institutions in New Brunswick and 13% in Saskatchewan had policies, and 0% of institutions in the territories had policies (Beveridge et al. 2015).

HEIs primarily understand and implement sustainability in an accommodative manner

Similar to the global terminology trends previously mentioned at the primary and secondary level (Aikens et al. 2016), shifts in the terminology used in higher education policies from 'environment' to 'sustainable development' and 'sustainability' were found (Beveridge et al. 2015). When this terminology was used within higher education strategic plans, it was most commonly done in a shallow way (Bieler and McKenzie 2017).

Bieler and McKenzie (2017) conducted a content analysis of a representative sample of strategic plans from 50 HEIs, including in relation to the five domains of a whole institution approach. The study found that 18% of strategic plans did not discuss sustainability at all, 40% had accommodative responses (addressed sustainability as one of many policy priorities), 16% had reformative responses (institutional priorities were aligned with 3–5 whole institution domains discussed in varying depth), 26% had progressive responses (sustainability addressed in mission, goals, and policy in relation to most whole institution domains), and 0% had transformative responses (sustainability integrated in a way that supports decarbonization) (see Figure 4). Site visits showed that universities overwhelmingly focused on operational upgrades and other low hanging fruit (for instance, recycling and water bottle refilling stations) as opposed to more transformative, decarbonized approaches, such as, for instance, divestment and social justice (McKenzie and Chopin 2023a).

Transformative sustainability responses require HEIs to re-think existing educational paradigms and re-conceptualize the purpose of higher education to support the transition to decarbonized societies (Bieler and McKenzie 2017). One potential transformative approach is fossil fuel divestment. However, SEPN found widespread resistance to fossil fuel divestment among HEIs, despite increasing pressure from students and faculty (Maina et al. 2020). In 2019, across the 220 accredited universities and colleges in Canada, there were 38 active divestment campaigns, but only six institutions had committed to divesting their financial portfolio to some degree (Maina et al. 2020).

In Canada, transformative sustainability approaches also mean respectfully engaging with Indigenous knowledge to move beyond superficial sustainability integration (Vizina 2018, McKenzie and Wilson 2022). A sample of 10 Indigenous higher education programmes indicated that more work was also needed in this area in relation to policy and practice (Vizina 2018). For instance, participants viewed facilities management at their institutions as also encompassing the type of curriculum offered, and suggested the need for more land-based learning (Vizina 2018).

SEPN found that conducting a sustainability assessment played a significant role in encouraging HEIs to develop more holistic sustainability responses. For example,

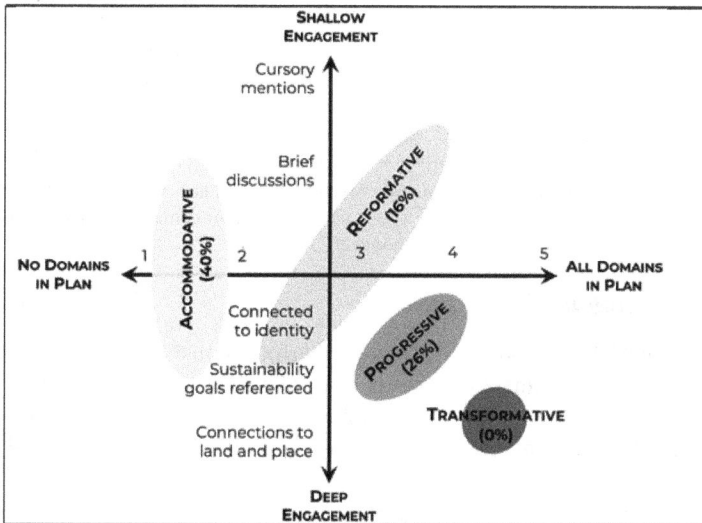

Figure 4. Types of sustainability approaches in higher education strategic plans, ranging from accommodative to transformative responses. Source adapted from Bieler and McKenzie 2017.

an analysis of institutional strategic plans found that institutions rated as STARS[7] by Association for the Advancement of Sustainability in Higher Education (AASHE) were more likely to have reformative and progressive responses to sustainability in their strategic plans (Bieler and McKenzie 2017). An analysis of AASHE-affiliated institutions also found they were more likely to have sustainability or environmental policies (Lidstone et al. 2015). That is, 67% of ASSHE-affiliated institutions had a sustainability or environmental policy (Lidstone et al. 2015), compared to 50% of the 220 HEIs in Canada (Beveridge et al. 2015). Additionally, SEPN's analysis of climate-related policy found that over half (63%) of institutions with a climate-focused policy were AASHE members, suggesting that sustainability assessments played a role in supporting climate action in institutions (Henderson et al. 2017). During site visits, it was found that participants rated STARS institutions higher across policy and practice domains relative to those in non-STARS institutions (see Figure 5).

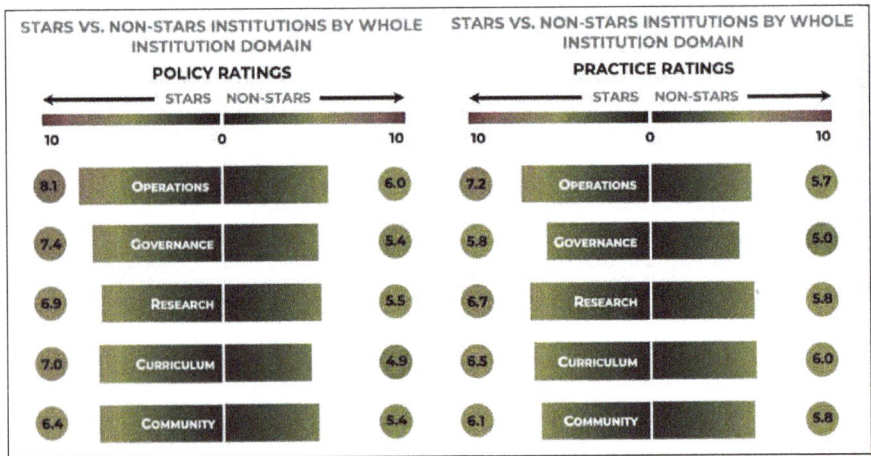

Figure 5. Comparison of Heat Diagram ratings between STARS and non-STARS institutions in relation to sustainability policy and practice uptake across whole institution domains. Source: McKenzie and Chopin 2023a.

HEIs are lagging in climate action

SEPN found that HEIs can advance climate action more holistically and systematically. A content analysis of climate change-related policies from a representative sample of 50 HEIs in Canada found that only 22 institutions (44%) had climate-specific policies (Henderson et al. 2017). Of those 22, only 11 institutions had official

---

[7] AASHE's STARS program is a self-reporting assessment framework that colleges and universities can use to track their sustainability progress (AASHE n.d.). Institutions report on their activities within the areas of: (1) academics (includes curriculum and research), (2) engagement (includes on and off campus), (3) operations, (4) planning and administration, as well as (5) innovation and leadership (an open-ended category that allows institutions to report on innovative practices not covered in the other domains) (AASHE n.d.).

climate change policies, while the remaining 11 were energy or emissions plans (Henderson et al. 2017). The most common words referenced in the policies were 'energy' and 'building.' Institutional climate policies were rare, and, when present, disproportionately focused on campus operations (Henderson et al. 2017).

In relation to climate action on the ground, perceptions of institutional engagement were often higher in non-operational domains (McKenzie and Chopin 2023a). For example, fewer than one in five national survey participants indicated they were aware of their institutions tracking institutional carbon footprints or greenhouse gas emissions (McKenzie and Chopin 2023a). SEPN's national survey also indicated room for improvement regarding institutional prioritization of climate change research. While the participants ranked climate change research third highest of the research topics asked about in the national survey, only 25% of the respondents indicated their institution prioritized climate change research to a large extent (McKenzie and Chopin 2023a). Only 20% of the survey participants indicated that their institutions were prioritizing research related to alternative energy to a large extent, and only 13% of participants indicated that environmental justice was prioritized by their institutions to a large extent (see Figure 6).

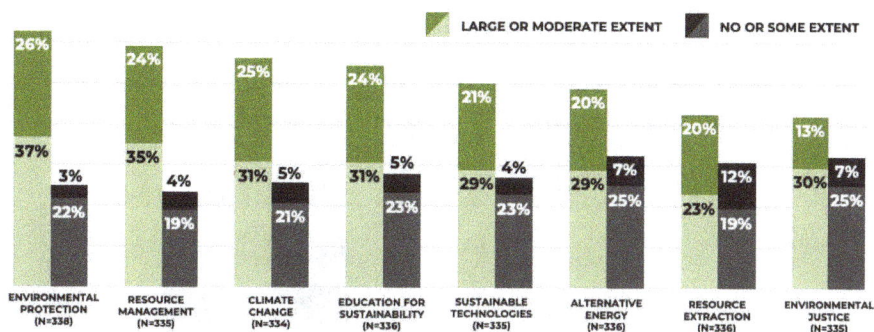

Figure 6. Prioritization of research topics at HEIs from SEPN survey results. Source: McKenzie and Chopin 2023a.

Unbalanced uptake of a whole institution approach in HEIs

SEPN found that HEIs were consistently engaging more deeply with sustainability in the domains of operations and governance as compared to other sustainability domains, especially in relation to policy (McKenzie and Chopin 2023a). For example, Vaughter and colleagues (2016) analyzed sustainability policies and plans from a representative national sample of 50 HEIs. Of the 40 institutions with a sustainability-related policy and/or plan, all of them referenced operations initiatives and most (32/40) incorporated sustainability into governance (Vaughter et al. 2016). Most policies, however, lacked detail on curriculum, research, and community outreach (Vaughter et al. 2016). An examination of sustainability uptake in relation to Indigenous knowledge in a sample of 10 HEIs also indicated that Indigenous knowledge was not well integrated across sustainability domains (Vizina 2018).

On the other hand, at universities in the northern Canadian provinces, Indigenous knowledge and practice was embedded across the whole institution, leading their campus community to act in more sustainable ways (McKenzie and Chopin 2023a). This integration was facilitated, in part, through practices, such as the integration of Indigenous knowledge in all institutional planning documents, the inclusion of Indigenous content in all courses, and the involvement of a Council of Elders within institutional governance structures (e.g., in Nunavut Arctic College, McKenzie and Wilson 2022).

SEPN also found that many HEIs had an unbalanced whole institution approach on the ground, engaging more with the domains of curriculum and operations depending on institutional size. Site visit participants rated curriculum as the 'hottest' practice domain at small institutions on SEPN's Heat Diagram surveys, whereas participants at large institutions rated operations as the 'hottest' practice domain (see Figure 7).

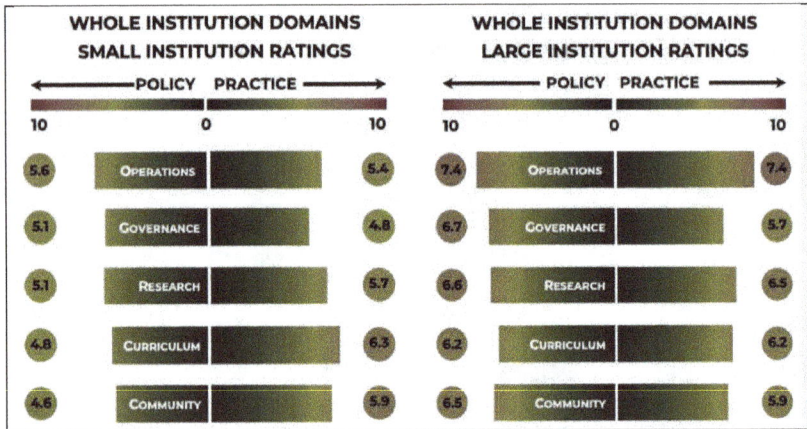

WHOLE INSTITUTION DOMAINS SMALL INSTITUTION RATINGS			WHOLE INSTITUTION DOMAINS LARGE INSTITUTION RATINGS		
←———POLICY PRACTICE———→			←——— POLICY PRACTICE ———→		
10	0	10	10	0	10
5.6	Operations	5.4	7.4	Operations	7.4
5.1	Governance	4.8	6.7	Governance	5.7
5.1	Research	5.7	6.6	Research	6.5
4.8	Curriculum	6.3	6.2	Curriculum	6.2
4.6	Community	5.9	6.5	Community	5.9

Figure 7. Comparison of sustainability practice and policy Heat Diagram ratings at small and large HEIs. Source: McKenzie and Chopin 2023a.

Policy-practice relationships are supportive, inhibitive, and iterative in HEIs

Participants described policy (or the lack thereof) as both a driver and a barrier to successful sustainability uptake at their institution (McKenzie and Chopin 2023a). During SEPN's site visits, participants at larger institutions described top-down relationships where sustainability-related policies at their institution drove practice (McKenzie and Chopin 2023a). Participants at these larger institutions often described policy-practice relationships as iterative and mutually supportive, with policy driving practice and vice versa (McKenzie and Chopin 2023a). At one institution, the absence of sustainability-related policy drove the student's union to take action to develop its own policy (McKenzie and Chopin 2023a). Instances of policy acting as a barrier during site visits were also found; for example, at one institution, supplier contracts blocked attempts to bring local food on campus (McKenzie and Chopin 2023a).

During site visits, municipal, provincial, and federal government policies, as well as cultural policies were found to play important roles in sustainability uptake in some institutions (McKenzie and Chopin 2023a). Several participants mentioned instances of non-education sustainability-related governmental policies trickling down to the institution (McKenzie and Chopin 2023a). An analysis of the drivers of embedding Indigenous knowledge into sustainability education within 10 Indigenous higher education programmes[8] found that provincial, national, and international policies and declarations supporting Indigenous self-determination were crucial (Vizina 2018). Cultural policies in relation to Indigenous knowledge and worldviews (e.g., oral policies from Elders) were also key to sustainability uptake at HEIs in the northern territories (McKenzie and Wilson 2022).

Emerging trends as well as current and future needs of ESE in Canada

This section discusses several future key priority areas for ESE as well as emerging trends in Canada, including Indigenous engagement and alignment with global metrics such as the SDGs. In relation to future needs, education systems at the primary, secondary, and higher education levels in Canada can do more to meaningfully engage with Indigenous education. A major emerging trend across primary, secondary, and higher education systems is represented through attempts to align with global competencies and goals.

Countries with colonial histories must engage with the settler-colonial contexts in which they are embedded to meaningfully address sustainability (Maina-Okori et al. 2018). In relation to Indigenous education, HEIs in Canada can engage more meaningfully with Indigenous knowledge systems. In some settings, the inclusion of Indigenous knowledge was primarily seen as for Indigenous learners (Vizina 2018). In contexts where Indigenous students, faculty, and staff were in the minority, there was much more resistance to inclusivity, even though broader society and other higher education learners are important to reconciliation efforts (Vizina 2018). On the other hand, participants at Nunavut Arctic College described their policy-practice context as one where policy drove practice, with their comments highlighting how the local Inuit culture guided the development of sustainability-related policies (McKenzie and Wilson 2022).

At the primary and secondary level, Indigenous-managed schools continue to face barriers related to Canada's colonial context (Bentham et al. 2019). An examination of sustainability uptake in Indigenous-managed schools identified severe underfunding, small population sizes, distance from urban centres, lack of community leadership, restricted employment opportunities, and historical contexts all acted as barriers to sustainability uptake (Bentham et al. 2019). Evidence of systemic racism within Indigenous education systems was also found, whereby outside priorities and pedagogies were favoured over locally determined priorities (Bentham et al. 2019). Moreover, several gaps in teacher education programmes

---

[8] This included HEIs with an Indigenous focus and HEIs with Indigenous programming.

were described by participants (Bentham et al. 2019). On the other hand, promising practices were found in Nunavut where Indigenous land-based knowledge was central to how sustainability was engaged with in schools (Aikens and McKenzie 2021, McKenzie and Aikens 2021).

At the primary, secondary, and higher education levels in Canada there are efforts to align with global initiatives such as the SDGs. At the primary and secondary level, the Council of Ministers of Education Canada has begun shifting focus away from ESD and towards 'global competencies', with sustainability now embedded within one of six, CMEC priorities (CMEC 2016, CMEC 2018). This commitment was illustrated through a recent Pan-Canadian Global Competencies Framework, which was endorsed by all ministers of education (CMEC 2018, UNESCO-IBE 2020).[9] This shift is, at least in part, driven by broader trends within the Programme for International Student Assessment (PISA) to assess for global citizenship competency (OECD 2018, 4).

At the higher education level, institutions in Canada have particularly taken on commitments to achieve the global SDGs, which is illustrated in international rankings and supported through key publications. In April 2019, the *Times Higher Education* released its inaugural University Impact Ratings, which measures the contribution of HEIs to 11 out 17 SDGs (Halliday 2019). More than 450 institutions were ranked, with 10 institutions in Canada making the list, most of which were in the top 100, and three of which ranked in the top 10 (THE 2019). These rankings illustrate a general trend to commit to the SDGs within HEIs in Canada. Efforts to move in this direction are also supported by publications, such as the *SDG Toolkit for Canadian* Colleges and Institutes (Colleges and Institutes Canada n.d.).

While the SDGs are notable in that they bring nations together to work towards common goals, they are not without critique. The SDGs come packaged with a number of overlaps and contradictions as well as a long list of targets and indicators, some of which have evaded attempts at measurement (ISC n.d., Swain 2018). The SDGs have also been critiqued for operating within neo-liberal frameworks that prioritise economic growth over achieving justice for all (Struckman 2017). Moreover, the editors of a recent special issue within *Higher Education*, which focused on the uptake of the SDGs within HEIs, noted an alarming focus on the human dimensions of sustainable development within the issue's submissions, versus also considering environmental aspects, such as climate change (Chankseliani and McCowan 2021). The extent to which the development focus within the SDGs may water down ESE approaches is a question that merits further investigation.

As SEPN conducted its comparative research in Canada, it became clear there is a substantial global need for more and better climate change education across a range of sectors. With only 60% of Canadians believing that climate change is caused mostly by human activities (e.g., Mildenberger et al. 2016), and continuing climate

---

[9] SDG 4 (quality education for all) is a key pillar upon which the framework is based (CMEC 2018, UNESCO-IBE 2020). SDG indicator 4.7 focuses on ESD and global citizenship education (United Nations General Assembly resolution 70/1 2015).

change denial, extensive calls for education and action have been made by youth and allies around the world. However, a clear understanding of effective climate change education (its quality), as well as the global benchmarks and targets to support inter-governmental processes to encourage the world's nations to increase climate change education (its quantity) remain missing.

As a result, SEPN began the Monitoring and Evaluating Climate Communication and Education (MECCE) Project in 2020. MECCE is a six-year Partnership Grant largely funded by the Canadian Social Sciences and Humanities Research Council. The MECCE Project involves over 100 leading experts and international agencies, including the IPCC, UNFCCC, UNESCO, and Environment and Climate Change Canada, collaborating to increase the impact of communication and education in addressing the climate crisis.

Conclusion

The findings presented above provide an overview of sustainability-related policies and practices, as well as policy-practice dynamics, in the formal education system in Canada. Approaches to climate change education and whole institution approaches, as engaged within Canada, are also presented. Canada's unique social, political, economic, and historical contexts in relation to global ESE trends are also considered.

In relation to policy presence, it was found that around 50% of the primary and secondary ministries of education and school divisions as well as HEIs have sustainability-specific policies (Beveridge et al. 2015, Beveridge et al. 2019). Across primary, secondary, and higher education levels, shifts away from 'environment' and towards 'sustainable development' and 'sustainability' in policy titles were noted (Beveridge et al. 2015, Aikens et al. 2016). At the primary and secondary level, it was found that conceptualizations of sustainability within ministry of education policies fall into three clusters (i.e., ESD, EE, and Indigenous Education) each of which have implications for ESE pedagogy and action (McKenzie and Aikens 2021). At the higher education level, a lack of transformative approaches within an institution's strategic plans was found (Bieler and McKenzie 2017). Within HEIs, having conducted a sustainability assessment played a significant role in more holistic responses to sustainability (Lidstone et al. 2015, Bieler and McKenzie 2017).

Across primary, secondary, and higher education, sustainability-specific policies rarely focused on climate change. When they did, it was usually in relation to operational energy upgrades (Henderson et al. 2017, Beveridge et al. 2019). At the primary and secondary level, climate policies were more likely to mention the need for climate change education than education policies (Bieler et al. 2018).

The use of a whole institution approach was found to contribute to increases in sustainability uptake but was under-utilized in institutions at primary, secondary, and higher education levels in Canada (Vaughter et al. 2016, Beveridge et al. 2019, McKenzie and Aikens 2021). A whole institution approach in Indigenous contexts differs in that the domains are seen as enmeshed, with schools seen as a focal point for community activities at the primary and secondary level (Bentham et al. 2019). At the higher education level, Indigenous worldviews and practices were seen as embedded across institutions within northern and Indigenous contexts, such as through institutional governance structures that included a Council of Elders, which oversaw the incorporation of Indigenous knowledge throughout the institution (McKenzie and Chopin 2023a).

In relation to policy-practice relationships at the primary and secondary level, it was found that a high ESE policy presence makes sustainability practices more likely across all levels (McKenzie and Aikens 2021). At the higher education level, evidence of policy-practice relationships that were supportive, inhibitive, and iterative was found (McKenzie and Chopin 2023a). At the primary, secondary, and higher education levels, evidence was found of non-education-related policy influencing sustainability uptake (Vizina 2018, McKenzie and Aikens 2021). At Indigenous HEIs, cultural oral policies and Indigenous knowledge were also key to sustainability uptake (McKenzie and Chopin 2023a).

In relation to future directions, primary, secondary, and higher education continue to deal with challenges associated with Canada's settler-colonial context (Vizina 2018, Bentham et al. 2019). Education systems across all levels in Canada have also signaled a trend to align with global competencies and goals. We are watching these trends with great interest to see what they mean for the future of ESE in Canada and encourage future research in this area.

Acknowledgements

SEPN's research was made possible by 2,600 research participants across Canada. Central to the project were the collaboration and contributions of nearly 100 partners and collaborators as well as numerous staff and graduate student research assistants. We extend our sincere appreciation to all those who contributed to the research in various ways.

We gratefully acknowledge the financial and in-kind support of our partners and contributors. The project was funded by a Partnership Grant from the Social Sciences and Humanities Research Council (SSHRC 2012–2020, Grant No. 895-2011-1025), as well as funding from the Canadian Foundation for Innovation Leader's Opportunity Fund. We also acknowledge the University of Saskatchewan, College of Education, and Sustainability Education Research Institute (SERI) for their financial and in-kind support of this project.

References

Aikens, K., M. McKenzie and P. Vaughter. 2016. Environmental and sustainability policy research: a systematic review of methodological and thematic trends. Environmental Education Research 22(3): 333–359.

Aikens, K. and M. McKenzie. 2021. A comparative analysis of environment and sustainability in policy across subnational education systems. The Journal of Environmental Education 52(2): 69–82.

Association for the Advancement of Sustainability in Higher Education (AASHE). n.d. Getting started with STARS. https://www.aashe.org/wp-content/uploads/2017/04/Getting-Started-with-STARS. pdf.

Bai, H. 2009. Reanimating the universe: Environmental education and philosophical animism. *In*: McKenzie, M., P. Hart, H. Bai and B. Jickling (eds.). Fields of Green: Restorying Culture, Environment, and Education. Hampton Press, New Jersey.

Bentham, D., A. Wilson, M. McKenzie and L. Bradford. 2019. Sustainability education in First Nations schools: A multi-site study and implications for education policy. Canadian Journal of Educational Administration and Policy 191: 22–42.

Bieler, A. and M. McKenzie. 2017. Strategic planning for sustainability in Canadian higher education, Sustainability 9: 161. Doi:10.3390/su9020161.

Bieler, A., R. Haluza-DeLay, A. Dale and M. McKenzie. 2018. A national overview of climate change education policy: Policy coherence between subnational climate and education policies in Canada (K-12). Journal of Education for Sustainable Development 11(2): 63–85. Doi: 10.1177/0973408218754625.

Beveridge, D., M. McKenzie, P. Vaughter and T. Wright. 2015. Sustainability in Canadian post-secondary institutions: The interrelationships among sustainability initiatives and geographic and institutional characteristics. International Journal of Sustainability in Higher Education 16(5): 611–638.

Beveridge, D., M. McKenzie, K. Aikens and K.M. Strobbe. 2019. A national census of sustainability in K-12 education policy: implications for international monitoring, evaluation, and research. Canadian Journal of Educational Administration and Policy 188: 36–52.

Breunig, M. 2013. Food for thought: An analysis of pro-environmental behaviours and food choices in Ontario environmental studies programmes. Canadian Journal of Environmental Education 18: 155–172.

Burch, S.L. and S.E. Harris. 2014. Understanding Climate Change: Science, Policy, and Practice. University of Toronto Press, Toronto.

Canadian Charter of Rights and Freedoms, s 16, Part 1 of the Constitution Act, 1982, being Schedule B to the Canada Act 1982 (UK), 1982, c 11. Retrieved from https://laws-lois.justice.gc.ca/eng/Const/page-15.html.

CBC News. 2019, September 24. Vancouver School Board will allow students to skip class to attend global climate strikes. Retrieved from https://www.cbc.ca/news/canada/british-columbia/vsb-climate-strike-vote1.5294962.

CEGN. 2006. Environmental Education in Canada: An Overview for Grant-makers. CEGN.

Chankseliani, M. and T. McCowan. 2021. Higher education and the sustainable development goals. Higher Education 81: 1–8.

CMEC. 2010. Report to UNECE and UNESCO on indicators of education for sustainable development: Report for Canada. Council of Ministers of Education, Canada, in collaboration with Environment Canada and The Canadian Commission for UNESCO. CMEC.

CMEC. 2012. Education for sustainable development in Canadian faculties of education. Council of Ministers of Education, Canada, in collaboration with Manitoba Education, the International Institute for Sustainable Development, and Learning for a Sustainable Future. CMEC.

CMEC. 2016. Pan-Canadian Systems-level Framework on Global Competencies. Council of Ministers of Education, Canada. CMEC.

CMEC. 2018. Global Competencies. Retrieved from https://www.globalcompetencies.cmec.ca/global-competencies.

Colleges and Institutes Canada. n.d. SDG toolkit for Canadian colleges and institutes. Retrieved from https://sdgcicanguide.pressbooks.com/front-matter/introduction/.

EEPSA. 2022. History of EEPSA. https://eepsa.org/history-of-eepsa/.

EECOM. 2019. History. Retrieved from https://eecom.org/about/history/.

Erickson, P. and M. Lazarus. 2014. Impact of the Keystone XL Pipeline on global oil markets and greenhouse gas emissions. Nature Climate Change 4(9): 778–780.

Fawcett, L. 2009. Feral sociality and (un)natural histories: On nomadic ethics and embodied learning. *In*: McKenzie, M., P. Hart, H. Bai and B. Jickling (eds.). Fields of Green: Restorying Culture, Environment, and Education. Hampton Press, New Jersey.

Fisher, D., K. Rubenson, G. Jones and T. Shanahan. 2009. The political economy of post-secondary education: a comparison of British Columbia, Ontario and Québec. Higher Education 57: 549–566.

Gebhard, A. 2017. Reconciliation or racialization? Contemporary discourses about residential schools in the Canadian prairies. Canadian Journal of Education 40(1): 1–30.

Government of Canada. 2002. A framework for environmental learning and sustainability in Canada. Retrieved from http://publications.gc.ca/collections/Collection/En40-664-2002E. pdf.

Government of Canada. 2019, July 15. Canada takes action on the 2030 Agenda for Sustainable Development. Retrieved from https://www.canada.ca/en/employment-social-development/programs/agenda-2030.html.

Government of Canada. 2020, April 27. United Nations Framework Convention on Climate Change and the Paris Agreement. Retrieved from https://www.canada.ca/en/environment-climate-change/corporate/international-affairs/partnerships-organizations/united-nations-framework-climate-change.html.

Halliday, M. 2019, April 29. Canadian universities score big in new sustainable development rankings. University Affairs. Retrieved from https://www.universityaffairs.ca/news/news-article/canadian-universities-score-big-in-new-sustainable-development-rankings/.

Harden, J. 2017. The case for renewal in post-secondary education. Canadian Centre for Policy Alternatives. Retrieved from https://www.policyalternatives.ca/publications/reports/case-renewal-post-secondary-education.

Hargis, K. and M. McKenzie 2020. Responding to Climate Change Education: A Primer for K-12 Education. The Sustainability and Education Policy Network, Saskatoon, Canada.

Hart, P. 1990. Environmental education in Canada: Contemporary issues & future possibilities. Australian Journal of Environmental Education 6: 45–65.

Hart, P. 1996. Problematizing enquiry in environmental education: Issues of method in a study of teacher thinking and practice. Canadian Journal of Environmental Education 1: 56–88.

Hart, P. and C. Hart. 2014. It's not that simple anymore: Engaging the politics of culture and identity within environmental education/education for sustainable development (EE/ESD). *In*: Lee, J.C. and R. Efird (eds.). School for Sustainable Development Across the Pacific. Springer, Heidelberg.

Henderson, J., A. Bieler and M. McKenzie. 2017. Climate change and Canada's higher education system: An institutional policy analysis. Canadian Journal of Higher Education 47(1): 1–26.

Hopkins, C. 2013. Education for sustainable development in formal education in Canada. pp. 23–36. *In*: McKeown, R. and V. Nolet (eds.). Schooling for Sustainable Development in Canada and the United States. Springer, London.

International Science Council (ISC). n.d. A guide to SDG interactions: From science to implementation: Executive summary. Retrieved from https://council.science/wp-content/uploads/2017/05/SDGs-interactions-executive-summary.pdf.

IPCC. 2018. Global warming of 1.5°C. An IPCC Special Report on the impacts of global warming of 1.5°C above pre-industrial levels and related global greenhouse gas emission pathways, in the context of strengthening the global response to the threat of climate change, sustainable development, and efforts to eradicate poverty [Masson-Delmotte, V., P. Zhai, H.O. Pörtner, D. Roberts, J. Skea, P.R. Shukla, A. Pirani, W. Moufouma-Okia, C. Péan, R. Pidcock, S. Connors, J.B.R. Matthews, Y. Chen, X. Zhou, M.I. Gomis, E. Lonnoy, T. Maycock, M. Tignor and T. Waterfield (eds.)]. In Press.

Jeppesen, S. and H. Nazar. 2012. Beyond academic freedom: Canadian neoliberal universities in the global context. Topia. 28: 87–113.

Jickling, B. 1992. Why I don't want my children to be educated for sustainable development. Journal of Environmental Education 23(4): 5–8.

Jones, G.A. 2014. An introduction to higher education in Canada. pp. 1–38. *In*: Joshi, K.M. and S. Paivandi (eds.). Higher Education Across Nations (vol. 1). B.R. Publishing, Delhi.

Lidstone, L., T. Wright and K. Sherren. 2015. An analysis of Canadian STARS-rated higher education sustainability policies. Environment, Development, Sustainability 17(2): 259–278.

Maina, N.M., J. Murray and M. McKenzie. 2020. Climate change and the fossil fuel divestment movement in Canadian higher education: The mobilities of actions, actors, and tactics. Journal of Cleaner Production 253: 1–10.

Maina-Okori, N.M., J.R. Koushik and A. Wilson. 2018. Reimaging intersectionality in environmental and sustainability education: A critical literature review. The Journal of Environmental Education 49(4): 286–296.

McClaren, M. 2009. The place of the city in environmental education. *In*: McKenzie, M., P. Hart, H. Bai and B. Jickling (eds.). Fields of Green: Restorying Culture, Environment, and Education. Hampton Press, New Jersey.

McIvor, O. and J. Ball. 2019. Language-in-education policies and Indigenous language revitalization efforts in Canada: considerations for non-dominant language education in the global south. Fire: Forum for International Research in Education 5(3): 12–28.

McKenzie, M., A. Bieler and R. McNeil. 2015. Education policy mobility: Reimagining sustainability in neoliberal times. Environmental Education Research 21: 319–337.

McKenzie, M., K. Aikens and N.S. Chopin. 2017. Scale Matters in Policy Flows: A Comparative Case Study of Sustainability in K-12 Education. Sustainability and Education Policy Network, University of Saskatchewan, Saskatoon, Canada.

McKenzie, M. and K. Aikens. 2021. Global education policy mobilities and subnational policy practice. Globalisation, Societies and Education 1–15. Doi: 10.1080/14767724.2020.1821612.

McKenzie, M. and N. Chopin. 2023a. Sustainability in the Canadian post-secondary education system. Sustainability and Education Policy Network, University of Saskatchewan, Saskatoon, Canada (in press).

McKenzie, M. and N. Chopin. 2023b. Tracking sustainability in K-12 education: A national comparative study. Sustainability and Education Policy Network, University of Saskatchewan, Saskatoon, Canada (in press).

McKenzie, M., P. Hart, H. Bai and B. Jickling (eds.). 2009. Fields of Green: Restorying Culture, Environment, and Education. Hampton Press, New Jersey.

McKenzie, M. and A. Wilson. 2022. Sustainability as wild policy: Mobile SDG interventions and land-informed policy in education. Discourse: Studies in the Cultural Politics of Education 1–16.

Metcalfe, A.S. 2010. Revisiting academic capitalism in Canada: No longer the exception. Journal of Higher Education 81: 489–514.

Mildenberger, M., P.D. Howe, E. Lachapelle, L.C. Stokes, J.R. Marlon and T. Gravelle. 2016. The distribution of climate change public opinion in Canada. Retrieved from https://ssrn.com/abstract=2732935.

OECD. 2018. Preparing our youth for an inclusive and sustainable world: The OECD PISA global competence framework. OECD.

OECD. 2020. Linking Indigenous communities with regional development in Canada, OECD Rural Policy Reviews, OECD Publishing, Paris, https://doi.org/10.1787/fa0f60c6-en.

Olssen, M. and M.A. Peters. 2005. Neoliberalism, higher education and the knowledge economy: From the free market to knowledge capitalism. Journal of Education Policy 20: 313–345.

Parker, L. 2017. Creating a crisis: Selling neoliberal policy through the rebranding of education. Canadian Journal of Educational Administration and Policy 183: 44–60.

Patrick, D. 2017. Language policy and education in Canada. pp. 401–441. In: May, S. and N.H. Hornberger (eds.). Language Policy and Political Issues in Education: Encyclopedia of Language and Education (Vol. 1). Springer, Cham, Switzerland.

Russell, C.L., A.C. Bell and L. Fawcett. 2000. Navigating the waters of Canadian environmental education. pp. 196–216. In: Goldstein, T. and D. Selby (eds.). Weaving Connections: Education for Peace, Social and Environmental Justice. Sumach Press, Toronto.

Sauvé, L., R. Brunelle and T. Berryman. 2005. Influence of the globalized and globalizing sustainable development framework on national policies related to environmental education. Policy Futures in Education 3(3): 271–283.

Schuetze, H.G., L. Kuehn, A. Davidson-Harden, D. Schugurensky and N. Weber. 2011. Globalization, neoliberalism and schools: The Canadian story. pp. 62–84. In: Olmos, L., C.A. Torres and R.V. Heetum (eds.). Educating the Global Citizen in the Shadow of Neoliberalism: Thirty Years of Educational Reform in North America. Bentham eBooks.

Statistics Canada. 2015, November 27. Academic outcomes of public and private high school students: What lies behind the differences? Retrieved from https://www150.statcan.gc.ca/n1/pub/11f0019m/11f0019m2015367-eng.htm.

Statistics Canada. 2017, March 30. Population size and growth in Canada: Key results from the 2016 Census. Retrieved from https://www150.statcan.gc.ca/n1/daily-quotidien/170208/dq170208a-eng.htm.

Statistics Canada. 2021a, February 1. Table 17-10-0009-01. Population estimates, quarterly. https://doi.org/10.25318/1710000901-eng.

Statistics Canada. 2021b, February 1. Table 37-10-0011-01. Post-secondary enrolments, by field of study, registration status, program type, credential type and gender. https://doi.org/10.25318/3710001101-eng.

Struckmann, C. 2017. A Post-colonial Feminist Critique of the 2030 Agenda for Sustainable Development: A South African Application (Master's thesis, Stellenbosch University).

Swain, R.B. 2018. A critical analysis of the Sustainable Development Goals. pp. 341–356. *In*: Filho, W.L. (ed.). Handbook of Sustainability Science and Research. Springer International Publishing, Cham, Switzerland.

The Canadian Press. 2019, September 18. Some Canadian schools, colleges move to accommodate climate strikes. Retrieved from https://www.cbc.ca/news/canada/toronto/schools-climate-rally-1.5288179.

Times Higher Education (THE). 2019. Impact Ratings 2019. Retrieved from https://www.timeshighereducation.com/rankings/impact/2019/overall#!/page/0/length/25/sort_by/rank/sort_order/asc/cols/undefined.

UN. 2015. Paris Agreement. Retrieved from https://unfccc.int/sites/default/files/english_paris_agreement.pdf.

UNDP. 2021. What are the Sustainable Development Goals? Retrieved from https://www.undp.org/content/undp/en/home/sustainable-development-goals.html#:~:text=The%20Sustainable%20Development%20Goals%20(SDGs,peace%20and%20prosperity%20by%202030.

UNESCO. 2014. UNESCO roadmap for implementing the Global Action Programme on Education for Sustainable Development. UNESCO.

UNESCO-IBE. 2020, February 27. Canada establishes a pan-Canadian global competencies framework for education. Retrieved from http://www.ibe.unesco.org/en/news/canada-establishes-pan-canadian-global-competencies-framework-education.

United Nations General Assembly resolution 70/1. 21 October 2015. Transforming Our World: The 2030 Agenda for Sustainable Development, A/RES/70/1.

United Nations General Assembly resolution 74/223. 24 January 2020. Sustainable development: Education for sustainable development, A/RES/74/223.

Vaughter, P., M. McKenzie, L. Lidstone and T. Wright. 2016. Campus sustainability governance in Canada: A content analysis of post-secondary institutions' sustainability policies. International Journal of Sustainability in Higher Education 17(1): 16–39.

Virone, R. 2016. Neoliberalism and the current educational climate: The effect on school culture and at-risk students. Master's thesis, Concordia University. Retrieved from https://spectrum.library.concordia.ca/981017/.

Vizina, Y.N. 2018. Indigenous Knowledge and Sustainability in Post-secondary Education (Doctoral dissertation, University of Saskatchewan).

Zhao, Y. 2020. Two decades of havoc: A synthesis of criticism against PISA. Journal of Educational Change 21: 2450266.

Chapter 16

Pathways to Achieving the Sustainable Development Goals through Environmental and Sustainability Education in the USA

Michaela Zint,* Jessica Ostrow Michel, Sarah R. Collins, Emma C. Sloan, Maite Elizondo Piñeiro, Lauren Balotin, Veronica Correa, María Isabel Dabrowski, Bita Davoodi, Joseph Dierdorf, Daniela Fernández Méndez Jiménez, Jessica Miller, Isabel Nakisher, Connor Roessler, Paige Schurr, Peter Siciliano and *Joshua Thompson*

Introduction

The United States of America (USA), with a population approaching 330 million, has one of the largest economies in the world (Organization for Economic Co-operation and Development, OECD 2020). The country is the second greatest contributor to global carbon emissions (Climate Action Tracker 2020) and is experiencing a range

School for Environment & Sustainability, School of Education, College of Literature, Science & the Arts, University of Michigan, 440 Church St, Ann Arbor, MI 48109, USA.

Emails: michjess@umich.edu; srsc@umich.edu; ecsloan@umich.edu; mepmaite@umich.edu; lbalotin@umich.edu; vcorrea@umich.edu; mariadab@umich.edu; davoodib@umich.edu; dierdorf@umich.edu; danielaf@umich.edu; jessfm@umich.edu; inakishe@umich.edu; croessle@umich.edu; pschurr@umich.edu; psicili@umich.edu; thomjosh@umich.edu

* Corresponding author: zintmich@umich.edu

The first five authors are listed based on their contributions to the chapter, with the remaining authors listed alphabetically.

of climate change impacts (such as severe storms, wildfires, flooding) (National Aeronautics and Space Administration 2020). Although the USA has not adopted the *2030 Agenda for Sustainable Development* (the country is not a UNESCO member), a range of actors and organization have committed to addressing climate change and achieving the Sustainable Development Goals (SDGs). Many, for example, joined movements like the *US Climate Alliance* and *We are Still In*, to uphold the objectives of the Paris Climate Agreement. Moreover, states like New York, cities like Los Angeles, and companies like Verizon are all working toward the SDGs (Brookings-UN Foundation 2020).

This is, however, not the case for the majority of the USA's education system, which includes more than 50.7 million students in pre-kindergarten to 12th grade (PK-12) public schools alone (National Center for Education Statistics (NCES) 2020). This is particularly unfortunate because the USA's education system is affected by a range of sustainability challenges. For example, with regard to:

- The environment: The proximity of schools to pollution sources (e.g., major highways and industrial facilities) negatively impacts American students' performance on academic tests (Kweon et al. 2016, Mohai et al. 2011). This issue is especially problematic because racially minoritized students disproportionately attend schools in more polluted areas. For example, in Michigan, 82 percent of black students compared with 44 percent of white students attend schools that are in the top 10 percent of areas in the state most burdened by industrial air pollution (Mohai et al. 2011).

- Society, nearly 75 percent of five billion cafeteria lunches in 2019 were free or reduced-price due to recipients living in poverty (Department of Agriculture Economic Research Service 2020).

- Economy, $27,357 per year is the average cost of a bachelor's degree (NCES 2019), partly as a result of declining federal and state funding for Higher Education Institutions (HEIs).

In contrast to President Clinton who established the President's Council on Sustainable Development resulting in *Education for Sustainability (EfS): An Agenda for Action* (1996) as well as President Obama who spoke in support of the SDGs (inter)nationally (The White House, Office of the Press Secretary 2015), President Trump rolled back over 100 sustainability-related regulations (Regulatory Rollback Tracker 2020) and exited the USA from the Paris Climate Accord in 2020. Although the USA formally rejoined the agreement under President Biden just a few months later (The White House, The Briefing Room 2021), the country has not formally committed to the SDGs. This is despite 85 percent of American voters agreeing that it is important for the USA to maintain an active role with the UN (Better World Campaign 2020).

Although political leadership has largely been lacking, particularly at the federal level, numerous environmental and sustainability education (ESE) champions and organizations across the country have taken it upon themselves, often with little or no funding, to work toward the goals of the UN Decade of Education for Sustainable Development (ESD) (Smith et al. 2015) and more recently, the SDGs.

For example, Shelburne Farms has a *Cultivating Pathways to Sustainability* program that partners youth and adult teams with experts to address one of the SDGs over the course of a year. The Girl Scouts have a badge to recognize girls for their leadership contributions to the SDGs. Arizona State University's Teacher College requires a pre-service course for K-8 educators that prepares teachers to incorporate the SDGs into their curricula. The University of Minnesota launched a grant program to fund efforts by students, faculty and staff to advance the achievement of the SDGs (Global Programs and Strategy Alliance: University of Minnesota 2020). Carnegie Mellon University conducted a self-assessment based on the SDGs (Sustainability Initiative 2020). The University of Michigan offers the Massive Open Online Course (MOOC) *Act on Climate: Steps to Individual, Community and Political Action* to contribute to SDG 13 as well as an online certificate titled *Sustainability and Development Master Track* focused on the SDGs. And *Learning For the Future*, a coalition of Vermont educators, utilizes the SDGs to teach about resilience and sustainability. An additional coalition goal is to encourage the state of Vermont to adopt its equity centered *Framework for Change*.

To enable readers to learn about ESE in the USA and its contributions to the SDGs based on a range of actors and organizations, we begin by providing background about the country and its ESE history since the late 1960s. This historic information is followed by a section to strengthen readers' understanding of current ESE in the country, supported by a section describing educational programs and scholarship illustrative of work that can contribute to the SDGs nationally and internationally. We conclude with recommendations for advancing ESE in the USA and beyond.

Because a relatively short chapter like this cannot possibly capture all of the past nor current ESE efforts across a large country like the USA, we acknowledge that significant contributions and perspectives across the country may have been missed. Most importantly, we are not able to represent Indigenous peoples' ESE contributions, in part due to our own limited WEIRD (i.e., "western, educated, industrialized, rich and democratic") histories.

Note that we deliberately focus on programs and draw on scholarship that includes USA authors, to highlight efforts throughout the country that are already, or we believe could be contributing to achieving the SDGs nationally and internationally. While we fully recognize and have personally benefited from the international foundation in ESE (e.g., Bengtsson et al. 2018), we deliberately chose a USA focus for the purpose of this particular chapter and book. While we decided to address ESE in both PK-12 and higher education settings in the USA, their respective contexts, efforts, challenges, and associated scholarship have been quite distinct. We therefore again acknowledge that important national contributions have been missed.

About the USA' political, economic, cultural and social contexts and conditions, including its education system

The USA is facing a range of societal challenges including poverty, racial and income inequality, and lack of wide-spread access to healthcare. Environmentally, former President Trump's agenda put the USA even further behind, especially within the context of national climate action. Not surprisingly, therefore, the USA is ranked 35th out of 162 countries on progress toward achieving the Sustainable Development Goals (SDGs), the worst among the OECD (Sachs et al. 2019).

As illustrated by the most recent presidential election, the USA is in a "culture war" in which social groups are engaging in polarized conflict to establish dominance based on different worldviews (Hunter 1991). The growing divides between socially constructed groups (e.g., republican-democrat, laissez faire-regulated capitalism, religious-agnostic, ethnic-racial, old-young) are centered around issues such as abortion, gun regulations, health care, and climate change (Pew Research Center 2017). For more about how these divides have been fueled by powerful interests that stand to gain from them, see Hoffman (2015) and Oreskes and Conway (2010). For how the country's democracy may be rebuilt, see Orr et al. (2020). As social institutions, the country's education systems are heavily influenced by these culture wars, among other conditions and factors (Nolet 2017).

In the USA, public-funded education begins in most states for four to five year old children in pre-kindergarten (PK). PK is not mandatory and often requires families to pay out of pocket if they want early education for their children. Youth attend public or private schools from about age five to 18, often referred to as "K-12." Afterward, about 19.7 million attend higher education in colleges focused on vocational and technical associates degrees or universities offering four year bachelor's as well as master's and PhD degrees (Education Data Initiative, unpublished data).

In recent years, reduced federal, state, and local funding for education (The Education Trust 2018) has been identified as responsible for decreased rates of (scientific) literacy across the country (OECD 2021), and higher education is becoming increasingly unaffordable (NCES 2019). In K-12 education, there are national educational standards across subjects, but federal funding is tied to national assessments that focus primarily on reading and math through the Every Student Succeeds Act (Plans 2015). The latter often results in school districts feeling pressure to "teach to the test," prioritizing student learning in mathematics and literacy. While educational policies are determined at the state and local education district levels, and these are strongly influenced by national standards, policies, textbooks, and curriculum developers, among others.

There are select states (e.g., Maryland, Vermont), school districts (e.g., Fairfax County, Virginia, Portland, Oregon), and model schools (e.g., Arthur Morgan School, North Carolina, Lowell School, Washington D.C., Teton Science Schools, Wyoming, Willow School, New Jersey) where ESE receives significant attention. However, the focus on ESE in formal education overall is limited.

This is attributable to a variety of causes. At the school level, for example, there is limited space in already "full" curricula centered on the aforementioned national assessments. At the societal level, the politicized nature of subject matter like climate change, has resulted in disagreements among members of the public and thus, among policymakers as to what, if anything, should be taught. Nonetheless, there is public support for strengthening ESE. For example, 78 percent of registered voters across the USA support "schools teaching children the causes, consequences, and potential solutions to global warming" (Leiserowitz et al. 2020).

Non-governmental organizations (NGOs) with environmental or sustainability missions (e.g., conservation, nature centers, museums, zoos) seek to fill these gaps and are a main source of non-formal ESE in the USA, along with select federal, state, and local government agencies. Additionally, HEIs are increasingly embracing ESE in a myriad of forms (Michel 2020c). The country has over 2,500 environmental science and studies programs (Vincent et al. 2017) and many HEIs are infusing sustainability to expose all of their students to associated topics (Michel 2020b).

About the USA's ESE history

Although ESE in the USA today has many shortcomings, the country was an early leader in environmental education (EE). In 1970, for example, environmental literacy was promoted through Earth Day teach-ins and the National Environmental Education Act (NEEA) was signed into law. Just a year earlier, Stapp (1969) published what would become the seminal definition of EE driving (inter)national policies and practices, such as the guidelines and principles of the Tbilisi declaration (UNESCO 1977) and the US. Environmental Protection Agency's (EPA) definition of the EE (EPA, unpublished data). Stapp's (1969) work led to GREEN (now Earthforce) promoting action research as a means to protect the environment and promote peace (Simmons et al. 2001) and was accompanied by others who introduced similar pedagogies to translate knowledge *about* the environment into action *for* the environment (e.g., Hungerford and Volk 1990). Today, NEEA (funded at about $8 million annually) remains the only national EE policy and the North American Association for Environmental Education (NAAEE) is the country's primary professional association for environment and sustainability educators and scholars. Although about 35 USA states have environmental literacy plans, ESE continues to be driven almost exclusively by grassroots NGO efforts (Braus 2020).

Sustainability education (SE) in the USA was inspired by (inter)national events like the National Council for Science and the Environment's (NCSE) *Education for a Sustainable and Secure Future* (2003) conference and has been advanced through the leadership of select individuals and organizations, such as the Association for the Advancement of Sustainability in Higher Education (AASHE) and the Association for Environmental Studies and Sciences (AESS). While President Clinton's Council on Sustainable Development (1996) defined Education for Sustainability as *"… a lifelong learning process that leads to an informed and involved citizenry having the creative problem-solving skills, scientific and social literacy, and commitment to engage in responsible individual and cooperative actions"* and researchers like one of the authors have proposed others [e.g., *"… is the process of developing students' sustainability knowledge, attitudes, and behaviors in favor of the environment and its economic and social implications"* (Michel 2020c, 25)], we are not aware of a single definition of SE that has broad acceptance in the USA. Moreover, there are on-going debates about the extent to which EE and SE differ in practice.

On the SE resource front, McKeown et al. (2002) prepared one of the first toolkits for more sustainable communities, now translated into over 60 languages. Rowe created the U.S. Partnership for Sustainability Education and associated networks to advance SE across the country, including numerous SE resources (Rowe et al. 2015).

AASHE has played a major role in HEIs' sustainability efforts through tools like the Sustainability Tracking Assessment and Rating System (STARS) and promotion of The Sustainability Literacy Test (Sulitest). And climate change educators have collaborated to scale fragmented educational efforts (e.g., Bowman and Morrison 2020).

In terms of advancing SE scholarship, McKeown and Nolet's (2013) book *Schooling for Sustainable Development in Canada and the United States* contains chapters by SE leaders such as Santone, an advocate for teaching ecological versus traditional economics (Santone 2013) and *Facing the Future*, a source for K-12 lessons on a range of sustainability topics (Church and Skelton 2013). Additionally, Smith et al.'s (2015) report provides an inventory of SE organizations across the country.

It is important to note that HEI scholars, like the ones who contributed to the above book have been, and continue to be, among the main advocates for SE in the country by, for example, researching sustainability competencies (Brundiers et al. 2021) and effective pedagogies (Michel 2020a), bringing to light Indigenous peoples' contributions (Bull and Hat 2019, Pierotti and Wildcat 2000), and preparing national reviews to advance SE (e.g., National Academies of Sciences, Engineering, and Medicine 2020).

Understanding ESE in the USA

As already alluded through the examples shared so far, a range of ESE programs exist across the USA, despite limited political leadership, policies, and funding. Although EE has historically had more support in the USA than SE, even EE has not gained a foothold in PK-12 education. Primary reasons include mandated education standards and assessments that are tied to public education funding and used to assess school performance (Hopkins 2013). These standards, with a few exceptions, do not address the environment or sustainability and even those that do (e.g., National Science Education Standards: climate change) tend to focus on natural science and engineering content rather than on underlying social drivers or key normative questions (Feinstein and Carlton 2013). This may be due in part to the siloing of subjects such as science and social studies, treating them as isolated rather than connected, as is key to ESE. It is therefore not surprising that many states get poor grades on their climate education standards (Madson 2021). With a few exceptions, teacher ESE pre- and in-service training is not mandated (Kirk et al. 1997), and thus, teachers tend not to experience a range of barriers due to a lack of relevant ESE knowledge (Ennes et al. 2021). The limited administrative and financial support for ESE professional development is unfortunate, especially in light of the hundreds of sustainability science and ESE scholars who have relevant content and pedagogical expertise.

We concur with Hopkins (2013) that ESE in the USA's K-12 schools can best be described as having (1) little to no coverage, (2) a student and teacher co-led extracurricular club/project, (3) a school-wide project on recycling or another environmental topic, (4) environmental topics incorporated into core subjects, primarily middle school science subjects, (5) a whole school or system approach, ranging from a focus on green school certification (e.g., Green Ribbon Schools,

E S E in the U S A is shifting (from → to)			Sample Reference
The environment	→	Other 'E's (i.e. economy, equity) and all three 'E's	Chang et al. (2016)
Individual behaviour	→	Leadership to collectively transform systems, communities, and societies	NASEM (2020)
Instrumental (develop knowledge, skills and attitudes for contributing to society) goal	→	Emancipatory (critical, autonomous, self-determined agents who can bring about transformative change) goal	Papenfuss et al. (2019)
Disciplinary (primarily natural science) and inter-disciplinary approaches	→	Approaches based on natural/social science and humanities, toward holistic and trans-disciplinary ways	Bartlett et al. (2020)
White, western (e.g. individualist, nature as separate from humans) perspectives	→	Diverse and indigenous perspectives (e.g., collectivist, nature as mother)	Frandy (2018)
Fear appeals and the past	→	Promoting hope through envisioning alternative futures	Trott (2019)
Apolitical	→	Political engagement and leadership	Orr (2020)

Figure 1. Shifts in foci underway in ESE in the USA.

The Center for Green Schools) and with less frequency, to holistic infusion (e.g., Higgs and McMillan 2006, Hurd and Ormsby 2020), or in isolated cases (6) a whole community approach facilitated by, for example, Regional Centre of Expertise like Greater Portland's Sustainability Education Network.

One of the more positive ESE trends can be found in HEIs. A recent Princeton Review (2022) survey highlighted what is likely to be the main reason behind that trend; i.e., the 75% of students whose decision of which HEI to attend is influenced by the latter's environmental commitment. Driven by this market demand in addition to student activism (e.g., 350.org), HEIs rankings (e.g., Princeton Review Guide to Green Colleges) and accreditation (e.g., Desha et al. 2019), as well as faculty interest and collaborations (e.g., Barlett and Chase 2013, Hong 2020), we anticipate a growing focus on whole institution, and other ESE initiatives (e.g., Hess and Maki 2019, Michel 2019, Bartlett et al. 2020) in the USA.

In contrast, adult ESE has been relatively limited in the USA. Notable exceptions have focused on climate change (e.g., Climate Reality Leadership Corps, Project Drawdown) with the goal of fostering mitigation and adaptation behaviors (e.g., Zint and Wolske 2014, Carman 2020). A relatively new trend for adult professional and young learners are the growing number of HEIs Massive Open Online Courses (MOOCs) on environmental and sustainability topics, available for free or small fees through Coursera and edX, as are a range of organizations' digital ESE learning opportunities as a result of COVID-19 (e.g., Soper 2020).

Importantly, promising trends are underway in how ESE is evolving across the USA across a range of learning settings (Figure 1).

ESE and the SDGs in the USA

The *UN Decade of ESD* (2005-2014) inspired a proliferation of grassroots SE initiatives in the USA and the *Global Action Programme on Education for Sustainable Development* (2015–2019) planted the seed for educational attention to the SDGs, despite neither being formally adopted by the federal USA government. As shared in the introduction and earlier sections of this chapter, there are some champions and organizations that offer ESE in the USA *explicitly* to contribute to the SDGs. Most ESE efforts, however, do not mention ESD for 2030 or the SDGs, although they can clearly have an impact on associated targets. Examples of such programs and scholarship are described in Table 1.

The purpose of compiling this table was to show that the USA offers a range of programs and scholarship that can advance the SDGs (often in combination) nationally and internationally, even if not explicitly focused on these particular goals. The programs and research featured in Table 1 were identified through internet and database searches using combinations of keywords inspired by the SDG indicators (U.S. National Statistics for the U.N. Sustainable Development Goals, unpublished data) and ESD learning outcomes (UNESCO 2017) (e.g., "sustainable development", "climate change education", "quality education") and others (e.g., "K-12", "higher education", "program", and "USA"). From the list of programs and research that were identified through these searches, we selected ones that appeared most promising with regard to contributing to the respective SDGs—and—to collectively, cut across a range of educational settings/approaches as well as geographic locations. The list in Table 1 is therefore illustrative and not intended to be comprehensive.

As indicated by the examples in Table 1 and throughout the chapter, there are a diverse set of formal, non-formal, and informal ESE *programs* in the USA that are already or can contribute to the SDGs for audiences including PK-12 and HEI students as well as adult professionals and citizens more broadly. These programs

Table 1. Sample USA ESE programmes and scholarship with potential to contribute to SDGs.

SDG	Sample ESE Programme	Sample ESE Research
1: No Poverty	*Fair Trade Campaigns* (FCP), an NGO, connects high schools and HEIs with curricula and activities to raise awareness of global fair-trade practices. With nearly 50 campaigns across the USA, FCP educates students about fair trade practices, wealth inequality, and child labour. FCP empowers students to address global inequalities, thus, working toward SDG 1 to eliminate poverty.	A commentary by Van Lancker and Parolin (2020) described the anticipated effects of COVID-19 school closures. With 14% of American children facing food insecurity and 2–5% in public schools living with instability, including homelessness, they suggest it is critical to provide food, learning resources, and income support for low income households to avoid an increase in children in poverty which SDG 1 seeks to address.
2: Zero Hunger	*Growing Hope* is a community organization in Ypsilanti, Michigan, that provides education and outreach programmes to teach PK-12 students and adult community members how to grow healthy and sustainable food. Growing Hope addresses SDG 2 by aiming to improve nutrition and promote sustainable agriculture, including through resources for home and community gardening as well as local food entrepreneurship.	Twill et al. (2016) studied the creation and maintenance of a food pantry for HEI students by documenting and analyzing the process of establishing the pantry. They found that the pantry supported food-insecure students and assisted in retaining students. The support this programme provides aids in ending hunger (SDG 2) through ensuring food security.
3: Good Health and Well-Being	The CDC's Whole School, Whole Community, Whole Child (WSCC) model frames the implementation of health education in K-12 schools. The framework, comprising 10 components that span the health continuum, can be tailored to individual school needs and is collaborative among community agencies, school leaders, and other groups. Contributing to SDG 3, WSCC provides a holistic, integrative approach to promote good health and well-being practices.	Philipsborn et al. (2020) aimed to fill a gap in the education of physicians through a framework for integrating the health impacts of climate change into medical school curricula. The suggested learning objectives can support adaptation to climate change through health care delivery, ensuring the SDG 3 goal of healthy lives for all.
4: Quality Education	In the *Birding Communities Program*, Brown University students lead birding walks for local elementary school students, to support them in learning about bird biodiversity. Many participating elementary students come from underserved backgrounds, and benefit from this learning experience. Additionally, as a result of their participation, HEI students benefit from the opportunity to apply their knowledge in real-life settings, supporting lifelong learning, and inclusive and equitable education as envisioned by SDG 4.	Dockry et al. (2016) introduced the College of Menominee Nation Sustainable Development Institute's model of sustainable development and described applications of this model to course design and community planning. They found that use of the model improved sustainable thinking in HEI students, and facilitated diverse, participatory community research. As such, the model provides a means to educate students about the SDG 4 goals of quality, inclusive and equitable education.

Table 1 contd. ...

...*Table 1 contd.*

SDG	Sample ESE Programme	Sample ESE Research
5: Gender Equality	*Girls' Outdoor Adventure in Leadership and Science* (GOALS) provides a free 2-week immersion summer programme facilitated by the University of California at Davis, and Sequoia National Forest. This experience strengthens girls' gender-expansive youth leadership and science skills in an outdoor setting. The programme supports SDG 5 by closing the opportunity gap between genders in the science, technology, engineering, and mathematical (STEM) fields.	Thébaud and Charles (2018) examined gender segregation across STEM fields in the USA based on a literature review. They found that western cultural stereotypes are driving STEM education segregation. Recommendations are offered for improved USA policies to desegregate STEM education, thus advancing the SDG 5 goal of gender equality.
6: Clean Water and Sanitation	The National Oceanic and Atmospheric Administration's (NOAA) *Meaningful Watershed Educational Experiences* (MWEE) are national K-12 student-directed projects in which students investigate a locally-relevant water-related issue (such as schoolyard erosion or stream habitat quality) throughout an academic year. Through data collection in and out of the classroom and an action project, MWEEs increase student environmental literacy and stewardship (Zint et al. 2014). MWEEs help to educate students on a range of water issues and support local community participation, thus supporting SDG 6.	Zint et al. (2014) explored the relationship between predictors of environmentally-responsible behaviour and MWEEs through multi-level qualitative analysis of student questionnaires results. They found that MWEEs increased the environmental stewardship characteristics of students. For example, MWEEs positively impact students' understanding of clean water and their intentions to engage in water stewardship actions, matching the SDG 6 goal of ensuring clean water for all.
7: Affordable and Clean Energy	The *National Energy Education Development Project* that offers K-12 curricula to educators to engage students in learning about renewable energy (e.g., solar panels). This programme furthers SDG 7 through creating a new generation of energy leaders.	Based on studying Wayne County, Michigan, Reames et al. (2018) learned that energy-efficient bulbs were less available in high poverty areas and smaller stores, and that their costs were higher. Education efforts through informing staff and creating store displays about energy savings through more efficient lighting could assist residents, along with other interventions to achieve SDG7.
8: Decent Work and Economic Growth	Susan Santone's 'Sustainability and Economics 101: A Primer for Elementary Educators' is a guide for K-2 teachers to engage their students with ecological economics. This guide, available through the United States Society for Ecological Economics, addresses how to use a sustainability lens to teach economics to youth. By providing accessible economics education to young students, this programme contributes to the SDG 8 goal of inclusive economic growth.	Venkatesan (2020) explored the results of teaching about sustainability in an introductory HEI economics course. Students' assignment reflections suggested that they thought more critically about the impact of their personal consumption and values. The findings have implications for how to support students' conceptions of the SDG 8 economic growth goal.

9: Industry, Innovation, and Infrastructure	The *U.S. Green Building Council* offers professional education courses in sustainable building design, urban development design, architecture, and construction planning. Courses cover a range of topics such as the Leadership in Energy and Environmental Design (LEED) building rating system and the intersection of building design and social equity, thereby contributing to the SDG 9 goals of inclusive and sustainable industrialization and infrastructure.	Cole (2013) explored the functional capacity of buildings to serve as teaching tools for sustainability. By analyzing human interactions with information and the physical environment, the author concluded that built structures have the ability to utilize design for EE. The proposed methods for teaching about green buildings can foster the innovations needed to meet SDG 9.
10: Reduced Inequalities	*Doris Duke Fellowship Programs* provides funding to support and mentor HEI students from diverse groups to aid their pursuit of environmental careers. Consistent with SDG 10, the programme aims to reduce inequalities in this context by ensuring that the environmental field becomes more representative of the USA's diverse population.	Garibay and Vincent (2018) analyzed the impact of diversity initiatives on the enrollment and retention of diverse students in HEI sustainability programmes. They found that programmes focused on environmental justice and community engagement with more enrolled diverse students were more likely to enroll and retain new diverse students. By exploring ways to increase the representation of diverse students in HEI sustainability programmes, this study contributes to SDG 10.
11: Sustainable Cities and Communities	The *International Council for Local Environmental Initiatives* (ICLEI USA) provides programming to reduce greenhouse gases. Through its Higher Education Affiliate Programme, ICLEI provides HEIs access to resources (such as guest lecturers, software) and assistance to create climate action plans. ICLEI offers internships for students to acquire skills for reducing GHG emissions and creating climate action plans. ICLEI helps make cities safe, resilient, and sustainable, which defines SDG 11.	Zint and Wolske (2014) conducted a literature review of educational programmes and initiatives for engaging urbanites in climate actions. They describe effective interventions designed based on social science research including information provision, social marketing, diffusion of innovation, asset-based community development, and participatory democracy. Their research identifies interventions that cities and communities can implement to bring about SDG 11 through sustainable behaviours and transformations.
12: Responsible Consumption and Production	The *Story of Stuff/Facing the Future* is a two-week curriculum that teaches high school students about production and consumption. Students learn about the impact of consumption on people, the economy, and environment. Becoming informed and empowered consumers allows students to visualize what sustainable production and consumption can look like, fulfilling SDG 12.	Armstrong et al. (2016) interviewed fashion-conscious HEI students to identify how various factors influence their consumption choices. They learned that individual, social, and cultural norms play into a student's desire to consume, such as envy of others' purchases or concern over fitting in. Understanding what factors influence students' choices is crucial to achieving the SDG 12 goal of sustainable consumption behaviours.

Table 1 contd. ...

...Table 1 contd.

SDG	Sample ESE Programme	Sample ESE Research
13: Climate Action	The Massive Open Online Course (MOOC) *Act on Climate: Steps to Individual, Community, and Political Action* was co-designed by a University of Michigan faculty member and HEI students. Within the context of food, energy, transportation, and the built environment, learners are encouraged to engage in climate mitigation and adaptation actions, and develop a personal climate action plan. Options include behaviours that individuals can engage in on their own, as well as in collaboration with their communities or politically. This MOOC equips learners with the knowledge and confidence to act on SDG 13 and take climate action.	Lawson et al. (2019) conducted survey research to investigate the impact of a climate change curriculum on the parents of middle-school students. They found that their intervention increased climate change concern among both students and their parents. Students (particularly girls) were able to convince their conservative fathers to change their perspectives on climate change. Findings demonstrate the effectiveness of intergenerational learning as a means to raise awareness and support for climate action, and therefore SDG 13.
14: Life Below Water	The *Northwest Indian College's Salish Sea Research Center* has a research lab and offers a marine education programme. Both promote understanding of the marine environment and focus on place-based learning, traditional ecological knowledge, and Western science. Their emphasis on education, sustainability, and the ocean engages with SGD 14 by supporting local youth and communities' learning about marine environments.	Kelly and Kast's (2020) case study analyzed interactive learning experiences between fifth graders and marine scientists. Based on their study, the authors created a Next Generation Science Standards-approved lesson plan for studying plankton. The findings illustrate how hands-on experiences with experts can help children learn about ocean ecosystems, essential to SDG 14.
15: Life on Land	*Monarch Butterfly Teacher and Student Resources* were developed by an American government agency in collaboration with a Mexican NGO. The resource manuals provide information and activities for PK–12 students about Monarch butterflies and their conservation. The manuals teach students about local and global ecosystem interactions and conservation strategies, increasing students' understanding of biodiversity, crucial for SDG 15.	Lewandowski and Oberhauser (2017) analyzed survey data to examine if educational citizen science projects about butterflies resulted in volunteers' involvement in conservation. Most volunteers increased their efforts, ranging from planting native plants to engaging in informal conservation conversations. This research illustrates that citizen science can contribute to SDG 15 by increasing adults' awareness of, and commitment to, sustainable habits and biodiversity.

16: Peace, Justice, and Strong Institutions	*#BlackBirdersWeek* is a week-long social media campaign that normalizes and celebrates the presence of non-white people in nature. It creates and promotes community-based and accessible SE through detailed, research-based social media posts. The promotion of inclusive and peaceful nature education is central to SDG 16, because it focuses SE on justice.	Andrzejewski (2005) contributed towards the creation of social justice, peace, and EE standards based on a review of relevant (inter)national and indigenous educational research. The standards can guide educators and curricula central to SDG 16 because they draw on a diversity of voices to evaluate peace, justice, and EE.
17: Partnerships for the Goals	*The US Partnership for Education for Sustainable Development* connects schools, businesses, and NGOs to advance SE. Examples of initiatives include recommendations for integrating sustainability into curricula, resources for promoting sustainability in HEIs, and networks between institutions to promote carbon neutrality. This programme establishes the multi-stakeholder partnerships needed for SDG 17, to share knowledge and resources for SE.	Daneri et al. in a 2015 case study, examined the Oberlin Project, a multi-stakeholder collaboration between Oberlin College, local NGOs, governments, and K-12 schools for sustainability. The study found that experiential learning through community-student partnerships is critical to SE, reflective of SDG 17.

are offered primarily by NGOs and HEIs and to a lesser extent government agencies, with partnerships/networks providing critical support.

The scholarly publications in Table 1 provide a glimpse into the ESE *research* conducted in the USA. This research is published in a range of journals focused on ESE as well as the respective (inter)disciplinary topics associated with the SDGs. The USA's ESE research tends to be empirical, drawing on quantitative and qualitative research methods, including case studies and syntheses. Findings from this significant body of scholarship provide insight into needed competencies, pedagogies, and assessments with implications for designing, implementing, and evaluating ESE, including for the SDGs (inter)nationally.

Recommendations for advancing ESE in the USA

Because UNESCO's (2020) *Education for Sustainable Development (ESD) for 2030 program* identifies five priority actions to ensure ESD can contribute to the SDGs by 2030, our initial recommendations for advancing ESE in the USA focus on these areas:

Advancing policy

President Biden, after re-joining the Paris Agreement, had Special Presidential Envoy for Climate, John Kerry, announce that the USA will contribute 50 million USD to the Global Climate Adaptation Fund (UNFCCC 2021). This marks the USA's first contribution to this fund and is evidence for recognition of the need to tackle the climate crisis. In contrast, there has been no indication that the USA will re-join UNESCO and thus, adopt ESD for 2030. This will unfortunately limit international ESE collaborations, through for example, the UNESCO Chair and University Twinning and Networking Programme (UNITWIN) and ESD-net. As such, other federal and state policies are urgently needed to inspire, support, and fund the scaling of ESE across the country. Although the *Every Student Succeeds Act* marks the first time a national education statute makes EE eligible for federal education funding, it does not require or even encourage states to do so (Itza 2017). Encouraging states to support their environmental literacy plans, increasing funding available through the National Environmental Education Act, and a Higher Education Sustainability Act, could be among the first steps toward advancing policy, along with other ESE mandates. HEI policies to advance ESE have the potential for particularly strong impact by preparing the next generation of sustainability leaders, through whole institution approaches including living labs, generating usable sustainability science research through boundary organizations, and disseminating innovations through re-purposed extension networks. While we have shown that some HEIs, companies, cities and states in the USA are explicitly targeting the SDGs, ESE policies framed in terms of resilience and health, a clean energy economy and workforce, and well-being (i.e., a better life for all) will likely be better received across the divided American public. To inform and enact new ESE policies, we encourage a cross-sector effort that engages practitioners from NGOs, government, corporations, youth, as well as (Indigenous) scholars, similar to Germany's process (Greiner 2020, Kopnina 2020).

Transforming learning environments

Despite the country's current lack of appetite to invest significantly in education, it may be possible to transform the USA's learning environments in response to growing concerns over public PK-12 schools as well as higher education. Such sources of dissatisfaction include not meeting workforce, citizenship, or sustainability needs, inequities in the quality of education diverse students receive, insufficient pay for teachers, and rising HEI tuition costs (Orr et al. 2020). One possible way to transform the country's learning environments is through reorienting PK-12 as well as accreditation standards and assessments (Kwauk and Casey 2021). Because K-12 education is mandatory in the United States, formal education provides a space to have a great deal of influence on education related to the SDGs, reaching a large audience who may not self-select into or have access to non-formal ESE programs. At the HEI level, changes can be facilitated by ranking systems that measure contributions to the SDGs (e.g., Times Higher Education Impact Rankings), ideally resulting in reorganization focused on identifying solutions for achieving SDG targets. All of the above could be influenced by greater grassroots calls for change to education systems, such as the UK's *Teach the Future*.

Building capacities of educators

Studies consistently suggest that the majority of K-12 and HEI educators in the USA agree that students should learn about sustainability, but few actually teach about the topic (e.g., Braus 2020, Michel 2020b). When they do, they tend to focus on the problems, typically from a natural science and engineering perspective, rather than solutions and associated (in)justices (Feinstein and Carlton 2013). ESE is still not prominent in teacher preparation programs in the United States (McKeown and USTESD 2013), partly because there are few relevant state sanctioned certifications (i.e., required accreditation to teach in K-12 schools) in ESE. Additionally, educators certified to teach in K-8 classrooms are often not confident in their capacities to teach science, including within the context of ESE. Nonetheless, as we have shown in this chapter many ESE resources as well as pre- and in-service opportunities exist for teachers in the USA, including for the purposes of contributing toward the achievement of the SDGs. Unfortunately, many educators are not aware of, or able to take advantage of them due to limited administrative, financial, and other support (Merritt et al. 2019, Brandt et al. 2019). Based on research conducted for this chapter, these professional development programs are more likely to focus on some SDG challenges more so than others (e.g., climate change versus poverty, income-inequality, and/or consumerism) and tend not to confront the underlying values and worldviews responsible for their existence. Current resources and professional development programs are therefore not sufficient in terms of providing educators with the content and pedagogical knowledge they will need to facilitate self and societal transformation. Communities of practice are one way to help educators overcome ESE challenges such as addressing controversial topics in their classes (Hong 2020). One such community of practice is The Global Council for Science and the Environment's Education for Energy Transformation, which, as its title suggests, strives to bring together educators to enable and empower them to bring about the necessary energy transition. Instituting communities of practice focused on ESE for the SDGs across the country's education system and institutionalizing ESE in teacher education programs at all levels are likely critical for instructors to facilitate the authentic, real-world, multi-actor, solution-oriented learning experiences necessary.

Empowering and mobilizing youth

Engagement by the country's youth and young adults has (e.g., First Earth Day in 1970) and will continue to be critical to bringing about transformation in the USA. Student activism, for example, has resulted in HEIs divesting from fossil fuels and committing to carbon neutrality (e.g., 350.org) and young adults' social media campaigns are in the process of changing powerful organizations like the Chamber of Commerce (https://www.changethechamber.org/). In the USA middle school girls have been particularly effective in raising conservative fathers' concern about climate change (Lawson et al. 2019). Providing youth with SDG leadership opportunities in their schools, as initiated in France, as well as engaging youth in action research, action civics, and civic science (e.g., Earth Force, Alliance for Climate Education, Schusler et al. 2018) can be effective in achieving the SDGs in local communities. When youth have opportunities to develop the content and procedural knowledge necessary to influence societal change (e.g., power analysis, coalition creation), they are often able to achieve policy changes (Kirshner 2015, Mirra et al. 2015). New technologies such as connectivist MOOCs could serve to capture and build such collective knowledge, and a climate corps (building on the Ameri- and Peace corps models) as proposed by the incoming administration (Biden 2020) could help to translate this knowledge into practice.

Accelerating local level actions

In addition to the recommendations for how ESE may contribute to achieving the SDGs in local USA communities, and consistent with SDG 17, we have no doubt that ESE partnerships will be critical to accelerating local impact. The latter can be achieved by adopting the collective impact framework (Kania and Kramer 2011, Raderstrong and Boyea-Robinson 2016). This parsimonious framework for effective partnerships suggests the need for a common agenda, shared measurement system, mutually reinforcing activities, continuous communication, and backbone support. Collaborations for collective impact have brought about transformative local education, health, environmental, and other outcomes (ORS Impact and Spark Policy Institute 2018). PK-12 schools and HEI can serve as learning hubs, facilitating positive future sustainability visions, as well as convene NGOs, community and religious groups, sovereign Indigenous and other governments, and businesses toward achieving them (Braus 2020, Weymouth et al. 2020). In light of the decline in civic education as well as general political disillusionment, such partnerships should pay particular attention to motivating and supporting political leadership and participation (Levy and Zint 2013). We acknowledge that this kind of work will require the creation of innovative power structures to cultivate large-scale systemic change (Ryan 2014).

We now turn to additional recommendation advancing ESE for the SDGs in the USA and beyond.

Infusing diversity, equity, and inclusion

It is critical that diversity, equity, and inclusion (DEI) considerations be infused throughout the above priority areas, if ESE is to achieve the SDGs. Although ESE programs in the USA have been striving to serve a greater number of diverse audiences based on, for example, race and socioeconomic status, to provide more equal access to ESE, and to be more welcoming of, for example, cultural differences (e.g., NAAEE' equitable and inclusive EE initiative, unpublished data), this goal has not yet been achieved (Stapleton 2020). Indeed, if the ESE programs and scholarship described in this chapter were examined from a DEI lens, we suspect that they would reveal primarily dominant WEIRD perspectives. While there are many reasons for the lack of DEI in ESE in the USA, a key one is that the majority of staff and board members of ESE organizations have been white (Taylor 2018), which must change in order for ESE to truly be more diverse, equitable and inclusive. Fellowship programs like those funded by the Doris Duke Foundation are seeking to change this landscape but more funding and support for ESE DEI initiatives is needed. Embracing practices of place-based and culturally responsive pedagogies, such as incorporating the diversity of students' cultural and community knowledge (Ladson-Billings 2009) are additional ways that ESE can center DEI in SGD priority areas.

Within this context, it must also be noted that there have been particularly few attempts by environment and sustainability educators in the USA to collaborate with, support or center the experiences and voices of Indigenous peoples. Indigenous people in the USA face especially difficult challenges due to the country's colonialist history and have much to offer ESE, including in terms of their traditional knowledge, skills, and values (e.g., College of Menominee Nation Sustainable Development Institute, unpublished data, Bang et al. 2014, Calderon 2014, Giegerich 2021). Efforts to decolonize education (Shahjahan et al. 2021), like at the authors' HEI, are needed to ensure that colonial influences and inequalities do not continue to be reproduced and that marginalized peoples, their knowledge, and ways of being are re-centered.

Addressing injustices and mental health

In addition to infusing DEI and decolonizing education it is critical to center justice in ESE because sustainability cannot be achieved without equity. The limited focus on environmental and social justice in ESE in the USA to date is particularly troubling because associated inequities disproportionately affect the country's most diverse communities (Adams et al. 2019, Nolet 2017). Fortunately, there is a growing awareness of the relationships between environmental and human health—and—racial/ethnic and class inequities, particularly among the country's youth (Oglesby 2021, Quiroz-Martinez et al. 2005). This awareness has also reached the highest level of governance. The Biden Administration is currently working to make justice a central focus in the country's efforts toward a more sustainable future with guidance from The White House Environmental Justice Advisory Council (EPA 2021). The first of its kind, this Council will provide insight on current and historic environmental injustices and offer a range of recommendations to ensure clean energy (SDG 7) and climate action (SDG 13), among others, are achieved in just ways.

As just alluded to, one of the main drivers underlying the Biden Administrations' efforts is the need for climate justice. This is because environmental hazards like climate change pose the greatest physical and mental health threats to the vulnerable communities and populations like children (Chalupka et al. 2020). While Albrecht et al. 2005 coined 'solastalgia' to describe the emotional distress experienced by the degradation of one's home environment, eco-grief (Cunsolo and Ellis 2018) and eco-anxiety (Clayton 2020) are the now more commonly recognized terms for the emotional distress caused by climate change. Fortunately, there are a growing number of opportunities for environmental and sustainability educators to learn how to help adults and youth cope (e.g., Ray 2020). Without positive, constructive coping mechanisms, it is unlikely we will be able to face the threats associated with the SDGs, let alone achieve them. We therefore encourage practitioners and scholars in the USA and beyond to focus on this affective dimension of ESE, especially within the context of environmental and climate injustices.

Learning from other countries

Finally, we urge the USA to learn from other countries who are more invested in the SDGs. One way to learn from other countries is through international networks and partnerships. While there are many examples, one is the Sustainable Development Solutions Network (SDSN) a "global network of knowledge-generating institutions [including educational institutions] advancing the SDGs," which situates work in the USA into the global context, and allows for informal learning from others globally (SDSN, unpublished data). Another example, the Global Environmental Education Partnership (GEEP) is a learning network that promotes shared learning around the SDGs through such as case studies and other resources (GEEP, unpublished data).

Conclusion

The USA has expertise and resources to bring about positive societal transformations to achieve the SDGs through ESE. To what extent the USA will do so will depend on many factors such as the political leadership's ability to reunite this deeply divided country into one willing to invest in transformative education to achieve well-being for all. Such education needs to develop leaders and citizens empowered to revolutionize the country's approach to the environment, economy, and equity. Given the powerful interests that benefit from the USA's divisions and their influence on the country's politics, as well as their faith in unfettered capitalism and consumer culture, there are many reasons to be pessimistic. At the same time, we remain optimistic about the possibility of a sustainable future. For one, the country's youth and young adults are highly motivated to bring about alternative ways for protecting the planet, organizing economic and social life, and overcoming racial and income inequalities. We are also encouraged by the thousands of USA educators and scholars dedicated to ESE, including their growing focus on the SDGs, along with the increasing number of K-12 schools and HEIs modeling the way to sustainability. It is our sincere hope that the majority of K-12 schools and HEIs will embrace their community and societal roles as conveners, facilitators, and partners to achieve the SDGs. The education system, in great part through its youth and young adults, has brought about revolutions in the past. The recommendations we propose, building on the ESE transformations already underway, have great potential for ushering in a peaceful sustainability revolution in the USA and beyond.

Acknowledgements

We acknowledge the contributions of our reviewers, Peggy Barlett, Julian Dautremont, and Melissa Hopkins Taggart, who provided constructive insight as we developed this chapter. We thank them for assisting in strengthening our chapter.

References

Adams, M., S. Klinsky and N. Chhetri. 2019. Barriers to sustainability in poor marginalized communities in the United States: The criminal justice, the prison-industrial complex and foster care systems. Sustain 12(1): 220.

Albrecht, G. 2005. 'Solastalgia'. A new concept in health and identity. PAN: Philosophy Activism Nature 3: 41–55.

Andrzejewski, J. 2005. The social justice, peace, and environmental education standards project. Multicult. Perspect. 7(1): 8–16.

Armstrong, C.M.J., K.Y.H. Connell, C. Lang, M. Ruppert-Stroescu and M.L. LeHew. 2016. Educating for sustainable fashion: Using clothing acquisition abstinence to explore sustainable consumption and life beyond growth. J. Consum. Policy 39(4): 417–439.

Bang, M., L. Curley, A. Kessel, A. Marin, E.S. Suzukovich and G. Strack. 2014. Muskrat theories, tobacco in the streets, and living Chicago as Indigenous land. Environ. Educ. Res. 20(1): 37–55.

Barlett, P.F. and G.W. Chase (eds.). 2013. Sustainability in Higher Education: Stories and Strategies for Transformation. Urban and Industrial Environments. The MIT Press, Cambridge, MA, USA.

Bartlett, P.W., M. Popov and J. Ruppert. 2020. Integrating core sustainability meta-competencies and SDGs across the silos in curriculum and professional development. pp. 71–85. *In*: Nhamo, G. and V. Mjimba (eds.). Sustainable Development Goals and Institutions of Higher Education. Springer International Publishing, Switzerland, AG.

Bengtsson, S.E.L., B. Barakat and R. Muttarak. 2018. The Role of Education in Enabling the Sustainable Development Agenda. Routledge, London.

Better World Campaign. 2020. Nationwide Poll on U.S.-UN Relations. Retrieved Dec. 28, 2020, from https://betterworldcampaign.org/us-un-partnership/public-opinion-polling/us-un-poll/.

Biden, J. 2020. The Biden Plan for a Clean Energy Revolution and Environmental Justice. Joe Biden for President: Official Campaign Website. Retrieved Dec. 28, 2020, from https://joebiden.com/climate-plan/.

Bowman, T. and D. Morrison. 2020. An ACE National Strategic Planning Framework for the United States [Online]. Created in collaborative reflection with the U.S. ACE Community. Retrieved, May 31, 2022, from http://aceframework.us.

Brandt, J.O., L. Bürgener, M. Barth and A. Redman. 2019. Becoming a competent teacher in education for sustainable development: Learning outcomes and processes in teacher education. Int. J. Sustain. High. Educ. 20(4): 630–653.

Braus, J. 2020. Civic and environmental education: protecting the planet and our democracy. pp. 183–195. *In*: Orr, D.W., A. Gumbel, B. Kitwana and W.S. Becker (eds.). Democracy Unchained: How to Rebuild Government for the People. The New Press, New York, NY, USA.

Brookings-UN Foundation. 2020, Sept. 16. American Leadership in Advancing the Sustainable Development Goals. Retrieved Dec. 28, 2020, from https://www.brookings.edu/events/american-leadership-in-advancing-the-sustainable-development-goals-2/.

Brundiers, K., M. Barth, G. Cebrián, M. Cohen, L. Diaz, S. Doucette-Remington et al. 2021. Key competencies in sustainability in higher education—toward an agreed-upon reference framework. Sustain Sci. 16: 13–29.

Bull, C.C. and E.R.W. Hat. 2019. Cangleska Wakan: The ecology of the sacred circle and the role of tribal colleges and universities. Int. Rev. Educ. 65(1): 117–141.

Calderon, D. 2014. Speaking back to Manifest Destinies: a land education-based approach to critical curriculum inquiry. Environ. Educ. Res. 20(1): 24–36.

Carman, J. 2020. Understanding and Expanding the Role of Personal and Household Behavior in Climate Change Adaptation. Ph.D. Dissertation, University of Michigan, Ann Arbor, MI, U.S.A.

Centers for Disease Control and Prevention (CDC). 2020. Whole School, Whole Community, Whole Child (WSCC). Retrieved Feb. 10, 2021, from https://www.cdc.gov/healthyschools/wscc/index.htm.

Chalupka, S., L. Anderko and E. Pennea. 2020. Climate change, climate justice, and children's mental health: A generation at risk? Environ Justice 13(1): 10–14.

Chang, H.C., R.M. Kelly and E.P. Metzger. 2016. A qualitative study of teachers' understanding of sustainability: Education for Sustainable Development (ESD), dimensions of sustainability, environmental protection. pp. 206–234. *In*: Urban, M.J. and D.A. Falvo (eds.). Improving K-12 STEM Education Outcomes through Technological Integration. IGI Global, Hershey, PA, USA.

Church, W and L. Skelton. 2013. Infusing sustainability across the curriculum. pp. 183–195. *In*: McKeown, R. and V. Nolet (eds.). Schooling for Sustainable Development in Canada and the United States. Schooling for Sustainable Development, vol 4. Springer, Dordrecht, Netherlands.

Clayton, S. 2020. Climate anxiety: Psychological responses to climate change. J. Anxiety Disord. 74: 102263.

Climate Action Tracker. 2020. Climate Action Tracker Warming Projections Global Update. Retrieved Dec. 28, 2020, from https://climateactiontracker.org/countries/usa/.

Cole, L.B. 2013. The Teaching Green School Building: a framework for linking architecture and environmental education. Environ. Educ. Res. 20(6): 836–857.

College of Menominee Nation. unpublished data. Sustainable Development Institute. Retrieved Dec. 28, 2020, from http://sustainabledevelopmentinstitute.org/.

Consolo, A. and N.R. Ellis. 2018. Ecological grief as a mental health response to climate change-related loss. Nat. Clim. Chang. 8(4): 275–281.

Daneri, D.R., G. Trencher and J. Petersen. 2015. Students as change agents in a town-wide sustainability transformation: The Oberlin Project at Oberlin College. Curr. Opin. Environ. Sustain 16: 14–21.

Department of Agriculture Economic Research Service (USDA ERS). 2020, Oct. 1. National School Lunch Program. Retrieved Dec. 28, 2020, from https://www.ers.usda.gov/topics/food-nutrition-assistance/child-nutrition-programs/national-school-lunch-program/.

Desha, C., D. Rowe and D. Hargreaves. 2019. A review of progress and opportunities to foster development of sustainability-related competencies in engineering education, Australas. J. Eng. Educ. 24(2): 61–73.

Dockry, M.J., K. Hall, W. Van Lopik and C.M. Caldwell. 2016. Sustainable development education, practice, and research: An indigenous model of sustainable development at the College of Menominee Nation, Keshena, WI, USA. Sustain Sci. 11(1): 127–138.

Education Data Initiative. Unpublished data. College enrollment statistics [2022]: Total + by demographic. Retrieved May 25, 2022, from https://educationdata.org/college-enrollment-statistics.

Ennes, M., D.F. Lawson, K.T. Stevenson, M.N. Peterson and M.G. Jones. 2021. It's about time: perceived barriers to in-service teacher climate change professional development. Environ. Educ. Res. 27(5): 762–778.

Environmental Protection Agency (EPA). 2021. White House Environmental Justice Advisory Council. Retrieved May 31, 2022, from https://www.epa.gov/environmentaljustice/white-house-environmental-justice-advisory-council.

Environmental Protection Agency (EPA). unpublished data. What is Environmental Education? Retrieved May 25, 2022, from https://www.epa.gov/education/what-environmental-education#:~:text=Environmental%20education%20is%20a%20process,make%20informed%20 and%20responsible%20decisions.

Feinstein N.W. and G. Carlton. 2013. Education for sustainability in the K-12 educational system of the United States. *In*: McKeown, R. and V. Nolet (eds.). Schooling for Sustainable Development in Canada and the United States. Schooling for Sustainable Development, vol 4. Springer, Dordrecht, Netherlands.

Frandy, T. 2018. Indigenizing sustainabilities, sustaining indigeneities: decolonization, sustainability, and education. J. Sustain Educ. 18.

Garibay, J.C. and S. Vincent. 2018. Racially inclusive climates within degree programs and increasing student of color enrollment: An examination of environmental/sustainability programs. J. Divers High Educ. 11(2): 201–220.

Giegerich, S. 2021. TC to Partner with Columbia on Climate Change Education Initiatives. TC Columbia University, New York, NY, USA. Retrieved Nov. 17, 2021, from https://www.tc.columbia.edu/ articles/2021/october/tc-to-partner-with-columbia-on-climate-change-education-initiatives/.

Global Environmental Education Partnership (GEEP). unpublished data. Case Studies. Retrieved May 31, 2022, from https://thegeep.org/learn/case-studies.

Global Programs and Strategy Alliance: University of Minnesota. 2020, Sept. 3. UMN Sustainable Development Goals Initiative. Retrieved Dec. 28, 2020, from https://global.umn.edu/.

Greiner, T. 2020, Dec., 1. Regional online launch of the ESD for 2030 roadmap—Europe and North America—YouTube. Retrieved Dec. 28, 2020, from https://www.youtube.com/ watch?v=bJmdPXplvCQ.

Hess, D.J. and A. Maki. 2019. Climate change belief, sustainability education, and political values: Assessing the need for higher-education curriculum reform. J. Clean. Prod. 228: 1157–1166.

Higgs, A.L. and V.M. McMillan. 2006. Teaching through modeling: four schools' experiences in sustainability education. J. Environ. Educ. 38(1): 39–53.

Hoffman, A.J. 2015. How Culture Shapes the Climate Change Debate. Stanford, California: Stanford Briefs, an Imprint of Stanford University Press, Standford, CA, USA.

Hong, W. 2020. Build it and they will come: the faculty learning community approach to infusing the curriculum with sustainability content. pp. 49–60. *In*: Nhamo, G. and V. Mjimba (eds.). Sustainable Development Goals and Institutions of Higher Education. Sustainable Development Goals Series. Springer, Cham, Switzerland, AG.

Hopkins, C. 2013. National Institute for Educational Policy Research (NIERP). Research Bulletin No. 142. pp. 25–35. National Institute for Educational Policy Research, Japan.

Hungerford, H.R. and T.L. Volk. 1990. Changing learner behavior through environmental education. J. Environ. Educ. 21(3): 8–21.

Hunter, J.D. 1991. Culture Wars: The Struggle to Define America. Basic Books, New York, NY, USA.

Hurd, E. and A.A. Ormsby. 2020. Supporting K-12 teachers in the context of whole-school sustainability: four case studies. Appl. Environ. Educ. Commun. 20(4): 303–318.

Itza, F. 2017. Environmental education in the Every Student Succeeds Act and the role of advocates. Geo. Envt. L. Rev. 29(2): 417–434.

Kania, J. and M. Kramer. 2011. Collective Impact. Stanford Soc. Innov. Rev. 9(1): 36.

Kelly, K.J. and D.J. Kast. 2020. Planktonic Relationships! Authentic STEM connections through partnership between university of southern california's marine researchers and Los Angeles unified elementary school students. Current: The Journal of Marine Education 34(2): 12.

Kirk, M., R. Wilke and A. Ruskey. 1997. A survey of the status of state-level environmental education in the United States. J. Environ. Educ. 29(1): 9–16.

Kirshner, B. 2015. Youth Activism in an Era of Education Inequality. NYU Press, New York, NY, USA.

Kopnina, H. 2020. Education for the future? Critical evaluation of education for sustainable development goals. J. Environ. Educ. 51(4): 280–291.

Kwauk, C. and O. Casey. 2021. A New Green Learning Agenda: Approaches to Quality Education for Climate Action. Center for Universal Education at The Brookings Institution, Washington, D.C., USA.

Kweon, B.-S., P. Mohai, S. Lee and A.M. Sametshaw. 2016. Proximity of public schools to major highways and industrial facilities, and students' school performance and health hazards. Environment and Planning B: Urban Analytics and City Science 45(2): 312–329.

Ladson-Billings, G. 2009. The Dreamkeepers: Successful Teachers of African American Children. Jossey-Bass Publishers, Inc., San Francisco, CA, USA.

Lawson, D.F., K.T. Stevenson, M.N. Peterson, S.J. Carrier, R.L. Strnad and E. Seekamp. 2019. Children can foster climate change concern among their parents. Nat. Clim. Chang. 9(6): 458–462. Springer Science and Business Media LLC, London.

Leiserowitz, A., E. Maibach, S. Rosenthal, J. Kotcher, J. Carman, X. Wanget et al. 2020. Politics and Global Warming, December 2020. Yale University and George Mason University, New Haven, CT, USA: Yale Program on Climate Change Communication.

Levy, B.L.M. and M.T. Zint. 2013. Toward fostering environmental political participation: framing an agenda for environmental education research. Environ. Educ. Res. 19(5): 553–576.

Lewandowski, E.J. and K.S. Oberhauser. 2017. Butterfly citizen scientists in the United States increase their engagement in conservation. Biol. Conserv. 208: 106–112.

Madson, D. 2021, Jan. 8. Many states get poor grades on their climate education standards. Yale Climate Connections. Retrieved Jan. 13, 2021, from https://yaleclimateconnections.org/2021/01/many-states-get-poor-grades-on-their-climate-education-standards/.

McKeown, R., C. Hopkins, R. Rizzi and M. Chrystalbridge. 2002. Education for Sustainable Development Toolkit, Version 2.0. Retrieved Dec. 28, 2020, from http://www.esdtoolkit.org.

McKeown, R. and V. Nolet. (eds.). 2013. Schooling for Sustainable Development in Canada and the United States. Schooling for Sustainable Development, vol. 4. Springer, Dordrecht, Netherlands.

McKeown, R., with USTESD Network. 2013. Reorienting teacher education to address sustainability: The U.S. context, White Paper Series, No. 1. Indianapolis, IN, USA: United States Teacher Education for Sustainable Development Network.

Merritt, E., A. Hale and L. Archambault. 2019. Changes in pre-service teachers' values, sense of agency, motivation and consumption practices: A case study of an education for sustainability course. Sustain 11(1): 155.

Michel, J.O. 2019. Sustainability in higher education. pp. 369–376. In: Filho, W.L. (ed.). Encyclopedia of Sustainability in Higher Education. Springer, Charm, Switzerland, AG.

Michel, J.O. 2020a. Charting students' exposure to promising practices of teaching about sustainability across the higher education curriculum. Teach. High. Educ. 1–27.

Michel, J.O. 2020b. Mapping out students' opportunity to learn about sustainability across the higher education curriculum. Innov. High Educ. 45(5): 355–371.

Michel, J.O. 2020c. Toward conceptualizing education for sustainability in higher education. New Dir. Teach. Learn. 161: 23–33.

Mirra, N., A. Garcia and E. Morrell. 2015. Doing Youth Participatory Action Research: Transforming Inquiry with Researchers, Educators, and Students. Routledge, London.

Mohai, P., B.-S. Kweon, S. Lee and K. Ard. 2011. Air pollution around schools is linked to poorer student health and academic performance. Health Affairs (Project Hope) 30(5): 852–862.

National Academies of Sciences, Engineering, and Medicine (NASEM). 2020. Developing a sustainability workforce. pp. 156–173. In: Strengthening Sustainability Programs and Curricula at the Undergraduate and Graduate Levels. The National Academies Press, Washington D.C., USA.

National Aeronautics and Space Administration (NASA). 2020. The Effects of Climate Change 2020. Retrieved Dec. 28, 2020, from https://climate.nasa.gov/effects/.

National Center for Education Statistics (NCES). 2019. The NCES Fast Facts Tool Tuition costs of colleges and universities. Digest of Education Statistics, 2018 (NCES 2020-009), Table 330.10. Retrieved Dec. 28, 2020, from https://nces.ed.gov/fastfacts/display.asp?id=76.

National Center for Education Statistics (NCES). 2020. The NCES Fast Facts Tool. Back to school statistics. Retrieved Dec. 28, 2020, from https://nces.ed.gov/fastfacts/display.asp?id=372.

National Council for Science and the Environment (NCSE). 2003. Education for a Sustainable and Secure Future: A report to the third National conference science, policy, and the environment. NCSE, Washington, D.C., USA. Retrieved from, https://www.scribd.com/document/34813572/National-Council-for-Science-and-the-Environment-2003.

Nolet, V. 2017. Quality education: cultural competence and a sustainability worldview. Kappa Delta Pi Rec 53(4): 162–167.

North American Association for Environmental Education (NAAEE). unpublished data. Equitable and Inclusive EE. Retrived May 29, 2022, from, https://naaee.org/eepro/learning/eelearn/equity-inclusion.

Oglesby, C. 2021. "The Generational Rift over 'Intersectional Environmentalism.'" Grist. Retrieved Feb. 10, 2021 from, https://grist.org/justice/intersectional-environmentalism-justice-language/.

Oreskes, N. and E.M. Conway. 2010. Defeating the merchants of doubt. Nature 465(7299): 686–687.

Organization for Economic Co-operation and Development. 2020. Population (Indicator). Retrieved Jan. 4, 2021, from, https://data-oecd-org.proxy.lib.umich.edu/united-states.htm.

Organisation for Economic Co-operation and Development (OECD). 2021. Science performance (PISA) (indicator), Retrieved Jan. 4, 2021, from https://doi.org/10.1787/91952204-en.

Orr, D.W. 2020. Democracy and the (missing) politics in environmental education. J. Environ. Educ. 51(4): 270–279.

Orr, D., A. Gumbel, B. Kitwana and W. Becker. 2020. Democracy Unchained: How to Rebuild Government for the People. La Vergne: The New Press, New York, NY, USA.

ORS Impact and Spark Policy Institute. 2018. When Collective Impact Has an Impact: A Cross-Site Study of 25 Collective Impact Initiatives. Seattle, WA, USA. https://www.orsimpact.com/directory/ci-study-report.htm.

Papenfuss, J., E. Merritt, D. Manuel-Navarrete, S. Cloutier and B. Eckard. 2019. Interacting pedagogies: A review and framework for sustainability education. J. Sustain Educ. 20: 19+.

Pew Research Center. U.S. Politics and Policy. 2017, Oct. 5. The partisan divide on political values grows even wider. Retrieved Dec 28, 2020, from https://www.pewresearch.org/politics/2017/10/05/the-partisan-divide-on-political-values-grows-even-wider.

Philipsborn, R.P., P. Sheffield, A. White, A. Osta, M.S. Anderson and A. Bernstein. 2020. Climate change and the practice of medicine: Essentials for resident education. Acad. Med. 96(3): 355–367.

Pierotti, R. and D. Wildcat. 2000. Traditional ecological knowledge: The third alternative (Commentary). Ecol. Appl. 10(5): 1333–1340.

Plans, A. 2015. The every student succeeds act: Explained. Education Week 35(14): 16–17.

President's Council on Sustainable Development. 1996. Education for Sustainability: An Agenda for Action. President's Council on Sustainable Development, Washington D.C., USA.

Princeton Review. 2022. 2022 College Hopes & Worries Survey Report. Retrieved Mar. 14, 2022 from https://www.princetonreview.com/college-rankings/college-hopes-worries.

Quiroz-Martinez, J., D. Pei Wu and K. Zimmerman. 2005. ReGeneration: Young People Shaping Environmental Justice. Movement Strategy Center, commissioned by the Ford Foundation, Oakland, CA, USA. https://movementstrategy.org/b/wp-content/uploads/2015/08/MSC-ReGeneration_Young_People_Shaping_EJ_Movement.pdf.

Raderstrong, J. and T. Boyea-Robinson. 2016. The why and how of working with communities through collective impact. Community Dev J. 47(2) 181–193.

Ray, S.J. 2020. A Field Guide to Climate Anxiety: How to Keep your Cool on a Warming Planet. University of California Press, California.

Reames, T.G., M.A. Reiner and M.B. Stacey. 2018. An incandescent truth: Disparities in energy-efficient lighting availability and prices in an urban U.S. county. Appl. Energy 218: 95–103.

Regulatory Rollback Tracker. 2020, Oct. 20. Environmental and Energy Law Program. Retrieved Dec. 28, 2020, from https://eelp.law.harvard.edu/regulatory-rollback-tracker/.

Rowe, D., S.J. Gentile and L. Clevey. 2015. The U.S. partnership for education for sustainable development: progress and challenges ahead. Appl. Environ. Educ. Commun. 14(2): 112–120.

Ryan, M.J. 2014. Power Dynamics in Collective Impact. Standford Social Innovation Rev.

Sachs, J., G. Schmidt-Traub, C. Kroll, G. Lafortune and G. Fuller. 2019. Sustainable Development Report 2019. New York: Bertelsmann Stiftung and Sustainable Development Solutions Network (SDSN). Cambridge University Press, Cambridge.

Santone, S. 2013. Ecological economics education. pp. 153–167. *In*: McKeown, R. and V. Nolet (eds.). Schooling for Sustainable Development in Canada and the United States. Schooling for Sustainable Development, vol 4. Springer, Dordrecht, Netherlands.

Schusler, T.M., J. Davis-Manigaulte and A. Cutter-Mackenzie. 2018. Positive youth development. pp. 165–174. *In*: Russ, A. and M.E. Kransy (eds.). Urban Environmental Education Review. Comstock Publishing Associates, Cornell University Press, Ithaca, NY, USA and London.

Sustainable Development Solutions Network (SDSN). unpublished data. Join the SDSN. Retrieved May 31, 2022 from, https://www.unsdsn.org/join-the-sdsn?gclid=Cj0KCQjw-daUBhCIARIsALbkjSYJ1uG9jw-w7ZF0z82nLsckYeOZHYuwSXRluUO7b1F4oSW-pUG7C20aAvhiEALw_wcB.

Shahjahan, R., A.L. Estera, K.L. Surla and K.T. Edwards. 2021. "Decolonizing" Curriculum and Pedagogy: A comparative review across disciplines and global higher education contexts. Rev. Educ. Res. 92(1): 73–113.

Simmons, B., P. Hart and H. Hungerford. 2001. A tribute to William B. Stapp: 1930–2001. J. Environ. Educ. 32(4): 4.

Smith, K., P. Adriance, A. Mueller, R. McKeown, V. Nolet, D. Rowe et al. 2015. The Status of Education for Sustainable Development (ESD) in the United States: A 2015 Report to the U.S. Department of State, Washington D.C., USA.

Soper, L. 2020, Sept. 11. Environmental Education Goes Virtual: Creating Meaningful Learning Opportunities at Home. Retrieved Dec. 28, 2020, from https://blog.nwf.org/2020/09/environmental-education-goes-virtual-creating-meaningful-learning-opportunities-at-home/.

Stapleton, S.R. 2020. Toward critical environmental education: a standpoint analysis of race in the American environmental context. Environ. Educ. Res. 26(2): 155–170.

Stapp, W.B. 1969. The concept of environmental education. Environmental Education 1(1): 30–31.

Sustainability Initiative. 2020. 2020 Voluntary University Review of the Sustainable Development Goals. Carnegie Mellon University. Retrieved Dec. 28, 2020, from https://www.cmu.edu/leadership/the-provost/provost-initiatives/cmu-vur-2020.

Taylor, D. 2018. Diversity in Environmental Organizations: Reporting and Transparency, Report 1. School for Environment and Sustainability, University of Michigan, Ann Arbor, MI, USA. Retrieved May 31, 2022, from https://www.researchgate.net/publication/322698951_Diversity_in_Environmental_Organizations_Reporting_and_Transparency.

The Education Trust. 2018. Funding Gaps. An Analysis of School funding equity across the U.S. and within each State. Retrieved Dec. 28, 2020, from https://edtrust.org/resource/funding-gaps-2018/.

The United Nations Framework Convention on Climate Change (UNFCCC). 2021, Nov. 9. Adaptation Fund Raises Record US$ 356 Million in New Pledges at COP26 for its Concrete Actions to Most Vulnerable. Retrieved Mar. 14, 2022, from https://unfccc.int/news/adaptation-fund-raises-record-us-356-million-in-new-pledges-at-cop26-for-its-concrete-actions-to.

The United States Government. 2021, Dec. 3. The build back better framework. The White House. Retrieved Mar. 14, 2022, from https://www.whitehouse.gov/build-back-better/.

The White House, Office of the Press Secretary. 2015, Sept. 27. Remarks by the President on Sustainable Development Goals. [Press release]. Retrieved from Dec. 28, 2020, from https://obamawhitehouse.archives.gov/the-press-office/2015/09/27/remarks-president-sustainable-development-goals.

The White House, The Briefing Room. 2021, Jan. 20. Paris Climate Agreement. Retrieved Mar. 14, 2022, from https://www.whitehouse.gov/briefing-room/statements-releases/2021/01/20/paris-climate-agreement/.

Thébaud, S. and M. Charles. 2018. Segregation, Stereotypes, and STEM. Social Sciences 7(7): 111.

Trott, C.D. 2019. Reshaping our world: Collaborating with children for community-based climate change action. Action Research (Lond) 17(1): 42–62.

Twill, S.E., J. Bergdahl and R. Fensler. 2016. Partnering to build a pantry: a university campus responds to student food insecurity. Journal of Poverty 20(3): 340–358.

U.S. National Statistics for the U.N. Sustainable Development Goals. unpublished data. Retrieved Dec. 28, 2020, from https://sdg.data.gov/.

UNESCO. 1977. Intergovernmental Conference on Environmental Education: organized by UNESCO in co-operation with UNEP (Tbilsi, USSR, 14–26 October 1977) Final Report. unesdoc.unesco.org. Retrieved May 25, 2022, from https://unesdoc.unesco.org/ark:/48223/pf000003276.3.

UNESCO. Rieckmann, M. (ed.). 2017. Education for Sustainable Development Goals: Learning Objectives. UNESCO Publishing, Belgium.

UNESCO. 2020. Education for Sustainable Development A Roadmap. unesdoc.unesco.org. Retrieved Dec. 28, 2020, from https://unesdoc.unesco.org/ark:/48223/pf0000374802.

Van Lancker, W. and Z. Parolin. 2020. COVID-19, School Closures, and Child Poverty: A Social Crisis in the Making. Lancet Public Health 5(5): e243–e244.

Venkatesan, M. 2020. Teaching introductory economics to promote sustainability. New Dir. Teach. Learn. 161: 53–71.

Vincent, S., S. Rao, Q. Fu, K. Gu, X. Huang, K. Lindaman et al. 2017. Scope of Interdisciplinary Environmental, Sustainability, and Energy Baccalaureate and Graduate Education in the United States. National Council for Science and the Environment, Washington D.C., USA.

We Are Still In. 2020. Retrieved December 28, 2020, from https://www.wearestillin.com/.

Weymouth, R., J. Hartz-Karp and D. Marinova. 2020. Repairing political trust for practical sustainability. Sustainability 12: 7055.

Zint, M. and K.S. Wolske. 2014. From information provision to participatory deliberation: engaging residents in the transition toward sustainable cities. pp. 188–209. *In*: Mazmanian, D.A. and H. Blanco (eds.). Elgar Companion to Sustainable Cities. Edward Elgar Publishing, Inc., Northampton, MA, USA.

Zint, M., A. Kraemer and G. Kolenic. 2014. Evaluating Meaningful Watershed Educational Experiences: An exploration into the effects on participating students' environmental stewardship characteristics and the relationships between these predictors of environmentally responsible behavior. Stud. Educ. Evaluation 41: 4–17.

Oceania

Chapter 17

Environmental and Sustainability Education in Australia

Annette Gough,[1,]* *Alan Reid*[2] and *Robert B. Stevenson*[3]

Introduction

The field of environmental education in Australia has a substantial and complex history. Its roots can be traced to the work of field naturalists and conservation education in the early parts of the twentieth century when concerns about the impact of humans on the environment were first being raised in the media. The environmental education movement grew in the 1970s apace with international developments, typically aligning with those initiatives associated with UN-level interests, conferences and programmes seeking to link education, environment, and later sustainable development, in formal, informal and non-formal education settings. However, the field's developmental progress and impacts in Australia can be likened to a game of snakes and ladders (Greenall 1987) with leaps forward at some times, and falls backward at others. This chapter illustrates why that is still the case, through a focus on environmental and sustainability education in formal education.

About Australia's political, economic, cultural and social contexts and conditions, including its education system

Australia is a federation of six states and two territories. First Nations peoples have lived in what is now called Australia for over 60,000 years. It was occupied

[1] School of Education, RMIT University, Australia.
[2] Faculty of Education, Monash University, Australia.
[3] The Cairns Institute, James Cook University, Australia; School of Education, University of Queensland, Australia.
Emails: alan.reid@monash.edu; bob.stevenson@jcu.edu.au
* Corresponding author: annette.gough@rmit.edu.au

by the British in January 1788 as a colony and has been a constitutional monarchy with a Westminster system of government since 1 January 1901. Further waves of substantial settlement and urbanization primarily to create a 'White Australia' have been replaced by multiculturalism and closer attention to Australia's contested and evolving place in the world (especially in relation to Asia) since the mid-late 20th century.

Another key feature of the context for environmental and sustainability education[1] in Australia is its population growth and changing composition. The official tally of citizens and permanent residents has increased from 4 million in 1900 to 7.5 million in 1945 and to over 25 million in 2019. Over 85% of the population lives on 0.22% of the land area, typically in large metropolitan areas associated with a capital city and within 75 km of the coast. Housing and industrial development are increasingly marked by a proliferation of sprawling suburbs and changes in land use. This growth has mainly come from migration to and within the states and territories, such that Australia is now a very multicultural country with diverse and competing expectations and experiences of people-planet relationships. These include those that, in part, reflect the various waves of migration post disasters and people seeking a 'better life' in an industrial and post-industrial setting, as well as indigenous and colonizer cosmologies. According to the Australian Human Rights Commission (n.d.), 'We are home to the world's oldest continuous cultures, as well as Australians who identify with more than 270 ancestries.'

Each wave of colonization and migration has had an impact on Australia's unique land-based, coastal and marine environments. Because Australia is an island continent, it has a large number of animal and plant species, and many fragile ecosystems, with one of the world's highest rates of species extinction (Australian Conservation Foundation 2018). For example, according to the University of New South Wales' Centre for Ecosystem Science, 'more than 10% of endemic terrestrial land mammal species have become extinct over the last 200 years, which represents 50% of the global mammal extinctions during that period' (The Senate 2019, 19).

According to the World Bank, in the early 2010s, the gross domestic product (GDP) per person reached a peak of US$ 68,000, and has dropped steadily towards $ 52,000 in the early 2020s, still putting Australia in the High Income Country category. The Australian economy is greatly dependent on the export of primary produce with coal, iron ore, natural gas as the top exports. The 'national purse' of $ 2 trillion GDP also relies on highly carbon intensive and carbon extractive industries per capita and per real dollar of GDP (DISER 2020). This includes many of Australia's most notable second tier industries: education services, agriculture, healthcare, construction and personal travel services, with Australian and visitor ecological footprints also strongly associated with transport, travel and tourism activities.

---

[1] Please note that in this chapter the terms environmental and sustainability education, education for sustainability (EfS) and education for sustainable development (ESD) are used interchangeably. The most used terms in Australia are EfS and sustainability education, compared with the international usage of ESD.

Historically, under the Australian Constitution, education (and environment matters) is a state responsibility not a national one. However, since the 1960s, the Australian Government has sought to influence state-level education through grants programmes (such as for school science laboratories and libraries) (Harrington 2011), and there has been an Australian Government Minister for the Environment since 1971. The Australian education system comprises government (71%), Catholic (18%) and independent (11%) schools. After much discussion more than 20 years 'emphasizing education's contribution to economic productivity' (Brennan 2011, 261), Australia now has a national curriculum—the Australian Curriculum—and the Foundation-Year 10 curriculum was endorsed Australia's education ministers in September 2015:

> The curriculum provides teachers, parents, students and the community with a clear understanding of what students should learn regardless of where they live or what school they attend. The national curriculum was introduced to improve the quality, equity and transparency of Australia's education system (ACARA 2016, n.p.).

Most states and territories follow the Australian Curriculum (ACARA 2021, 2022a), however, the states of New South Wales, Queensland, Victoria and Western Australia have their own local curriculum documents and these do, at times, differ from the Australian curriculum, but generally not to any great extent.

The content of teacher education programmes for early childhood, primary and secondary schooling is the responsibility of the institution offering the programme (usually a university), although these programmes now need to meet national standards (AITSL 2019) to be accredited.

Development of environmental and sustainability education in Australia

The first national conference specifically focused on environmental education in Australia was convened in April 1970 under the auspices of the Australian Academy of Science. Here the chair of the National Committee for the International Biological Program, Sir Otto Frankel, noted that the deterioration of the environment threatened to engulf the whole world and concluded that this "is now perhaps the most pressing and most important aspect of education for the coming decades" (1970, 8).

In the years following this conference, the Australian Government responded to various calls for action on environmental education by designating it as a priority area for curriculum materials development by the national Curriculum Development Centre in 1973. It also sponsored participation in the UNESCO and UNEP conferences and workshops on environmental education (such as those held in Belgrade in 1975 and Tbilisi in 1977) which helped shape the movement (Gough 1997). The Curriculum Development Centre published Australia's first national statement on environmental education for schools (Greenall 1980), which all state and territory education authorities endorsed. This statement promoted environmental education as 'an orientation in the curriculum,' which continues to the present day with sustainability as a cross-curriculum priority in the Australian Curriculum (ACARA 2021).

From the late 1970s through to the early 1990s, Australian states published environmental education policies and curriculum guides to support environmental education in schools, often in collaboration with local environmental education associations and with the involvement of environmental (educational) activists (see, for example, New South Wales c1989, Queensland 1977, 1993, South Australia 1987, Tasmania c1976, Victoria 1990, 1998, Western Australia c1977). The Curriculum Development Centre also published a Bicentennial Australian Studies Schools Project Bulletin related to environmental education (Fien 1988) but, as Greenall (1987) described, the political history of environmental education up to this time could be likened to a game of snakes and ladders.

The first national action plan for environmental education, titled *Environmental Education for a Sustainable Future,* was released in 2000 by Environment Australia. As part of this plan, the Department of the Environment and Heritage published the second national statement on environmental education (Gough and Sharpley 2005). This suggested a different, 'whole school approach,' for environmental and sustainability education, consistent with the basis of the Australian Sustainable Schools Initiative, the funding of which was associated with the action plan (Gough 2011). This approach saw a curriculum-only focus as inadequate, and successful implementation of environmental education requiring action across the whole school. The action plan also established the short-lived Australian Research Institute in Education for Sustainability (ARIES) which conducted various reviews related to Education for Sustainability (EfS) (such as Tilbury et al. 2005).

The second national action plan was released in 2009 by the Department of the Environment, Water, Heritage and the Arts, this time titled a *National Action Plan for Education for Sustainability*. This plan was abandoned in 2010 (Larri and Colliver 2020). That these significant documents about environmental education were published by environment rather than education agencies is an on-going problem around Australian Government action in environmental education, which is diminishing—as evidenced by the changing wording in the national statements on schooling.

The need for Australian school students to learn about the environment was included in the first national education declaration, the 1989 *Hobart Declaration on Schooling* (MCEETYA 1989). This included as one of the 10 goals, 'an understanding of, and concern for, balanced development and the global environment and complex environmental and social challenges' (p.2). This rather weak goal was already taking on the political, cultural, and economic logic of neoliberalism (Hursh et al. 2015) consistent with a focus on treating the environment as a commodity in its reference to 'balanced development.' However, that the goal existed was sufficient to stimulate a range of environmental education activities by departments of education around Australia (see Gough 1997, 2002).

The 1999 *Adelaide Declaration on National Goals for Schooling in the Twenty-First Century* (MCEETYA 1999) moved recognition of complex environmental and social challenges to the preamble, and added a new goal, that students should have 'an understanding of, and concern for, stewardship of the natural environment, and the knowledge and skills to contribute to ecologically sustainable development' (p.3).

The 2008 *Melbourne Declaration on Educational Goals for Young Australians* (MCEETYA 2008) expanded on the environmental content of the Adelaide Declaration and recognized the unprecedented challenges posed by climate change. Here, the preamble noted new demands on Australian education, including 'complex environmental, social and economic pressures such as climate change that extend beyond national borders pose unprecedented challenges, requiring countries to work together in new ways' (p.5). The goal to accompany this statement was for students to become 'active and informed members of the community who ... have empathy for the circumstances of others and work for the common good, in particular sustaining and improving natural and social environments' (p.9), and the associated action was a resolution that 'a focus on environmental sustainability will be integrated across the curriculum' (p.14).

Although sustainability is primarily included as a cross-curriculum priority in the Australian Curriculum (ACARA 2021), its placement within the curriculum is haphazard, and generally develops a shallow understanding of sustainability, if it is taught to students at all (Nicholls 2017). Indeed, a 2014 review of the curriculum by Donnelly and Wiltshire recommended that 'ACARA re-conceptualize the cross-curriculum priorities and, instead, embed teaching and learning about ... sustainability explicitly, and only where educationally relevant, in the mandatory content of the curriculum' (p.247). This recommendation, we note, is highly conservative. Rather than foster a holistic, interdisciplinary, focused and whole-school approach to EfS—which is consistent with the research literature and best practice—this 'snake' (Greenall 1987) has proven particularly problematic. It entails, on the one hand, disengaging students and teachers from engaging mindsets, shifting dispositions and enhancing young people's abilities to both acknowledge contemporary problems. On the other, it foregoes developing a coherent and progressive educational approach for developing and applying knowledge and skills in schools, to remediate unsustainable practices in the present, as well as in the future (Wals 2011).

The *Alice Springs* (*Mparntwe*) *Education Declaration* (Department of Education, Skills and Employment 2019), which replaced the Melbourne Declaration in December 2019, fails in this regard, too. It simply sees education as preparing 'young people to thrive in a time of rapid social and technological change, and complex environmental, social and economic challenges' (p.2), and repeats the goal from the Melbourne Declaration. It is notably silent on climate change, and reduces consideration of sustainability in the curriculum to encouraging students to 'engage with complex ethical issues and concepts such as sustainability' (p.15).

Following the release of this Declaration, in June 2020 the education ministers announced another review of the Australian Curriculum with the terms of reference including to:

- Revisit, and improve if necessary, the organizing frameworks for the cross-curriculum priorities with reference to current research
- Declutter the content of the Australian Curriculum by improving the relationship of the cross-curriculum priorities to learning area content, removing any repetition of content between the cross-curriculum priorities and the learning

areas and replacing the current 'icon tagging' for cross-curriculum priorities on the Australian Curriculum website with a more user-orientated approach (ACARA 2020, 5).

The Review panel released a consultation paper in April 2021 (ACARA 2021b) which stated,

> revisions to the sustainability cross-curriculum priority reflect evolving understanding of the concepts that underpin sustainability and the features of effective sustainability education. In particular, the revisions position the priority with reference to the Alice Springs (Mparntwe) Education Declaration (2019) and the United Nations Sustainable Development Goals (p.2).

In particular, the Review panel claimed that these revisions:

- Broaden actions for sustainability to include the mitigation of human impacts and restoration of environments, in addition to preservation.
- Provide clearer support to explore how individuals and communities can take action and effect positive change.
- Ensure that organizing ideas fit naturally within learning areas and can be applied to content descriptions and elaborations (p.2).

The revised Australian Curriculum 9.0 was endorsed by the ministers of education in the Australian States and Territories in April 2022, and sustainability has been retained as a cross-curriculum priority (ACARA 2022b). The framing documents for the changes were the *Alice Springs Education Declaration* (DESE 2019) which downgraded references to the environment and climate change compared with its predecessor document (MCEETYA 2008). Given also that Australia's reporting to the United Nations on the implementation of the Sustainable Development Goals (SDGs) in Australia makes no mention of education for sustainable development (ESD) (DFAT 2018), it is not surprising that there has been no greater promotion of sustainability. Attention to sustainability remains piecemeal rather than adopting a holistic approach which would be consistent with indigenous approaches to sustainability and land care which is comprehensive in its nature, and is what is needed.

In sum, the environment has long been treated by policy-makers and politicians in Australia as a political football rather than an educational priority. Moreover, at a time when climate and other environmental emergencies are upon us, and Australia is a signatory to not just the SDGs but also the Paris Agreement (United Nations 2015)—which included ESD and climate change education as action areas in their associated work programmes—it would seem long overdue for ESD and climate change education to become education priorities in Australia.

Indeed, in the associated Paris Agreement work programme, Australia had agreed to develop extensive climate change education policies. Yet climate change has been ignored in our national education agenda at a time when it is most desperately needed, including in response to explicit calls to recognize this shortcoming in education by Australian youth and Aboriginal and Torres Strait Islanders (e.g., Kos 2020).

As Smith and Stevenson (2017, 79) argue, this lack of current government education policy support for environmental and sustainability education creates for educators who are concerned about sustainability

1) a policy environment that is hostile to Education for Sustainable Development (ESD) or Sustainability (EfS) (by omitting or not supporting internationally recognized ESD/EfS policies);

2) is in a constant state of flux because of frequent changes in governments with different ideologies and agendas regarding the role of education in fostering sustainability; or places EfS/ESD policies in competition with other educational policies that (overtly or covertly) receive greater priority.

Nevertheless, many teachers and schools are finding ways to implement environmental and sustainability education, such as through participating in Sustainable Schools programmes (Rickinson et al. 2016, Larri and Colliver 2020), or by having school administrations that want it to work (Smith and Stevenson 2017).

An important contributor to both environmental and sustainability education experiences for students and informal professional development for teachers in Australia in this regard has been the Environmental Education Centres established and maintained by some state Departments of Education. Both New South Wales and Queensland have extensive networks of now over 20 such centres that began in the 1970s and are staffed by specialist educators who offer generally one to four day programmes for visiting schools from throughout their state. Research conducted at the Queensland centres has provided evidence of powerful and distinctive experience-based teaching and learning through direct experience in authentic natural and community settings (Ballantyne and Packer 2008, 2009, Renshaw and Tooth 2018). Students were identified as learning: by '*doing,*' by '*being in the environment,*' by '*addressing authentic tasks,*' '*from sensory engagement,*' and '*in their own context* (backyard) by being encouraged to explore and investigate local problems and issues' (Renshaw and Tooth 2018, 10–11). The environmental educators' intimate embodied knowledge of both the biophysical, cultural and historical places in which they work with students and the unique pedagogical affordances of these particular places—a linking together of which can be characterized as pedagogical content knowledge (Stevenson and Smith 2018)—are represented in the stories they tell about specific places. In fact, these stories often include elements *in, about* and *for* 'country,' to ensure that indigenous perspectives and knowledge are included and respected. Observations and student interviews revealed how these stories, perceived as the environmental educators being knowledgeable of and passionate for authentic, complex, and honoured places, increased visiting student engagement (Renshaw and Tooth 2018).

Understanding of environmental and sustainability education in Australia

Early understanding of environmental education in Australia was grounded in the 1970 IUCN definition of the field, the 1975 Belgrade Charter and the 1977 Tbilisi

Declaration. According to the latter, the goals for environmental education are (UNESCO 1978, 26–27):

a) to foster clear awareness of, and concern about, economic, social, political and ecological interdependence in urban and rural areas;
b) to provide every person with opportunities to acquire the knowledge, values, attitudes, commitment and skills needed to protect and improve the environment;
c) to create new patterns of behaviour of individuals, groups and society as a whole towards the environment.

These definitions were referenced in the previously-mentioned state education department statements on environmental education in the 1970s (Gough 1997). Based on Arthur Lucas' (1979) research, the notion of environmental education being education *in, about* and *for* the environment was popularized in Australia by Linke (1980) and Greenall (1980), among others. Local environmental educators continued to work with these definitions through the 1980s, as they continued to be endorsed internationally, for example, at the 1987 Tbilisi+10 conference (UNESCO-UNEP 1988). As noted previously, the 1989 *Hobart Declaration on Schooling* (Education Council 2014) included as one of its 10 goals, 'an understanding of, and concern for, balanced development and the global environment and complex environmental and social challenges.'

The term 'sustainable development' joined the lexicon from the *World Conservation Strategy* (IUCN 1980) and the *National Conservation Strategy for Australia* (DHAE 1984, 17) as 'living resource conservation for sustainable development,' from the report of the World Commission on Environment and Development (1987), and from the Australian Government discussion paper on *Ecologically Sustainable Development* (DPMC 1990) and the subsequent *Draft National Strategy for Ecologically Sustainable Development* (DPMC 1992).

At the 1992 United Conference on Environment and Development, the focus changed from environmental education to 'Reorienting education towards sustainable development' (United Nations 1993, 36.2), and the renaming of environmental education to education for sustainable development began to happen in Australia with reference to 'ecologically sustainable development' in *A Statement on Studies of Society and Environment for Australian Schools* (AEC 1994, 3). However, the shift was slow. As noted above, the first national action plan was for environmental education (Environment Australia 2000), but the second referred to education for sustainability or EfS (DEWHA 2009). As noted in this document:

> Australia's approach to education for sustainability has come a long way since its origins in environmental education in the 1970s. It has evolved from a focus on awareness of natural ecosystems and their degradation to equipping all people with the knowledge, skills and understanding necessary to make decisions based upon a consideration of their full environmental, social and economic implications (DEWHA 2009, 3–4).

Understandings of EfS are also elaborated in the revised version of the Australian Curriculum (ACARA 2022a, 2022b). According to the sustainability overview statement,

> The Sustainability cross-curriculum priority explores the knowledge, skills, values and world views necessary for people to act in ways that contribute to a sustainable future. Designing solutions and actions for a sustainable future requires an understanding of the ways environmental, social and economic systems interact, and an ability to make balanced judgements based on present and future impacts.
>
> *The Sustainability cross-curriculum priority is futures-oriented and encourages students to reflect on how they interpret and engage with the world. It is designed to raise student awareness about informed action to create a more environmentally and socially just world* [Emphasis added] (ACARA 2022b, paras. 4 and 5).

Although this statement is consistent with the holistic approaches to environmental education and sustainability outlined earlier, the actual content of the subject areas does not enact the statement's intent, nor is there guidance for teachers in implementing the Organizing Ideas (OIs) for Sustainability (ACARA 2021a, 2021b, 2022b). Importantly, compared with the previous sustainability overview statement (ACARA 2021a), the key emphasis on 'protecting environments and creating a more ecologically and social just world' has been deleted. There has also been a shift from endorsing 'actions that support more sustainable patterns of living' to merely raising 'student awareness about informed action.' The earlier statement was

> Sustainability education is futures-oriented, *focusing on protecting environments and creating a more ecologically and socially just world* through informed action. Actions that support more sustainable patterns of living require consideration of environmental, social, cultural and economic systems and their interdependence [Emphasis added] (ACARA 2021a, para. 6).

Table 1 shows both the previous and the revised organizing ideas for sustainability.

These organizing ideas are interesting for their valuing of the environment for its own sake, rather than just as a resource for exploitation. However, the revised organizing ideas are much more human-focused. For example, the first IO ('The biosphere is a dynamic system providing conditions that sustain life on Earth') has been deleted from the revised version and replaced with an expanded version of the previous IO.2: 'All life forms, including human life, are connected through the earth's systems (geosphere, biosphere, hydrosphere and atmosphere) on which they depend for their well-being and survival' as SS1. The new SS2 and SS3 are stronger in saying 'influence the sustainability of Earth's systems' rather than 'sustainable patterns of living rely on.' Also, IO.5 reworded as SW2, and IO.6 reworded as SF1 have added 'business and political' to the actions for sustainability.

The introduction of a new organizing idea, Design, ('The role of innovation and creativity in sustainably-designed solutions, including products, environments and services, that aim to reduce present and future impacts or to restore the health

Table 1. Comparison of the current and revised organising ideas of the sustainability cross-curriculum priority (ACARA 2022b) with key differences highlighted in bold

Version 8.4 Organising Ideas	Version 9.0 Organising Ideas
Systems	*Systems*
The biosphere is a dynamic system providing conditions that sustain life on Earth. (OI.1)	All life forms, including human life, are connected through Earth's systems (geosphere, biosphere, hydrosphere and atmosphere) on which they depend for their wellbeing and survival. (SS1)
All life forms, including human life, are connected through ecosystems on which they depend for their well-being and survival. (OI.2)	**Sustainable patterns of living require the responsible use of resources, maintenance of clean air, water and soils, and preservation or restoration of healthy environments.** (SS2)
Sustainable patterns of living rely on the interdependence of healthy social, economic and ecological systems. (OI.3)	**Social, economic and political systems influence the sustainability of Earth's systems.** (SS3)
World Views	*World Views*
World views that recognise the dependence of living things on healthy ecosystems, and value diversity and social justice, are essential for achieving sustainability. (OI.4)	World views that recognise the interdependence of Earth's systems, and value diversity, equity and social justice, are essential for achieving sustainability. (SW1)
World views are formed by experiences at personal, local, national and global levels, and are linked to individual and community actions for sustainability. (OI.5)	World views are formed by experiences at personal, local, national and global levels, and are linked to individual, community, **business and political** actions for sustainability. (SW2)
[Futures]	*Design*
Sustainable futures result from actions designed to preserve and/or restore the quality and uniqueness of environments. (OI.9)	Sustainably designed products, environments and services aim to minimise the impact on or restore the quality and diversity of environmental, social and economic systems. (SD1)
	Creative and innovative design is integral to the identification of new ways of sustainable living. (SD2)
Designing action for sustainability requires an evaluation of past practices, the assessment of scientific and technological developments, and balanced judgements based on projected future economic, social and environmental impacts. (OI.8)	Sustainable design requires an **awareness** of place, past practices, **research** and technological developments, and balanced judgements based on projected environmental, social and economic impacts. (SD3)
Future	*Future*
The sustainability of ecological, social and economic systems is achieved through informed individual and community action that values local and global equity and fairness across generations into the future. (OI.6)	A sustainable future is achieved through informed individual, community, **business and political** action that values local, national and global equity and fairness across generations into the future. (SF1)
Actions for a more sustainable future reflect values of care, respect and responsibility, and require us to explore and understand environments. (OI.7)	A sustainable future requires individuals to seek information, identify solutions, reflect on and evaluate past actions, **and collaborate with and influence others as they work towards a desired change.** (SF2)

or diversity of environmental, social and economic systems,' ACARA 2022b, para. 9) reinforces the Government's focus on STEM and technological solutions to environmental problems. However, as indicated in Table 1, two of the Design organizing ideas are merely the re-labelling of two previous Future organizing ideas.

There was also a mismatch between the previous versions of the Sustainability cross-curriculum priority statement and the OIs. For example, ethical and aesthetic concerns are prioritized for Foundation year students, and not revisited in late secondary schooling.

To elaborate, in the current version there is a specific symbol (⌁) to indicate particular content where the Sustainability cross-curriculum priority is appropriate in each of the four core curriculum statements. The symbol has been updated to ⊕ in the revised curriculum which is closer to the global focus of sustainability. In both versions, this symbol rarely occurs across the eleven years of schooling covered by those curriculum statements. It does not appear at all in English and Mathematics, while the six occurrences in the Science curriculum across eleven year levels focus on the Systems OIs. Arguably, the Futures OIs could be linked more strongly in science, and with the Humanities and Social Sciences (HASS), too. The year level descriptions in English from Year 5 onwards refer to students studying texts that explore themes of interpersonal relationships and ethical (and global in Years 9 and 10) dilemmas in real-world and fantasy/fictional settings, but there is no specific guidance to teachers for implementing the Sustainability OIs. The HASS learning areas (particularly Geography, History and Economics) offer the strongest links with sustainability. There are no links with Music, and sustainability is inconsistently visible in the Languages.

The lack of guidance is found particularly wanting on: (a) sequencing and synchronizing attention to the OIs in and across both cross-curricular work and all curriculum subjects, and (b) critically exploring the tensions that attention to sustainability can create with prevailing curriculum priorities: the prevailing and dominant curriculum of 'education for unsustainability' in many schools being the key case in point (Reid 2018). In comparison to the learning objectives offered in UNESCO's *Education for Sustainable Development Goals* (2017), many of the elaborations offered for the OIs are trite. The consultation paper on the Australian Curriculum Review (ACARA 2021b) noted that 'teachers, in particular, acknowledged the broad contexts that the cross-curriculum priority can support and emphasized the need for clear, descriptive content elaborations to assist teachers to identify opportunities to engage with sustainability across learning areas' (p.2). However, the revised Australian Curriculum does not seem to address this deficiency.

Furthermore, notably absent from the content statements are specific and urgent ecological issues facing Australia and the planet, such as the climate crisis and loss of biodiversity which do not appear in the 'high status' Science content, but are in Geography, in the current Australian Curriculum (Reid 2018). The revised

Science curriculum (ACARA 2021c), moves the content further away from an interdisciplinary understanding of concepts related to sustainability. For example,

- Content about effects of sudden geological changes and extreme weather has been removed from the science-understanding strand in Year 6, removing a duplication with HASS.
- Content about ecosystems has been removed from the science-understanding strand in Year 9, as students explore ecosystems with a focus on matter and energy flow at Year 7.
- Carbon cycle content has been moved from Year 10 to Year 9 Science.
- Models of climate change have been given greater emphasis in Year 10 with a focus on energy flow between earth's spheres, to enable a deeper exploration of the science underpinning climate models.

The only reference to climate change in the Science curriculum is at Year 10, and in the context of energy flow, which makes it clear how policies of the state simply emphasize safe knowledge that is unlikely to create community dissent or anxiety, even as it is unlikely to be assessed or evaluated as with standard curriculum areas (Buchanan 2021). This also means that knowledge is likely to be treated as apolitical and devoid of complexities, nuances, and uncertainties (Stevenson 2010). Put starkly, if EfS is about diverse communities learning within and beyond schooling in the face of risk, uncertainty and rapid change, it is neither clearly nor explicitly consistent with the vision for the 'learning domain' set out by ACARA (2022b, para 5), as in the emphasis added above, i.e.,

> The Sustainability cross-curriculum priority is futures-oriented and encourages students to reflect on how they interpret and engage with the world. It is designed to raise student awareness about informed action to create a more environmentally and socially just world.

It is all well and good for ACARA (2022c, 1) to assert that

> Cross-curriculum priorities are incorporated through learning area content; they are not separate learning areas or subjects. They provide opportunities to enrich the content of the learning areas, where most appropriate and authentic to do so, allowing students to engage with and better understand their world.

However, as long as teachers are not provided with guidance on how to implement such content, and such content is not part of any assessment regime, it could well go untaught. This is particularly the case regarding indigenous approaches to sustainability and land care which are silent in the discussions around sustainability in the Australian Curriculum. In addition, while the cross curriculum priority could be seen as promoting a holistic approach to sustainability, if the complex ecology of social, environmental and economic factors is not mentioned, let alone assessed, then it is unlikely to be taught.

In addition, there is the issue of global citizenship. While the SDG target 4.7 is about global citizenship as well as education for sustainable development and other areas, global citizenship isn't a cross curriculum priority in the Australian Curriculum, but it is in the Years 7–10 learning area Civics and Citizenship, which is part of Humanities and Social Sciences. This signals the need for more work to see how these could be better aligned and not siloed. Indeed, much more work is needed to develop a holistic vision of education, not a piecemeal hunt for a few keywords.

Implementation of environmental and sustainability education and Global Action Programmes on ESD in Australia

Over the past two decades there has been a diminishing interaction, perhaps best described as a deafening silence, between international initiatives in sustainability education and actions in Australia, to the point where there has been no national engagement with either the Global Action Program (UNESCO 2014) or the follow-up ESD for 2030 program (UNESCO 2020).

There were a number of activities related to EfS at the beginning of the UN Decade on Education for Sustainable Development (2005–2014) (DESD), including activities that had their origins in the first national action plan for environmental education (Environment Australia 2000), but only a few specifically focused on implementing the Decade in Australia (Gough 2011, Tilbury 2006). Tilbury (2006) cautioned that 'There is a tendency to adopt the UN DESD logo or label existing activities as DESD initiatives—when in reality those activities would have been developed regardless of whether there was a UN Decade or not' (p.80), and this has proved to be the case in Australia for the Decade, and since.

Although the aforementioned national action plan for EfS (DEWHA 2009) was produced during the Decade, and mapped an extensive sustainability curriculum framework to enact this (DEWHA 2010), the Decade was more snakes than ladders. This included the dismantling of the National Environmental Education Council, the National Environmental Education Network, and the national Australian Sustainable Schools Initiative among a range of sustainability education-related activities in early 2010, culminating in the wholesale abandonment of any Australian Government commitment to the national action plans (Larri and Colliver 2020). While EfS has been retained as a cross-curriculum priority in the Australian Curriculum—primarily because of the *Melbourne Declaration on Educational Goals for Young Australians* (MCEETYA 2008)—as noted above, there remains pressure from right-wing sources (most notably, mainstream (the 'Murdoch Press') and alternative media, think tanks, and pressure groups) to adopt the position that it be re-conceptualized and only taught 'where essential' (Donnelly and Wiltshire 2014).

As discussed previously, the recent reviews of the Australian Curriculum (ACARA 2020, 2021b, 2022a) have retained sustainability as a cross curriculum priority, albeit with a more resource and technology/design focus to its organizing ideas (see Table 1). In addition, the government at the time was increasingly championing science, technology, engineering and mathematics (STEM) education and research, and references to ecological concerns have been downgraded in the Alice Springs Declaration (DESE 2019).

Australia is a signatory to the United Nations 2030 Agenda for Sustainable Development and its 17 Sustainable Development Goals (SDGs). However, the government's 'voluntary' reports on progress towards implementing the SDGs (DFAT 2018a, 2018b) are significantly silent on ESD and EfS. For example, the report on tracking Australia's progress (DFAT 2018b) does not mention EfS or ESD. In the report on implementation of the SDGs (DFAT 2018a) where, related to SDG 4 on education, it is stated, in the only reference to ESD:

> Many Australian schools and universities have implemented sustainability programs to teach children and young people about resource sustainability and to improve resource management within their institutions. Sustainability is one of three national cross-curriculum priorities and has been incorporated in programs like Resource Smart Schools in Victoria. Many Australian universities are actively incorporating the SDGs into their curricula and student activities, including institutions that have signed up to the Principles for Responsible Management Education, which is working to embed the SDGs into management education (p.39).

SDG 4 is the responsibility of the Department of Education, Skills and Employment (DESE), but their website only links back to a generic page on the SDGs   (https://www.dese.gov.au/international-education/announcements/2030-agenda-sustainable-development): there are no related actions. Basically, there is nothing to report.

Emerging issues and trends as well as current and future needs of environmental and sustainability education in Australia

The greatest threat to the environment and sustainability and its associated education in Australia has been the former Australian Government. For example, Australia's key piece of national wildlife protection law, the *Environment Protection and Biodiversity Conservation Act 1999* (EPBC Act), was enacted in 2000. However, the previous right-wing coalition governments attempted to change this law to hand greater responsibility for development assessments to the states, which would increase risk and uncertainty over proposals and lead to more environmental destruction, not protection (Morton 2020). In addition, the previous Prime Minister and key cabinets of the Australian Government continued to resist international pressure to set emission targets for 2050, in accordance with the Paris Agreement on Climate Change (Bagshaw et al. 2020, Eckersley 2021) until shortly before the last federal election (held in May 2022) at which they were defeated.

This resistance to recognizing the importance of environmental protection and responding to climate change is reproduced in the current guiding document for education in Australia, the *Alice Springs (Mparntwe) Education Declaration* (Department of Education, Skills and Employment 2019). As noted earlier, this sees education as preparing 'young people to thrive in a time of rapid social and technological change, and complex environmental, social and economic challenges' (p.2), but there is still little in the Australian Curriculum to achieve this goal. Instead, the previous Australian Government invested in STEM education as part of its neoliberal future productivity agenda:

> The Australian Government regards high-quality science, technology, engineering and mathematics (STEM) education as critically important for our current and future productivity, as well as for informed personal decision making and effective community, national and global citizenship (Department of Education, Skills and Employment 2020, n.p.).

There are a number of National Innovation and Science Agenda (NISA) schools and early years initiatives being funded as part of the *National STEM School Education Strategy 2016–2026* (Education Council 2015) in the belief that 'a renewed national focus on STEM in school education is critical to ensuring that all young Australians are equipped with the necessary STEM skills and knowledge that they will need to succeed' (p.3). National concerns about the quality and relevance of STEM education, including in relation to EfS, have persisted for nearly a decade. As Gough (2015, 446) notes, within Australia, an 'avalanche of documents from a range of authoritative sources [is] calling for more STEM education to create smarter futures, economic competitiveness and growth as well as a more scientifically literate society.' Yet there is no mention of the environment or sustainability in the STEM agenda, even as the needs for schools-based and intergenerational environmental and sustainability education are huge (AESA 2014) and persist—all in the face of the past government's lack of prioritizing of environmental protection and responding to existential challenges such as climate change.

This is not to say that there is nothing substantial, authentic or valuable happening in schools under an umbrella of environmental and sustainability education. For example, Larri and Colliver (2020, p.71) note that, while there has not been a national evaluation of the Australian Sustainable Schools initiative since 2010, 'there is evidence of many schools implementing whole-school sustainability actions and education' across many states and territories. However, as Smith and Stevenson (2017) discuss, the disruptive policy environment—set in motion by neo-liberal reforms focused on standards, accountability, and international competitiveness—impacts on EfS initiatives in schools. Facilitating or focusing primarily on individualistic responses to sustainability challenges are inadequate as EfS; they do not to challenge dominant, systems-embedded practice that maintain 'cultures of unsustainability' (Reid 2018). Equally, in Australia, as found in many other minority world education systems, EfS is typically marginalized and positioned as a low priority, left to a few very dedicated teachers, or employed as an add-on to an already overcrowded curriculum (Reid 2018).

Nevertheless, there are schools who are implementing a wide range of enriching, progressive, and place-based environment-related educational programmes despite all of the political, curriculum and administrative barriers they face. These are typically programmes that foster a strong personal and socially-oriented ethical outlook that helps students and teachers manage context, conflict and uncertainty, and builds the skills, dispositions and capabilities to fully realize their capacities as active and informed democratic citizens. In contrast to neoliberal priorities for education, they are intended to actually promote cooperation and community in addressing sustainability challenges, rather than competition between individuals (Hursh et al. 2015).

A study in the state of Queensland of two primary schools that had established significant EfS programmes prior to the dismantling of policy support, revealed a number of factors distinguishing a school able to sustain a high-quality, balanced, critical and engaging program from one unable to do so (Smith and Stevenson 2017). In the context of Australia having fallen prey to the same preoccupation with raising student test scores and international competitiveness that now dominates educational discourse elsewhere, both schools were under steady and persistent pressure to implement new curriculum guidelines tied to state testing. What was striking is that at one school this pressure made it increasingly difficult for teachers to implement EfS initiatives, while at the other, its impact, although certainly in evidence, was not as inhibitory. Consistent with research on school reform, the latter school was characterized by a principal who cultivated a collaborative culture of trusting relationships and shared decision-making, created a common purpose and vision that reflected the school's values, and used the strengths and interests of staff and students to build on their assets.

Although research at … [these schools] did not surface insights about the way national or state policies can facilitate the implementation of EfS at the school level, it did reveal how educators in a turbulent policy environment can retain and even expand initiatives supportive of sustainability despite the less than favourable organizational and political context in which they work (ibid p.94).

Unfortunately, however, the inclusion of environmental and sustainability education in Australian teacher education programmes remains haphazard (Dyment and Hill 2015), and if included, it is often as an elective rather than part of the core programme (Evans et al. 2017, Ferreira et al. 2009, Gough 2016, Stevenson 2010). Achieving environmental and sustainability education in core teacher education and professional development offerings for classroom teachers and school leaders remains an issue, as does ensuring that systemic, lasting, whole-school change happens by embedding EfS across the school sector (AESA 2014).

Conclusion

Australia is facing major environmental challenges that have reached crisis proportions from climate change, disasters such as floods, bushfires, and the bleaching of the Great Barrier Reef, and extensive loss of biodiversity, to name just a few. We have had a government that was looking for a technological fix for climate change rather than acting to cut emissions, and which has a reputation for creatively 'gaming' the process of carbon accounting—'since 2018 Australia is also the only developed country to refuse to channel its climate finance through the multilateral Green Climate Fund' (Eckersley 2021, n.p.).

Just like the government's marginalization of environmental concerns in its policies and practices, its response to sustainability education was minimal. Environmental (and sustainability) education has been on the formal education agenda in Australia for 50 years. While there is much good that is happening at individual school levels, the past decade has not been prosperous, and currently, there is neither a strong platform nor momentum to achieve Agenda 2030 in Australia. As Buchanan (2021, 169) notes, there is 'a series of rifts between government rhetoric, government policy, education rhetoric, education policy, syllabuses, teacher practice and student engagement' related to sustainability education. However, whether it is the diversity and accessibility of land-based and other environments in Australia, the relatively high levels of support and robustness of its diverse education systems, the near-full enrolment of primary-aged students, and the continuing high levels of professionalism among teachers—including their willingness (often provoked by their students, many of whom have joined Strike4Climate marches) to address climate change and other environmental challenges—there are strong indicators that the situation could be turned around.

However, since the Australia Government de-funded the national action plan of education for sustainability (DEWHA 2009) in 2010 and sustainability became an unmeasured cross-curriculum priority in the Australian Curriculum, much activity has been left to the enthusiastic individual (who often burns out or leaves the field) rather than part of the mainstream. Just like environmental protection and responding to climate change are current challenges, so too is getting environmental and sustainability education formally recognized, administratively supported, and funded in any substantial way in the current political climate.

In May 2022 there was a change of government, and the Australian Labour Party was elected with a more environmentally-conscious agenda than its predecessor. We need this government to engage with and implement Agenda 2030, and the *Berlin Declaration on Education for Sustainable Development* (UNESCO 2021) which advocates for environmental education to be part of the core curriculum in schools by 2020 (UNESCO Harare Office 2021). We need the next Australian Curriculum to take the organizing ideas of sustainability seriously, particularly that 'Sustainable futures require individuals to seek information, identify solutions, reflect on and evaluate past actions, and collaborate with and influence others as they work towards a desired change.' (SF2) (ACARA 2022b). To date, at the time of writing, which is only a month after the new government assuming office, there has been no mention of changes to the positioning of sustainability in the Australian Curriculum. However there will be new pressures for the next version to deliver on international obligations and for there to be less silos or disconnection to achieve what is envisaged, including in relation to the SDGs. The recent change in government and the emergence of strong political support across the country for climate action leaves us with hope.

References

Australia. Department of Education, Skills and Employment. 2020. Support for Science, Technology, Engineering and Mathematics (STEM). https://www.education.gov.au/support-science-technology-engineering-and-mathematics.

Australia. Department of Foreign Affairs and Trade (DFAT). 2018a. Report on the Implementation of the Sustainable Development Goals 2018. https://www.dfat.gov.au/sites/default/files/sdg-voluntary-national-review.pdf.

Australia. Department of Home Affairs and Environment (DHAE). 1984. A National Conservation Strategy for Australia. Australian Government Publishing Service, Canberra.

Australia. DFAT. 2018b. Tracking Australia's Progress on the Sustainable Development Goals. https://www.dfat.gov.au/sites/default/files/sdgs-data-report-tracking-progress.pdf.

Australia. Department of the Environment, Water, Heritage and the Arts (DEWHA). 2009. Living Sustainably: The Australian Government's National Action Plan for Education for Sustainability. DEWHA, Canberra.

Australia. DEWHA. 2010. Sustainability curriculum framework: a guide for curriculum developers and policy makers. Commonwealth of Australia http://www.environment.gov.au/education/publications/pubs/curriculum-framework.pdf.

Australia. Department of Industry, Science, Energy and Resources (DISER). 2020. Quarterly Update of Australia's National Greenhouse Gas Inventory: June 2020. https://www.industry.gov.au/sites/default/files/2020-11/nggi-quarterly-update-june-2020.pdf.

Australia. Department of Prime Minister and Cabinet (DPMC). 1990. Ecologically Sustainable Development. A Commonwealth Discussion Paper. Australian Government Publishing Service, Canberra.

Australia. DPMC. 1992. Draft National Strategy for Ecologically Sustainable Development. A Discussion Paper. The Ecologically Sustainable Development Steering Committee. Australian Government Publishing Service, Canberra.

Australian Conservation Foundation. 2018. Fast-tracking extinction Australia's national environmental law. https://d3n8a8pro7vhmx.cloudfront.net/auscon/pages/6451/attachments/original/1536271571/08-2018_16pp_ACF_Fast-Tracking_Extinction_report_final_WEB.PDF?1536271571.

Australian Curriculum and Assessment and Reporting Authority. ACARA. 2016. About us. https://www.acara.edu.au/about-us.

ACARA. 2020. Review of the Australian Curriculum. Retrieved from https://www.acara.edu.au/curriculum/curriculum-review/.

ACARA. 2021a. The Australian Curriculum. 8.4: Sustainability. Retrieved from https://www.australiancurriculum.edu.au/f-10-curriculum/cross-curriculum-priorities/sustainability/.

ACARA. 2021b. Cross-Curriculum Priorities. Consultation – Introductory information and organising ideas. Retrieved from https://www.australiancurriculum.edu.au/media/7018/ccp_sustainability_consultation.pdf.

ACARA. 2021c. What has changed and why? Proposed revisions to the Foundation – Year 10 (F–10) Australian Curriculum: Science. Retrieved from https://www.australiancurriculum.edu.au/media/7122/ac_review_2021_science_whats_changed_and_why.pdf.

ACARA. 2022a. The Australian Curriculum version 9.0. Retrieved from https://v9.australiancurriculum.edu.au/.

ACARA 2022b. Understand this cross-curriculum priority: Sustainability. Retrieved from https://v9.australiancurriculum.edu.au/teacher-resources/understand-this-cross-curriculum-priority/sustainability.

ACARA. 2022c. Cross Curriculum Priorities. Retrieved from https://v9.australiancurriculum.edu.au/f-10-curriculum/f-10-curriculum-overview/cross-curriculum-priorities.

Australian Education Council (AEC). 1994. A Statement on Studies of Society and Environment for Australian Schools. Curriculum Corporation, Carlton.

Australian Education for Sustainability Alliance (AESA) 2014. Education for Sustainability and the Australian Curriculum Project: Final Report for Research Phases 1 to 3, AESA, Melbourne.

Australian Human Rights Commission. (n.d.) Face the facts. Retrieved from https://humanrights.gov.au/education/face-facts.

Australian Institute for Teaching and School Leadership (AITSL). 2019. Accreditation of initial teacher education programs in Australia: Standards and Procedures. AITSL, Melbourne. https://www.aitsl.edu.au/docs/default-source/national-policy-framework/accreditation-of-initial-teacher-education-programs-in-australia.pdf?sfvrsn=e87cff3c_28.

Bagshaw, E., N. O'Malley and M. Foley. 2020, 28 October. Australia defies international pressure to set emissions targets. The Age. https://www.theage.com.au/politics/federal/australia-defies-international-pressure-to-set-emissions-targets-20201028-p569ed.html.

Ballantyne, R. and J. Parker. 2008. Learning for Sustainability: The Role and Impact of Outdoor and Environmental Education Centres. School of Tourism, University of Queensland, St. Lucia.

Ballantyne, R. and J. Parker. 2009. Introducing a fifth pedagogy: Experience-based strategies for facilitating learning in natural environments. Environmental Education Research 15(2): 243–2623.

Brennan, M. 2011. National curriculum: A political-educational tangle. Australian Journal of Education 55(3): 259–280.

Buchanan, J. 2021. Environmental trust? Sustainability and renewables policy and practice in the school years. Curriculum Perspectives 41: 163–173.

Department of Education, Skills and Employment. 2019. Alice Springs (Mparntwe) Education Declaration. Education Council, Carlton South. Retrieved from https://www.dese.gov.au/alice-springs-mparntwe-education-declaration/resources/alice-springs-mparntwe-education-declaration.

Donnelly, K. and K. Wiltshire. 2014. Review of the Australian Curriculum. Final report. Australian Government Department of Education, Canberra.

Dyment, J.E. and A. Hill. 2015. You mean I have to teach sustainability too? Initial teacher education students' perspectives on the sustainability cross-curriculum priority. Australian Journal of Teacher Education 40(3) 21–35.

Eckersley, R. 2021, 12 November. 'The Australian way': How Morrison trashed brand Australia at COP26. The Conversation. Retrieved from https://theconversation.com/the-australian-way-how-morrison-trashed-brand-australia-at-cop26-171670.

Education Council. 2015. National STEM school education strategy, 2016–2026. Retrieved from https://www.dese.gov.au/australian-curriculum/support-science-technology-engineering-and-mathematics-stem/national-stem-school-education-strategy-2016-2026.

Environment Australia. 2000. Environmental Education for a Sustainable Future: National Action Plan. Environment Australia, Canberra.

Evans, N., R. Stevenson, M. Lasen, J-A. Ferreira and J. Davis. 2017. Approaches to embedding sustainability in teacher education: A synthesis of the literature. Teaching and Teacher Education 63: 405–417.

Ferreira, J-A., L. Ryan, J. Davis, M. Cavanagh and J. Thomas. 2009. Mainstreaming Sustainability into Pre-service Teacher Education in Australia. Australian Research Institute in Education for Sustainability, Sydney.

Fien, J. 1988. Education for the Australian Environment. Bicentennial Australian Studies Schools Project Bulletin 6. Curriculum Development Centre, Canberra.

Frankel, O. 1970. Chairman's remarks. pp. 7–8. *In*: Evans, J. and S. Boyden (eds.). Education and the Environmental Crisis. Australian Academy of Science, Canberra.

Gough, A. 1997. Education and the Environment: Policy, Trends and the Problems of Marginalisation. Australian Education Review Series No. 39. Australian Council for Educational Research, Melbourne, Victoria.

Gough, A. 2002. Mutualism: a different agenda for science and environmental education. International Journal of Science Education 24(11): 1201–1215.

Gough, A. and B. Sharpley. 2005. Educating for a Sustainable Future: A National Environmental Education Statement for Schools. Curriculum Corporation for the Department of the Environment and Heritage, Melbourne.

Gough, A. 2015. STEM Policy and science education: scientistic curriculum and sociopolitical silences. Cultural Studies of Science Education 10(2): 445–458. doi: 10.1007/s11422-014-9590-3.

Gough, A. 2016. Teacher education for sustainable development: past, present and future. pp. 109–122. *In*: Filho, W.L. and P. Pace (eds.). Teaching Education for Sustainable Development at University Level. Springer, Dordrecht.

Greenall, A. 1980. Environmental Education for Schools: Or How to Catch Environmental Education. Curriculum Development Centre, Canberra.

Greenall, A. 1987. A political history of environmental education in Australia: Snakes and ladders. pp. 3–21. *In*: Robottom, I. (ed.). Environmental Education: Practice and Possibility. Deakin University, Geelong.

Harrington, M. 2011. Australian Government funding for schools explained. Background Note. Parliament of Australia: Parliamentary Library. https://www.aph.gov.au/binaries/library/pubs/bn/sp/schoolsfunding.pdf.

Hursh, D., J. Henderson and D. Greenwood. 2015. Environmental education in a neoliberal climate. Environmental Education Research 21(3): 299–318.

International Union for the Conservation of Nature and Natural Resources (IUCN) in collaboration with the United Nations Environment Program (UNEP) and the World Wildlife Fund (WWF). 1980. World Conservation Strategy. IUCN, Gland.

Kos. 2020. Planting the SEED for Climate Action. 26 January. https://www.kosmagazine.com.au/stories/planting-the-seed-for-climate-action.

Larri, L. and A. Colliver. 2020. Moving Green to Mainstream: Schools as models of sustainability for their communities—The Australian Sustainable Schools Initiative (AuSSI). pp. 61–82. *In*: Gough, A., J.C.-K. Lee and E.P.K. Tsang (eds.). Green Schools Globally: Stories of Impact on Education for Sustainable Development. Springer, Cham.

Linke, R.D. 1980. Environmental Education in Australia. Allen and Unwin, Sydney.

Lucas, A.M. 1979. Environment and Environmental Education: Conceptual Issues and Curriculum Implications. Australian International Press and Publications, Melbourne.

Ministerial Council on Education Employment Training and Youth Affairs (MCEETYA). 1989. The Hobart Declaration on Schooling. In G. Spring. Australia's Common and Agreed Goals for Schooling in the Twenty First Century: A Review of the 1989 Common and Agreed Goals for Schooling in Australia (The Hobart Declaration). A Discussion Paper. MCEETYA, Carlton South.

Ministerial Council on Education Employment Training and Youth Affairs (MCEETYA). 1999. The Adelaide Declaration on National Goals for Schooling in the Twenty-first Century. MCEETYA, Carlton South.

Ministerial Council on Education Employment Training and Youth Affairs (MCEETYA). 2008. Melbourne Declaration on Educational Goals for Young Australians. MCEETYA, Carlton South.

Morton, A. 2020, 23 November. Changes to Australia's environment laws would risk return to 'confusion', inquiry told. The Guardian. https://www.theguardian.com/environment/2020/nov/23/changes-to-australias-environment-laws-would-risk-return-to-confusion-inquiry-told.

New South Wales. Department of Education. c1989. Environmental Education Curriculum Statement K-12. Department of Education, Sydney.

Nicholls, J. 2017. Understanding how Queensland teachers' views on climate change and climate change education shape their reported practices. Unpublished doctoral thesis, James Cook University, Cairns, Queensland, Australia.

Queensland. Department of Education. 1977. Environmental Education in Queensland Schools. Department of Education, Brisbane.

Queensland. Department of Education. 1993. P-12 Environmental Education Curriculum Guide. Department of Education, Brisbane.

Renshaw, P. and R. Tooth. (eds.). 2018. Diverse Pedagogies of Place: Educating Students in and for Local and Global Environments. Routledge, London.

Reid, A. (ed.). 2018. Curriculum and Environmental Education: Perspectives, Priorities and Challenges. Routledge, London.

Rickinson, M., M. Hall and A. Reid. 2016. Sustainable schools programmes: what influence on schools and how do we know? Environmental Education Research 22(3): 360–389. doi:10.1080/13504622.2015.1077505.

Smith, G.A. and R.B. Stevenson. 2017. Sustaining education for sustainability in turbulent times. The Journal of Environmental Education 48(2): 79–95. doi: 10.1080/00958964.2016.1264920.

South Australia. Education Department. 1987. Environmental Education. Education Department of South Australia, Adelaide.

Stevenson, R.B. 2010. Teacher preparation and environmental sustainability education: Motives, meanings and methods. Invited keynote address to the Australian Teacher Education Association (ATEA) Conference, Townsville, Queensland, 4–7 July.

Stevenson, R.B. and G.A. Smith. 2018. Environmental educators learning and theorizing place-responsive pedagogy. pp. 190–210. *In*: Renshaw, P. and R. Tooth (eds.). Diverse Pedagogies of Place: Educating Students in and for Local and Global Environments. London, Routledge.

Tasmania. Education Department. c1976. Environmental Education. Government Printer, Hobart.

The Senate, Environment and Communications References Committee. 2019, April. Australia's faunal extinction crisis: Interim report. Canberra: Australian Parliament House. https://www.aph.gov.au/Parliamentary_Business/Committees/Senate/Environment_and_Communications/Faunalextinction/Interim_report.

Tilbury, D., V. Coleman and D. Garlick. 2005. A National Review of Environmental Education and its Contribution to Sustainability in Australia: School Education. Department for the Environment and Heritage, and Australian Research Institute in Education for Sustainability, Canberra.

Tilbury, D. 2006. Australia's Response to a UN Decade in Education for Sustainable Development. Australian Journal of Environmental Education 22(1): 77–81.

United Nations. 1993. Agenda 21: Earth Summit: The United Nations Programme of Action from Rio. sustainabledevelopment.un.org/content/documents/Agenda21.pdf.

United Nations. 2015. Paris Agreement. Retrieved from https://unfccc.int/sites/default/files/english_paris_agreement.pdf.

UNESCO. 2014. Roadmap for implementing the Global Action Programme on Education for Sustainable Development. Retrieved from unesdoc.unesco.org/images/0023/002305/230514e.pdf.

UNESCO. 2017. Education for Sustainable Development Goals: learning objectives. https://unesdoc.unesco.org/ark:/48223/pf0000247444.

UNESCO. 2020. Education for sustainable development: A roadmap. Paris, France: UNESCO. https://unesdoc.unesco.org/ark:/48223/pf0000374802.

UNESCO. 2021. Berlin Declaration on Education for Sustainable Development. UNESCO World Conference on Education for Sustainable Development, 17–19 May 2021. Retrieved from https://en.unesco.org/sites/default/files/esdfor2030-berlin-declaration-en.pdf.

UNESCO. Harare office. 2021, 20 May. UNESCO declares environmental education must be a core curriculum component by 2025. Retrieved from https://en.unesco.org/news/unesco-declares-environmental-education-must-be-core-curriculum-component-2025-0.

UNESCO-UNEP. 1988. International Strategy for Action in the Field of Environmental Education and Training for the 1990s. UNESCO and UNEP, Paris/Nairobi.

Victoria. Ministry of Education. 1990. Ministerial Policy: Environmental Education. Ministry of Education, Melbourne.

Victoria. Department of Education. 1998. Investing in the Future: Environmental Education for Victoria's Schools. Department of Education, Melbourne.

Wals, A.E.J. 2011. Learning our way to sustainability. Journal of Education for Sustainable Development 5(2): 177–86.

Western Australia. Education Department. 1977. Environmental Education. Policy from the Director-General's Office. No. 8. Government Printer, Perth.

Chapter 18

Environmental and Sustainability Education in Aotearoa New Zealand in the Context of the Sustainable Development Goals

Kerry Shephard,[1,*] *Sally Birdsall,*[2] *Chris Eames*[3] and *Jenny Ritchie*[4]

Introduction

Aotearoa New Zealand has an international reputation for the quality of its natural environment, its cultural diversity, its democratic foundations and its educational processes. Yet the country also maintains a range of unsustainable practices which need to be addressed. This chapter describes and critically analyses Aotearoa New Zealand's historical engagement with environmental and sustainability education (ESE) and its preparedness for implementing ESE measures to contribute to the international achievement of the United Nations' Sustainable Development Goals (SDGs). In this chapter we attempt to distinguish between objectives and achievements, between commitments and evaluated progress, and between academic endeavour and peer-reviewed outcomes.

[1] University of Otago.
[2] University of Auckland.
[3] University of Waikato.
[4] Te Herenga Waka Victoria, University of Wellington.
Emails: s.birdsall@auckland.ac.nz; chris.eames@waikato.ac.nz; jenny.ritchie@vuw.ac.nz
* Corresponding author: kerry.shephard@otago.ac.nz

About Aotearoa New Zealand

In the context of their environment and progress towards sustainable living, the people of Aotearoa New Zealand have much to be proud of. After all, we headline our international presence with expressions such as '100% pure' (New Zealand Tourism 2020) and many readers will be familiar with our 'clean and green' international image. Readers of this book, no doubt, will use appropriate academic insights to interpret the information recorded in this section about Aotearoa New Zealand, and indeed in equivalent sections in other chapters about other countries. Much of the general information reported here about Aotearoa New Zealand has been sourced from the online encyclopaedia (Ministry of Culture and Heritage 2020).

Geologists suggest that Aotearoa New Zealand was once part of Gondwanaland. The islands of New Zealand separated from the rest about 85 million years ago and currently sit across two moving tectonic plates. New Zealanders and our visitors are familiar with earthquakes, volcanoes, hot springs and boiling mud. Being separate from the rest of the world for so long has also had a major impact on the plants and animals that have evolved here. Many are endemic and have evolved without the competitors and predators common in other countries. Some species of bird, for example, do not fly, and so are vulnerable to predators introduced from other countries. Sharing Aotearoa New Zealand with endemic species is both a privilege and a pressing responsibility, as 80% of our 168 remaining bird species are at risk of extinction (Wright 2017).

People also came to inhabit Aotearoa much later than they arrived in other places. Māori people came from other Pacific Islands approximately 800 years ago (Anderson 2016). Europeans first encountered Aotearoa, and the indigenous Māori, in 1642. British settlement proceeded at pace after the signing in 1840 of *Te Tiriti o Waitangi* | the Treaty of Waitangi. This treaty allowed for British settlement whilst making many commitments to Māori, including that they would have equal citizenship rights alongside the British new-comers. The settlers perpetrated significant breaches of the treaty, the impacts and injustices of which are only in recent decades being recognised in law and educational practice and with reparations via a treaty settlement process.

Today, Aotearoa New Zealand is home to a diverse population of approximately five million people, of whom the largest group (70.2%) are of European ancestry, with 16.5% Māori, 15.1% Asian and 8.1% of Pacific Islands descent and other small groups (Stats New Zealand 2019). Recent immigration policies have vastly increased New Zealand's social, cultural, ethnic and linguistic diversity resulting in the classification of 'super-diversity' (Royal Society of New Zealand 2013). Aotearoa New Zealand is governed by a democratically-elected sovereign parliament incorporating proportional representation. Aotearoa New Zealand is part of the British Commonwealth, a member of SEATO (South-East Asia Treaty Organisation), OECD (Organisation for Economic Co-operation and Development) and formerly of the ANZUS Pact (Australia, New Zealand, and United States) but retains its own distinctive stance on international relations. The departure from ANZUS was a result of the government adopting a nuclear-free stance in 1987. Prime Minister Jacinda

Ardern declared a climate change emergency in December 2021, describing this as 'one of the greatest challenges of our time' (Taylor 2020, para.1).

More than one-third of New Zealanders live in the greater Auckland city area and overall fewer than 15% of New Zealanders live in rural locations. So we have a lot of open spaces and this is no doubt beneficial for our two major foreign income earners: agriculture and tourism, the latter greatly impacted by the COVID-19 pandemic. Aotearoa New Zealand has diverse agricultural and horticultural industries, exporting timber, lamb, beef, dairy goods, wine, kiwifruit and apples amongst others. Tourists are primarily attracted to New Zealand (at least pre-Covid) because of our open spaces, mountains, natural areas and countryside, rather than to our cities. Our limited infrastructure (very few railway lines or motorways) contribute to tourists' appreciation for open spaces and access to wildlife. By reputation, New Zealanders are also passionate about the outdoors and regularly undertake activities to make the most of our spectacular landscape. On the international stage, New Zealanders are renowned for their love of climbing, walking, skiing, angling and sporting achievements in netball, cricket, rugby and sailing. Linked to our open spaces and mountains, we have access to abundant sources of relatively-green energy such as hydro-electricity and wind.

It is perhaps surprising that New Zealanders, although benefiting from relatively high per capita GDP, are amongst the highest per capita contributors to greenhouse gas emissions in the world, that our generosity in the form of overseas aid is not at the best end of the OECD spectrum, and that as a country we are far from eliminating poverty, ill-health and educational inequities (see for example, https://data.oecd.org). We would also note the negative social statistics pertaining to Māori, which are associated with the ongoing legacy of the impacts of colonisation and have been linked to recent neoliberal social and economic policies (Marriott and Sim 2014, Walker 2004).

The formal education system

The current Aotearoa New Zealand education system largely reflects our British colonial past. There are well-established structures of early childhood care and education (birth to 5 years of age), primary (5–12 years), secondary (13–18 years) and tertiary education (universities and vocational institutes). Significantly, there are also Māori language educational pathways at all of these levels, specifically *kōhanga reo, kura, wharekura* and *whare wānanga*.[1] Three and four-year-olds are entitled to 20 hours per week of government subsidy for early childhood education, whilst primary and secondary schooling is free and compulsory for all children to the age of at least 16. Māori education systems, along with the Māori language, were largely suppressed throughout much of Aotearoa New Zealand's history, but since the 1970s, as a result of ongoing Māori activism, a greater recognition of the intent of *Te Tiriti o Waitangi* | Treaty of Waitangi obligations has seen the development of

---

[1] *kōhanga reo, kura, wharekura* and *whare wānanga*, respectively early childhood, 'school', secondary and tertiary educational entities, generally where education is undertaken in *te reo* Māori (Māori-language).

Māori language immersion schools from early childhood to tertiary education. These remain, however, a small minority of schools and 97% of Māori children attend English-medium schooling (Education Review Office 2020). This recognition of *Te Tiriti* | Treaty obligations has extended into the current expectation of the inclusion of the Māori language and culture in all schools, and a greater recognition of these aspects in wider society.

Early childhood education - Te Whāriki

On its launch in 1996, the first national early childhood education curriculum, *Te Whāriki*, (Ministry of Education 1996) was widely lauded for its innovative bi-cultural, socio-cultural, holistic and integrative nature. The 2017 revision (Ministry of Education 2017a) unfortunately fails to acknowledge the current challenges of super-diversity, sustainability or the SDGs. It does, however, expect teachers to recognise *Te Tiriti* | the Treaty, the status of Māori as *tangata whenua* (indigenous to Aotearoa) and the relationship Māori have to their tribal lands and to *Papatūānuku*, the Earth Mother. Teachers are to share local Māori histories and place-based knowledge with children and to foster their connection to *Papatūānuku*. Furthermore, teachers are to encourage children's understandings of *kaitiakitanga* (stewardship of the environment) and provide them with opportunities to engage in the responsibilities of being *kaitiaki* (stewards) by, for example, 'caring for rivers, native forest and birds' (Ministry of Education 2017a, 33).

School education

Schools in Aotearoa New Zealand are expected to guide their work using The New Zealand Curriculum (NZC) in English-medium schools (Ministry of Education 2007) or *Te Marautanga o Aotearoa* in Māori-medium schools (Ministry of Education 2017b). These curricula are relatively non-prescriptive and schools are encouraged to undertake local curriculum development alongside their communities.

ESE is not a mandated learning area in the NZC. Instead, it is diffused throughout the curriculum in a thematic manner. For example, ecological sustainability is a suggested value to be modelled, explored and encouraged in teaching, and sustainability is a Future Focus Principle that can underpin teaching and learning. The NZC also includes a Vision statement, where the aspiration of students being connected to and actively involved in creating a sustainable social, cultural, economic and environmental future for their country is articulated. There are specific achievement objectives in the NZC that address environmental and sustainability aspects in the science, social science, technology and health learning areas. Notably, the most significant sustainability issue facing us today, climate change, does not explicitly feature at all in the NZC (Eames 2017), and its inclusion in schools is highly variable and often weak, despite schools believing it to be important (Bolstad et al. 2020).

According to the English translation of *Te Marautanga o Aotearoa*, teachers are to ensure that students 'are encouraged to practise sustainability of the land and the natural environment, and to consider the relationship between practice and the community. The spiritual and the physical aspects of people and the land are one' (p.62). This message about the inclusion of sustainability is infused through *Te Marautanga o Aotearoa*. It is inherent in its principle: '*ko te oranga taiao, he oranga tangata*' (the health of the environment is the health of the people) (p.7). As with the NZC, the *Putaiao* (science) and *Hauora* (health) subject areas refer to the environment and to human relationships with the environment.

Secondary qualifications

Student achievement in secondary schools is largely assessed by the National Certificate of Educational Achievement (NCEA) (New Zealand Qualifications Authority 2020). This is a standards-based assessment system and drives teaching and learning in the final three years of secondary school (15–18 years of age). Within broad constraints, schools choose which standards to offer and students select from within these. Since 2009, these have included a set of standards in sustainability which are widely regarded as appropriate and stimulating for student learning. However, for a variety of reasons including lack of teacher knowledge, challenges in subject timetabling, and interdisciplinary teaching and learning, the uptake of these standards by schools and students has remained low since their introduction.

Teacher education

We know that teacher education, both pre-service and in-service, is a powerful determinant of teacher capability and motivation to teach something. In Aotearoa New Zealand, pre-service teacher education opportunities in ESE are very restricted. Initial teacher education is predominately provided by the eight public universities with a lesser number of initial teacher education programmes in other tertiary institutes. Few of these tertiary entities offer courses with a clear ESE focus, with some notable exceptions at the undergraduate and postgraduate level (such as the University of Waikato, Te Herenga Waka Victoria University of Wellington, and Ara Institute of Technology). Other tertiary institutes strive to include this learning where they can. Teacher professional development in ESE was provided by the central government from 2002–2009 (Eames et al. 2010), but has not been provided since. However, non-government agencies (NGOs), sometimes funded indirectly by the central and local governments, have attempted to fill that gap, and although constantly under-resourced, have achieved some significant results, which are discussed later in this chapter.

Tertiary

For such a small country, Aotearoa New Zealand has a very complex tertiary education sector. There are more than 900 separate tertiary providers including universities, institutes of technology and polytechnics, industry training organisations, private training establishments and *whare wānanga* (providing tertiary education over a range of qualifications based on Māori principles and values), in addition to a limited number of community education providers and secondary schools involved in tertiary education. From our population of five million, at any one time, approximately 400,000 students are enrolled in tertiary education. Many tertiary education providers are empowered to teach and to award at both undergraduate and postgraduate levels, although notably the quality assurance processes that apply to universities are different from those for the other parts of the tertiary sector.

Although each of New Zealand's eight universities has significant independence, a range of government measures directs many of their actions; for example, processes directed at improving Māori and Pacific Islands student enrolment, retention and success. All universities and a number of other tertiary entities offer courses related to the environment and to sustainability, with some recent initiatives in sustainable practice and climate change, indicating a desire to include interdisciplinary teaching and learning that could address a number of the Sustainable Development Goals.

Non-formal and informal education

Beyond the formal education sector, ESE is a mix of provision by the central and local governments, and NGOs. The Department of Conservation (DOC, central government) manages the conservation estate of Aotearoa New Zealand, land and water areas that account for more than one-third of the country (DOC 2020). The DOC runs regular public education initiatives that focus on conservation of species and ecosystems. Local territorial authorities (regional and district councils) are required to look after the environment within their places, and work with industry, farmers, and communities to plan and manage natural resources and systems. NGOs (independent or sometimes supported by authorities) also offer opportunities to their communities to learn about and engage in environmental and sustainability initiatives.

Development of environmental and sustainability education in Aotearoa New Zealand

Since at least the 1970s, a network of environmentalists and educators have been active both in lobbying for the development of national ESE policy and curriculum, and in developing flaxroots[2] practice in Aotearoa New Zealand schools. This is evident in the convening of conferences; the formation of the New Zealand Association for Environmental Education (NZAEE); the design and teaching of optional pre-service environmental education courses (Law 1996); schools developing their own school-based EE curricula (Springett and Buchanan 1991); and in the development of a sequence of national environmental education strategies. At a policy level, from the ground-breaking policies of the Resource Management Act of 1991, which required local authorities to co-manage their local environments with their communities, to the partnerships between the governmental Department of Conservation and community-based NGOs, and between those organisations and schools, many New Zealanders of all ages have been engaged in ESE.

---

[2] Flaxroots is a Aotearoa New Zealand term for activities happening from the ground up (ca. grassroots). Flax (harakeke) is a *taonga* (treasure) plant for us.

From this early flaxroots activity around environmental science and outdoor education, notions of ESE first appeared in a national school curriculum in 1993 and now have a somewhat undefined place in the current curriculum (Ministry of Education 2007). Production of government guidelines for ESE (Ministry of Education 1999) and the development of a successful non-government programme, Enviroschools (Enviroschools 2020), have created a groundswell of activity that has seen around a third of all New Zealand schools and many early childhood education settings offering an ESE programme (Eames and Mardon 2020).

The enviroschools programme

The impact of attending the Earth Summit of 1992 in Rio de Janeiro led a group of Hamilton City Councillors to begin a pilot eco-school programme with the help of the University of Waikato in Hamilton. This pilot developed into the Enviroschools Programme. Early in its genesis, the programme reflected a *Tiriti o Waitangi* partnership model, with Te Mauri Tau, a community-based Māori educational, environmental and health organisation, working alongside local government in a facilitated developmental process. Underpinned by a strong commitment to *tikanga* and *te ao* Māori (Māori values and worldview), sustainable communities and whole school approaches, the programme aims to empower students to take action for real change. The programme now has a mix of central and local government funding.

Independent evaluations conducted in 2010, 2014 and 2017 showed that schools were undergoing organisational change 'in (the) development of more sustainable practices, in particular waste, energy and water use, more sustainability content in the curriculum, and improvements to the physical surroundings of the school' (Eames et al. 2010, iii). Importantly, the programme appeared to be developing student outcomes such as knowledge development, action-taking, increased engagement in learning, and transfer of learning from school to the home environment (Eames et al. 2010). More recently, '90% of Enviroschools that participated in the 2017 Census reported connecting with their communities through organisations such as local authorities, government agencies, ecological restoration groups, *iwi* (Māori tribes) and businesses.' Of all Enviroschools, 'over 90% reported being engaged in *kai* (food) production and distribution, while approximately 90% were involved in protecting/restoring biodiversity and 80% in water health and conservation' (Eames and Mardon 2020, 56–57). The long-term nature of a school's developmental journey as an Enviroschool and the support of a trained facilitator have been identified as key factors in these outcomes.

The New Zealand association for environmental education

The New Zealand Association for Environmental Education (NZAEE) was established in 1984 and is the 'peak body' for a diverse group of teachers, educators, policy-makers, academics, and service providers for whom ESE is core to their professional work. The NZAEE has run biennial conferences to support sharing and professional learning amongst its community, as well as providing advocacy and other services. It has also provided umbrella services for ESE initiatives, most notably providing this for Enviroschools in their early days, and the equally successful Seaweek (Seaweek 2020).

Requirements of the Teaching Council Aotearoa New Zealand

The Teaching Council of Aotearoa New Zealand is the body that oversees the teaching profession, including teacher education provision and the registration and certification of teachers. The current (2017) Education (now Teaching) Council's code of professional responsibility and standards for the teaching profession requires all teachers to commit to, for example, the following items:

- Promoting and protecting the principles of human rights, sustainability and social justice
- Demonstrating a commitment to a *Tiriti o Waitangi*-based Aotearoa New Zealand
- Fostering learners to be active participants in community life and engaged in issues important to the well-being of society (Education Council Aotearoa New Zealand 2017, 12)
- Under the standard 'design for learning', teachers are required to:
  - Design and plan culturally responsive, evidence-based approaches that reflect the local community and *Te Tiriti o Waitangi* partnership in Aotearoa New Zealand.
  - Harness the rich capital that learners bring by providing culturally responsive and engaging contexts for learners.
  - Design learning that is informed by national policies and priorities (2017, 20).

The extent to which this code and standards are enacted within New Zealand's schools is addressed further in this chapter. Given the recent parliamentary declaration of a climate emergency, it is to be hoped that environmental sustainability will become a strong focus in education in Aotearoa New Zealand via Teaching Council overview.

On indigenous knowledge and its links to ESE

Magni (2016) has suggested that the knowledge of indigenous people, being 'deeply rooted in their relationships with the environment as well as in cultural cohesion' … 'has allowed many communities to maintain a sustainable use and management of natural resources, to protect their environment and to enhance their resilience; their ability to observe, adapt and mitigate has helped many indigenous communities face new and complex circumstances that have often severely impacted their way of living and their territories' (p.3). Within Aotearoa New Zealand, *te ao* Māori (the Māori worldview) places a strong emphasis on the interconnectedness of people to planetary well-being in that it positions people along with trees, birds, insects and so on as the inter-related offspring of *Papatūānuku* and *Ranginui* (the Earth Mother and Sky Father). This inter-relationality and interconnectivity is recognised and sustained through the maintenance of spiritual forces that must be kept in balance through upholding core Māori values, respectful rituals and practices that sustain the well-being of all. Māori and other Pacific Islands peoples' onto-epistemologies have been described as being 'relational, functional, and contextualized,' grounded in principles of reciprocity and respect that provide 'balance and harmonious existence, a holistic type of sustainability' which 'advocates a spirit of preservation that Pacific peoples identify and align with in their ecologically anchored social/cultural practices' (Vaioleti and Morrison 2019, 654).

Sustainability's fourth pillar of cultural sustainability (social, economic and environmental) is thus of critical, integral importance in enabling the recognition, articulation and enactment of indigenous knowledges for sustainability.

Understanding of environmental and sustainability education in Aotearoa New Zealand

Our understanding of the trajectory and impact of ESE within Aotearoa stems from research and evaluation studies and commentaries. At the policy level, two important commentaries have highlighted the challenges of establishing guidelines for environmental education in New Zealand schools (Law and Baker 1997), and the influence of the political context on the development of ESE (Chapman 2011). Independent reports have highlighted the potential and directions for what was termed education for sustainability, but also the lack of progress in implementation and support by the central government (Parliamentary Commissioner for the Environment 2007). Major evaluative studies of school-based ESE were undertaken in 2002–3 and in 2009 which provided evidence for the development of ESE and its outcomes in Aotearoa. The first study concluded that 'some schools believe they are achieving significant educational and environmental outcomes. These achievements appear to be linked to strong leadership and whole school support, availability of professional development and strong links to environmental agencies. Yet many challenges remain for schools to develop EE practice through school-based curriculum development' (Eames et al. 2008, 46). The second report, which looked into three professional development programmes—the Enviroschools Programme, a Māori immersion programme (*Mātauranga Taiao*) and a national team of education for sustainability facilitators—found that 'each initiative was achieving greater inclusion of sustainability content and more integrative teaching across the curriculum; the development of facilitative teaching styles that were empowering students to become strongly engaged in their learning and to think critically about issues; and the development of sustainable practices in schools and their communities' (Eames et al. 2010, 2). However, challenges remain for integrating ESE in large primary and secondary schools and developing a coherent Aotearoa ESE strategy. Recent research on how schools are addressing climate change by Bolstad (2020) reinforces these concerns.

Aotearoa New Zealand's eight universities are very research active and are all ranked relatively highly in world ranking systems. Not surprisingly, these universities conduct a wealth of research on topics relating to the sustainable development goals, and much of our understanding about the pedagogy and outcomes of ESE in Aotearoa has emerged from academic and postgraduate-student thesis work in our universities. To illustrate the depth and breadth of some of these studies, they have shown us:

- *In teacher education*: The importance of the development of teacher pedagogical content knowledge for implementing ESE (Eames and Birdsall 2019); how teachers develop their ESE curriculum in their schools (John Lockley PhD) and develop a whole school approach to ESE from scratch (Tatiana Kalnins PhD); the preparedness of beginning teachers to implement ESE in their classrooms (Deborah Bandele PhD, Birdsall 2013); the potential for social media to assist in professional learning for dispersed ESE teachers (Fariba Mostafa PhD).

- *In student learning*: How mobile learning (Eames and Aguayo 2019) or permaculture (Nelson Lebo PhD) can be used to enhance student ecological literacy; how student environmental identity developed through the Enviroschools Programme endures through the teenage years (Eames et al. 2018); how youth leadership can be developed through ESE (Charlotte Blythe PhD); how secondary students perceive their role as agents of pro-environmental change in schools (Amber Pierce PhD); how students learn to take action (Birdsall 2010); the role of youth climate activism (Nairn 2019); and what features of a master's programme contributed to paradigm change in, and action competence of, its participants (Piasentin and Roberts 2018).

- *In tertiary education*: A model for university leadership for ESE (Pam Williams PhD); how to develop and use tools and processes for monitoring the development of sustainability-related values and attitudes in cohorts of undergraduate students (Shephard et al. 2015); that New Zealand academics have diverse opinions about what it is that they should be teaching in ESD contexts (see as examples, Brown et al. 2019, Shephard and Furnari 2012).

- *In indigenous knowledge*: The value of indigenous knowledge in ESE (Parsons et al. 2017).

Collectively this research emphasises the connections between our ESE and the features of our educational system in the context of our country. Perhaps most striking about Aotearoa New Zealand's understanding of ESE, in comparison with that of many other countries, is our increasing commitment to *Te Tiriti o Waitangi* | the Treaty of Waitangi and its integration within our ESE policies and strategies at all levels. As will be explored in subsequent sections of this chapter, we are far from systematically evaluating the impact of our educational efforts.

ESE, ESD, the GAP and 'ESD for the SDGs' in Aotearoa New Zealand

The international Decade of Education for Sustainable Development (2005–2014) and the following Global Action Programme (GAP) on ESD (2015 to 2019) are likely to have impacted educational developments in Aotearoa New Zealand at a strategic level. No doubt our strategies do impact what happens in our schools, universities and more widely. It is to be hoped that the 2030 UN Agenda for Sustainable Development, incorporating the 17 Sustainable Development Goals and

169 targets, will impact education in Aotearoa more visibly. But it could also be argued that the international drivers that have shaped these international endeavours have also had a direct impact on Aotearoa New Zealand themselves, rather than via UN-sponsored constructs.

For example, with clear links to international trends, Aotearoa New Zealand has structured its research funding in ways linked both to national productivity and to environmental and social justice gains. The National Science Challenges (Ministry of Business, Innovation & Employment 2016a), for example, demand measurable and visible contributions to environmental and social well-being, with emphasis on partnership and cooperation. There are also strong links between these ideas and the national strategic plan, 'A Nation of Curious Minds,' that promotes a scientifically and technologically engaged public and a publicly-engaged science sector (Ministry of Business, Innovation & Employment 2016b). Similarly, and also at a national level, the Environmental Education for Sustainability Strategy and Action Plan (2017–2021) could be seen as an umbrella educational strategy, mapping out 'how government agencies would work collaboratively with communities, local government, *whānau, hapū* and *iwi*,[3] research institutes, non-governmental organisations (NGOs), businesses, schools, *kura,* tertiary institutions and volunteers' (Department of Conservation 2017, n.p.). While some ministries and departments have particular responsibilities, the strategy applies to the whole of government.

Even so, it is challenging to confirm how these far-reaching, high-level strategies have directly influenced the policies, strategies and work of educational organisations. Most universities, for instance, have developed their own strategic frameworks in support of their own sustainability missions. The University of Otago, for example, operates its own Sustainability Strategic Framework 2017–2021 (University of Otago 2017) promising to (as examples) nurture a culture of sustainability and to increase sustainability literacy among students. The University of Auckland has over many years developed a wide range of structures and initiatives pertaining to sustainability, sufficient for it to be recognised as #1 globally in The Times Higher Education Impact Rankings in 2020, running alongside more conventional rankings, to address impact with respect to the SDGs. One quality measure, for example emphasising SDG17, is the proportion of academic publications with at least one co-author from a developing country (Times Higher Education 2020). The University of Auckland, and Aotearoa New Zealand universities generally, are highly networked internationally. In recognition of this work over many years, the University of Auckland has been designated by the United Nations as the international hub for SDG 4 (Academic Impact 2020).

PISA data and OECD comparisons

Data from the OECD's 2015 Programme for International Student Assessment (PISA) was analysed to explore the levels of environmental knowledge and awareness of Aotearoa New Zealand's English-medium secondary students. The results

---

[3] *whānau, hapū* and *iwi*, respectively family, cluster of related families and tribe.

were troubling as these 15-year-olds had low levels of environmental knowledge, environmental awareness and optimism about environmental issues improving. For example, they reported relatively low levels of ability to explain environmental issues such as the increase in greenhouse gases (ranked 58th out of 71 countries) or the extinction of plants and animals (68/71) when compared to their peers in other countries (Ministry of Education 2019). Moreover, their level of environmental awareness was very low, ranked 63rd out of 71 countries according to the 2015 PISA index. Added to these results, awareness of environmental issues among Aotearoa New Zealand students showed a moderate and statistically significant decrease between 2006 and 2015.

A possible reason for these results relates to scientific literacy scores. It seems that there is a relationship between science literacy and environmental awareness, as those students who showed higher environmental awareness also had higher scientific literacy scores (Ministry of Education 2019). Since the science literacy scores of Aotearoa New Zealand students have dropped 17 points between 2006 and 2015, the drop in levels of environmental awareness and knowledge may be consequential to the drop in scientific literacy. While this relationship could suggest that more environmental topics should be taught as part of science, more than knowledge is needed. Students also need to develop optimism and agency to empower them to tackle environmental issues.

ESE and the SDGs in Aotearoa New Zealand

Education is described as one of the 'fronts with high potential' for implementing efforts towards a more sustainable world (UNESCO 2005, 11). Due to this potential, teachers have a role as change agents as they shape the development of knowledge, attitudes and skills of future generations. Hence, the manner in which they incorporate SDGs into their teaching and learning activities is important. Teachers' incorporation of SDGs was explored through a survey of NZAEE members and other environmental educators at the end of 2018 by one of this chapter's authors (SB) and a preliminary analysis is presented here. Respondents were asked to rank the SDGs in order according to the amount of time they spent on each one in their education activities. Seventy Aotearoa New Zealand educators responded; 19 males and 51 females.

Table 1 presents the data with high and low response frequencies to illustrate clustering of response rates. The data shows that those Aotearoa New Zealand educators who responded spend the most time on SDGs 3, 4, 6, 7, and 10 as these SDGs have the highest response rates in the highest rank. As these SDGs relate to social justice, it would seem that Aotearoa New Zealand educators focus more on social sustainability. This focus on social sustainability is also evident in that, along with SDG 12, these educators prioritise time for learning on the first eight SDGs, and especially on SDGs 1, 2, 3, and 5. In contrast, less time is spent on SDGs 11, 13–17, with SDG 17 accorded the least amount of time.

There are interesting patterns to note in terms of individual SDGs. For example, for SDG 5, Gender Inequality, most of the responses were ranked in the top half with clusters in the 2nd and 3rd places. However, there was also a clustering in

Table 1. Frequency of Aotearoa NZ EE educators' incorporation of SDGs in ranked order (n = 70).

	SDG1	SDG2	SDG3	SDG4	SDG5	SDG6	SDG7	SDG8	SDG9	SDG10	SDG11	SDG12	SDG13	SDG14	SDG15	SDG16	SDG17
1st	5	3	7	8	6	9	5	1	4	7	4	3	2	2	3	0	0
2nd	7	3	2	5	10	7	8	4	4	7	3	4	1	1	1	1	2
3rd	10	5	11	3	11	7	7	7	1	3	0	2	0	5	2	0	0
4th	0	2	2	4	1	7	6	8	5	4	6	11	5	5	1	3	0
5th	1	5	6	0	2	3	6	9	6	4	5	4	7	6	4	2	0
6th	8	18	10	8	8	5	6	3	5	0	0	2	0	0	0	0	0
7th	1	7	7	8	1	6	9	9	8	6	4	0	2	1	0	1	0
8th	13	14	10	7	7	3	2	5	2	2	2	1	1	1	0	0	1
9th	0	2	4	2	2	1	2	2	6	9	6	10	10	5	8	0	4
10th	0	0	0	1	0	1	0	1	2	8	11	8	9	8	4	13	4
11th	18	5	3	7	1	3	1	1	3	1	8	10	2	3	3	1	0
12th	4	0	0	3	4	5	2	5	7	3	5	8	13	5	1	4	1
13th	0	0	0	1	0	1	0	0	2	0	0	2	9	11	12	8	24
14th	2	4	4	8	9	8	3	5	6	0	2	0	2	10	4	3	0
15th	0	1	1	1	5	2	4	4	5	3	4	2	3	6	18	8	3
16th	0	0	3	1	1	2	4	0	2	5	5	0	4	3	8	22	10
17th	0	2	1	3	2	1	5	6	3	7	5	3	0	2	1	4	25

Key:

Dark Grey – frequency of 10 or more

Medium Grey – frequency of 8-9

Light Grey – frequency of 1

Very Light Grey – frequency of 0

Note:

For each SDG, the relative positions of clusters of high rank responses suggests the degree to which educators are incorporating each SDG within their teaching. For SDG3 for example, most high rank responses are in the top half of the table so this SDG is highly incorporated. For SDG16, on the the other hand, most high rank responses appear in the lower half of the table, indicating that this SDG is less incorporated.

14th place, perhaps indicative that for some educators, this SDG was not as important. SDG 4 Quality Education also has widely spread rankings, with clustering points at 1st, 6th, 7th and 14th places, which is indicative of a wide variety of opinions about its importance.

Three SDGs seemed to be regarded as of some importance with clusters of higher responses in the middle of Table 1. For example, SDG 9 Industry, Innovation and Infrastructure, has one cluster in 7th place, SDG 10 Reduced Inequalities has clusters at 9th and 10th, and SDG 11, Sustainable Cities and Communities, has clustering at 10th and 11th.

SDG 12, Responsible Consumption, has a cluster in 4th position, but also a robust clustering at the 9th, 10th, 11th and 12th rankings, a cluster that makes up nearly 50% of responses. This finding suggests that for some educators, this SDG is important, but for nearly half of them, it is not as important as the social sustainability-focused SDGs. The Climate Action SDG 13 also has a strong clustering in the middle of the rankings with 61% of responses in the 9th to 13th rankings, perhaps also highlighting these educators' preferred focus on socially-related sustainability.

What is surprising, given New Zealanders' high use of their environment and valuing of its unique biodiversity of flora and fauna, is the ranking of SDG 13 Life Below Water and SDG 14 Life on Land. Both these SDGs had responses with high clustering rates near the bottom of Table 1, indicating that less time is spent on these SDGs than on others. With many environmental education programmes focused on conservation and restoration of the environment, this is surprising. For example, Bolstad et al. (2015) report that DOC programmes have a strong focus on community and place-based conservation projects. National organisations such as Experiencing Marine Reserves, and the Mountain to Sea Conservation Trust run education programmes related to protecting and conserving marine environments.

Both SDG 16 Peace, Justice and Strong Institutions and SDG 17 Partnerships had responses firmly placed in the bottom half of the rankings and very few in the top half. These two SDGs seemed to be the least important, particularly SDG 17 with 50% of responses in the bottom two rankings.

The data illustrates a 'tale of two halves,' with these educators prioritising the first eight SDGs and their focus on social sustainability. However, these findings are only a snapshot and the level of the respondents' familiarity with the nature of each SDG was not clear. Given these limitations, further exploration into how SDGs are being incorporated into environmental education activities is warranted.

Waikato RCE ESD 2014 and Otago RCE ESD 2020

Aotearoa New Zealand has two Regional Centres of Expertise (RCEs) on ESD (Education for Sustainable Development). The first was established at the University of Waikato in 2014 and more recently, Otago's was established in 2020. As with other RCEs internationally, these too are committed to the realisation of the Sustainable Development Goals through processes that mainstream ESD in formal, informal and non-formal educational settings. Otago's RCE is being developed in line with the principles of evaluative enquiry (RCE Network 2020).

Australasian campuses towards sustainability

Many campus-based institutions (primarily universities and polytechnics) in Aotearoa New Zealand (and Australia) are members of Australasian Campuses towards Sustainability (ACTS 2020). ACTS builds cross-sector partnerships, bringing together sustainability educators, practitioners and change-makers to promote and support change towards best practice sustainability. ACTS provides competitive awards, online guidelines, professional-development support and a forum within which sustainability-professional representatives of institutions can share best practices. It is closely affiliated with other national bodies, such as the UK's NUS (National Union of Students) and the USA's AASHE (Association for the Advancement of Sustainability in Higher Education). It would be challenging to evaluate the impact that membership in this organisation has on Aotearoa New Zealand's contribution to the SDGs, but clearly much of its impact would likely occur via its interaction with, and support for, formal, campus-based education institutions.

Although this article does focus on the implementation of ESE, partnership is key as per SDG17. This mission is broader than education and will involve education being integrated within a wide range of governmental and independent initiatives. Several initiatives in Aotearoa New Zealand are noteworthy in this context.

- At about the same time that the international community was deliberating on the SDGs, Aotearoa New Zealand was developing its own Living Standards Framework (LSF) (LSF, Treasury 2019), broadly based on the OECD Wellbeing Framework. The LSF was initially focused on supporting New Zealanders' opportunities, capabilities and incentives, but has developed into a structure with considerable overlap with the ideals of sustainability and with the achievement of the SDGs (Treasury 2018). Naturally, the LSF focuses on economic measures but argues that economics does provide powerful tools for thinking about policy possibilities and consequences. The LSF dashboard provides a wide range of data and commentary on capitals (natural, social, human, physical/financial) over a range of well-being domains. Education is a major element of the LSF, although, notably, ESD/ESE is not itself a measure or indicator within it as yet. Aotearoa New Zealand does benefit from a strong science base on which to build some parts of the LSF, having, for example, extensive data-collecting capacity with respect to climate, atmosphere and Aotearoa New Zealand's contribution to climate change (see for example Stats New Zealand 2020).

- More specifically focused on the SDGs, Aotearoa New Zealand's first national review of progress occurred in 2019 (Foreign Affairs and Trade 2020). Education is widely referenced in the resulting report, most strongly of course with respect to SDG4 (Quality Education), but there is one paragraph that specifically addresses teaching associated with sustainable development and global citizenship. Other than noting that our curriculum supports learning in the context of our educational strategies, no particular data are provided on relevant outcomes and outputs in this broad domain. Nevertheless, broadly-based national reviews do provide excellent opportunities not only to bring together data to support our national assertions but also to identify where the processes needed to gather such data may be inadequate or missing altogether.

- A range of other initiatives need to be identified. Aotearoa New Zealand works alongside Australia and Pacific Nations as an entity in the UN's Sustainable Development Solutions Network (SDSN 2020). This group has organised two summit meetings to promote communication and networking. Taking pride in its 'unofficial' status is SDG.org.nz, an organisation that created the People's Report (SDG.org.nz 2019) with an extensive commentary by many contributors on Aotearoa New Zealand's approach to achieving the SDGs based on a national survey of 187 participants.

Emerging issues, trends and futures for environmental and sustainability education in Aotearoa New Zealand

Although Aotearoa New Zealand clearly has some sophisticated policy, strategy and outcome-oriented educational frameworks in operation, some general trends and issues are notable.

Perhaps most apparent is the lack of organised and committed educational leadership for ESE, which appears for many as just another in a long list of worthy causes and educational enterprises. As a consequence, many developments occur with a pronounced bottom-up emphasis, based on the enthusiasm of individuals and small groups but naturally lacking integration, interconnectedness or unified oversight. From an ESE perspective, this is particularly worrying as we may not be adequately building on the enthusiasm of children and students, or on the insights that come from Aotearoa New Zealand's unique indigenous heritage, and may even be in danger of alienating those who we most wish to support. Aotearoa New Zealand clearly needs educational leaders in policy-making positions who are able to identify ESE as something fundamentally more important than many other pressing educational issues.

Despite the international prowess of our universities, there is a profound lack of research and especially of evaluation of the extent to which Aotearoa New Zealand is contributing to the international achievement of the SDGs via education, or even if we are moving in the right direction within our own institutions. In this context, international literature emphasises the importance of evaluative enquiry; this is something that our universities really do need to grasp if they are to contribute, not only to the achievement of SDG4, but also as they must to the achievement of all 17 SDGs. Similarly, there is a real danger in Aotearoa New Zealand that via the specific interests of individual scholars, our universities will contribute to the achievement of particular SDGs rather than to the achievement of all goals. Aotearoa New Zealand surely needs some effective integrating strategies, with funding, perhaps incorporating an expectation that many of our scholars develop interests and abilities in aspects of ESE and the SDGs, rather than depending on the good work and good nature of individuals with their own particular interests.

Similarly, an issue that often emerges from research, and particularly evaluative studies, is the balance of focus on particular environmental/sustainability outcomes compared to more general educational outcomes. This tension often stems from the researchers considering the particular sustainability-focused interests of the end-users of their work, be they funders, policy-makers or educators, rather than, or at least including, related, more broadly-based learning objectives within the institution of education itself. ESE must be seen as a partnership between making a difference and learning to make a difference, as the latter is more likely to develop learner motivation and transferable skills for new contexts.

The UNESCO Chair in Reorienting Teacher Education towards Sustainability has generated an international network, recognising that teacher education institutions are key in the process of fostering intergenerational change (Hopkins and McKeown 2005). More recently, and now that sustainability is required as a key commitment of teachers (as required by the Teaching Council Aotearoa New Zealand), it remains to be seen how well this particular aspect is taught and assessed in pre-service teaching qualifications, and whether, as part of its programme for alleviating climate change, for example, the government will provide in-service professional learning opportunities for teachers, to engage them in ways to deliver on this commitment. We encourage our Teaching Council and Government to evaluate the attainment of these important commitments to change.

In Conclusion

As environmental and sustainability educators and researchers, our purpose is to advocate for, research and implement an education that enables learners to work towards a sustainable future as imagined by the global achievement of the SDGs; and, in this chapter, to comment on our nation's progress. We give Aotearoa New Zealand a bare pass mark in this endeavour to date. We have some strong policies, strategies and initiatives already underway, and no shortage of activists, educators and role models with passion for ESE. But we emphasise the need for: organised and committed leadership towards environmental and sustainability education; research and evaluation of educational change and impact; and education and training to ensure that our institutions develop scholars capable of these things. We also emphasise the need, for the benefit of our planet, to plan for the achievement of all 17 SDGs, rather than to focus on those that address our particular nation's circumstances. To make real progress, we now need to commit to the infusion of the SDGs throughout our educational sectors, in a coordinated, meaningful way based on respect for future generations.

References

Academic Impact. 2020. SDG 4 (Quality Education). https://academicimpact.un.org/content/sdg-hubs.
ACTS. 2020. Australasian Campuses towards Sustainability. https://www.acts.asn.au.
Anderson, A. 2016. The First Migration: Māori Origins 3000BC–AD1450. Bridget Williams Books.
Birdsall, S. 2010. Empowering students to act: learning about, through and from the nature of action. Australian Journal of Environmental Education 26: 65–84.

Birdsall, S. 2013. Measuring student teachers' understandings and self-awareness of sustainability. Environmental Education Research 20: 814–835.

Bolstad, R., C. Joyce and R. Hipkins. 2015. Environmental education in New Zealand schools. Research update 2015. NZCER. Retrieved from https://www.nzcer.org.nz/system/files/EE%20Update%20 Report%20Final%202015_1.pdf.

Bolstad, R. 2020. Climate change and sustainability in primary and intermediate schools: Findings from the 2019 NZCER national survey of English-medium schools. New Zealand Council for Educational Research. https://www.nzcer.org.nz/research/publications/climate-change-and-sustainability-primary-and-intermediate-schools.

Brown, K., S. Connelly, B. Lovelock, L. Mainvil, D. Mather, H. Roberts, S. Skeaff and K. Shephard. 2019. Do we teach our students to share and to care? Research in Post-Compulsory Education 24: 462–481. https://doi.org/10.1080/13596748.2019.1654693.

Chapman, D.J. 2011. Environmental education and the politics of curriculum: a national case study. The Journal of Environmental Education 42: 193–202.

Department of Conservation. 2017. Environmental education for sustainability actions | Mātauranga Taiao mo nga mahi whakauka – Nga wehenga. https://www.doc.govt.nz/about-us/our-policies-and-plans/ education-strategies/environmental-education-for-sustainability-strategy-and-action-plan/.

Department of Conservation (DOC). 2020. Annual report for the year ended 2020. https://www.doc.govt. nz/about-us/our-role/corporate-publications/annual-reports-archive/annual-report-for-year-ended-30-june-2010/1-introducing-the-department-of-conservation/1_1-the-nature-and-scope-of-docs-functions/.

Eames, C., B. Cowie and R. Bolstad. 2008. An evaluation of characteristics of environmental education practice in New Zealand schools. Environment Education Research 14: 35–61. doi:10.1080/13504620701843343.

Eames, C., J. Roberts, G. Cooper and R. Hipkins. 2010. Education for sustainability in New Zealand schools: An evaluation of three professional development programmes. 1–302. Ministry of Education, Wellington, New Zealand.

Eames, C. 2017. Climate change education in New Zealand. Curriculum Perspectives 37: 99–102. Doi: 10.1007/s41297-017-0017-7.

Eames, C., M. Barker and C. Scarff. 2018. Priorities, identity and the environment: Negotiating the early teenage years. Journal of Environmental Education 49: 189–206. doi:10.1080/00958964.2017.141 5195.

Eames, C. and C. Aguayo. 2019. Using mobile learning in free-choice educational settings to enhance ecological literacy (Report to Teaching and Learning Research Initiative). http://www.tlri.org.nz/tlri-research/research-completed/cross-sector/using-mobile-learning-free-choice-educational-settings.

Eames, C. and S. Birdsall. 2019. Teachers' perceptions of a co-constructed tool to enhance their pedagogical content knowledge in environmental education. Environmental Education Research. doi: 10.1080/13504622.2019.1645445.

Eames, C. and H. Mardon. 2020. The Enviroschools Programme in Aotearoa New Zealand: Action-orientated, culturally-responsive, holistic learning. *In*: Gough, A., J.C. Lee and E.P.K. Tsang (eds.). Green Schools Globally: Stories of Impact on Education for Sustainable Development. Springer, Cham, Switzerland.

Education Council Aotearoa New Zealand. 2017. Our code, our standards. Code of professional responsibility and standards for the teaching profession. https://teachingcouncil.nz/professional-practice/our-code-our-standards/.

Education Review Office. 2020. Nihinihi Whenua – Valuing te reo Māori: Student and whānau aspirations. https://www.ero.govt.nz/publications/nihinihi-whenua/.

Enviroschools. 2020. The Enviroschools Programme. https://www.enviroschools.org.nz.

Foreign Affairs and Trade. 2020. New Zealand's first Voluntary National Review. https://www.mfat. govt.nz/en/peace-rights-and-security/work-with-the-un-and-other-partners/new-zealand-and-the-sustainable-development-goals-sdgs/nzunvnr2019/.

Hopkins, C. and R. McKeown. 2005. Guidelines and Recommendations for Reorienting Teacher Education to Address Sustainability. UNESCO. https://unesdoc.unesco.org/ark:/48223/pf0000143370.

Law, B. 1996. Learning for a Sustainable Environment – Innovations in Teacher Education Through Environmental Education. Christchurch: Christchurch College of Education.

Law, B. and R. Baker. 1997. A case study of dilemmas and tensions: The writing and consultation process involved in developing a national guideline document for environmental education. Environmental Education Research 3: 225–232.

Magni, G. 2016. Indigenous knowledge and implications for the sustainable development agenda. UNESCO. https://unesdoc.unesco.org/ark:/48223/pf0000245623.

Marriott, L. and D. Sim. 2015. Indicators of inequality for Māori and Pacific people. The Journal of New Zealand Studies, 20, June (np) https://doi.org/10.26686/jnzs.v0i20.3876.

Ministry of Business, Innovation and Employment. 2016a. National Science Challenges. http://www.mbie.govt.nz/info-services/science-innovation/national-science-challenges.

Ministry of Business, Innovation and Employment. 2016b. A Nation of curious minds – He Whenua Hihiri i te Mahara. http://www.curiousminds.nz.

Ministry of Culture and Heritage. 2020. Te Ara. Encyclopaedia of New Zealand. https://teara.govt.nz/en.

Ministry of Education. 1996. Te Whāriki. He whāriki mātauranga mō ngā mokopuna o Aotearoa: Early childhood curriculum. Learning Media, Wellington, New Zealand. https://www.education.govt.nz/assets/Documents/Early-Childhood/Te-Whariki-1996.pdf.

Ministry of Education. 1999. Guidelines for environmental education in New Zealand schools. Learning Media, Wellington, New Zealand. https://health.tki.org.nz/Key-collections/Action-for-well-being/References-and-resources/Ministry-of-Education-publications.

Ministry of Education. 2007. The New Zealand Curriculum. https://nzcurriculum.tki.org.nz/.

Ministry of Education. 2017a. Te Whāriki. He whāriki mātauranga mō ngā mokopuna o Aotearoa. Early childhood curriculum. https://www.education.govt.nz/assets/Documents/Early-Childhood/Te-Whariki-Early-Childhood-Curriculum-ENG-Web.pdf.

Ministry of Education. 2017b. Te Marautanga o Aotearoa. https://tmoa.tki.org.nz/Te-Marautanga-o-Aotearoa.

Ministry of Education. 2019. He Whakaaro Education Insights: How environmentally aware are New Zealand students? https://www.educationcounts.govt.nz/publications/series/he-whakaaro/he-whakaaro-how-enviornmentally-aware-are-new-zealand-students.

Nairn, K. 2019. Learning from young people engaged in climate activism: the potential of collectivizing despair and hope. Young 27: 435–450, DOI: 10.1177/1103308818817603.

New Zealand Tourism. 2020. 100% Pure New Zealand: Facts about New Zealand. https://www.newzealand.com/nz/facts/.

New Zealand Qualifications Authority. 2020. NCEA. https://www.nzqa.govt.nz/ncea/.

Parliamentary Commissioner for the Environment. 2007. Outcome Evaluation. See Change: Learning and Education for Sustainability. Parliamentary Commissioner for the Environment, Wellington, New Zealand.

Parsons, M., J. Nalau and K. Fisher. 2017. Alternative perspectives on sustainability: indigenous knowledge and methodologies. Challenges in Sustainability 5: 7–14.

Piasentin, F.B. and L. Roberts. 2018. What elements in a sustainability course contribute to paradigm change and action competence? A study at Lincoln University, New Zealand. Environmental Education Research 24: 694–715. https://doi.org/10.1080/13504622.2017.1321735.

RCE Network. 2020. Untitled. https://www.rcenetwork.org/portal/rce_evaluation.

Royal Society of New Zealand. 2013. Languages in Aotearoa New Zealand. https://royalsociety.org.nz/what-we-do/our-expert-advice/all-expert-advice-papers/languages-in-aotearoa-new-zealand/.

SDSN. 2020. Sustainable Development Solutions Network. https://www.unsdsn.org.

SDG.org.nz. 2019. People's Report. https://www.sdg.org.nz/peoples-report/.

Seaweek. 2020. Seaweek. https://www.seaweek.org.nz/about-us/.

Shephard, K. and M. Furnari. 2012. Exploring what university teachers think about education for sustainability. Studies in Higher Education 38(10): 1577–1590, DOI: 10.1080/03075079.2011.644784.

Shephard, K., J. Harraway, B. Lovelock, M. Mirosa, S. Skeaff, L. Slooten, M. Strack, M. Furnari, T. Jowett and L. Deaker. 2015. Seeking learning outcomes appropriate for 'education for sustainable development' and for higher education. Assessment & Evaluation in Higher Education 40: 855–866. doi:10.1080/02602938.2015.1009871.

Springett, D. and B. Buchanan. 1991. The eco-school. In: Springett, D. and C.M. Hall (eds.). Our Common Future: The Way Forward. Proceedings of the New Zealand Natural Heritage Foundation International Conference on Environmental Education, Palmerston North, 26–30 August 1991.

Stats New Zealand. 2019. New Zealand's population reflects growing diversity. https://www.stats.govt.nz/news/new-zealands-population-reflects-growing-diversity.

Stats New Zealand. 2020. New Zealand's Environmental Reporting Series: Our atmosphere and climate 2020. https://www.stats.govt.nz/information-releases/new-zealands-environmental-reporting-series-our-atmosphere-and-climate-2020.

Taylor, P. 2020. New Zealand declares a climate change emergency. The Guardian. https://www.theguardian.com/world/2020/dec/02/new-zealand-declares-a-climate-change-emergency.

Times Higher Education (THE). 2020. The Times Higher Education Impact Rankings (https://www.timeshighereducation.com/rankings/impact/2020/).

Treasury. 2018. The relationship between the sustainable development goals and the living standards framework. https://www.treasury.govt.nz/publications/dp/dp-18-06-html.

Treasury. 2019. Our living standards framework. https://www.treasury.govt.nz/information-and-services/nz-economy/higher-living-standards/our-living-standards-framework).

University of Otago. 2017. Sustainability Strategic Framework 2017–2021. https://www.otago.ac.nz/coo/otago664842.pdf.

United Nations Educational, Scientific and Cultural Organisation (UNESCO). 2005. Guidelines and Recommendations for Reorienting Teacher Education to Address Sustainability. Education for Sustainable Development in Action. Technical Paper No. 2. UNESCO, France.

Vaioleti, T.M. and S.L. Morrison. 2019. The value of indigenous knowledge to education for sustainable development and climate change education in the Pacific. pp. 651–670. *In*: McKinley, E.A. and L.T. Smith (eds.). Handbook of Indigenous Education. Springer Nature. https://doi.org/https://doi.org/10.1007/978-981-10-3899-0_8.

Walker, R. 2004. Ka Whawhai Tonu Matou. Struggle without end (revised ed.). Penguin.

Wright, J. 2017. Taonga of an island nation: saving New Zealand's birds. Retrieved from http://www.pce.parliament.nz/media/1695/taonga-of-a-island-nation-web-final-small.pdf.

Concluding Thoughts on the Global Delivery of Environmental and Sustainability Education in the Context of the Sustainable Development Goals

Rosalba Thomas Muñoz[1],* and *Marco Rieckmann*[2]

The crisis the planet is going through, which is jeopardizing the survival of humans and all living beings that inhabit it, is multidimensional and complex. The foregoing chapters illustrate the importance of these two characteristics, mentioning them in the context of practically all countries considered. Undoubtedly, analysis and alternative solutions must be developed with a view to motivating us to change the way we construct knowledge about the world, promoting collaboration and dialogue between disciplines. Sustainability remains a multidisciplinary, problem-driven, solution-oriented and transformational field.

In the face of the crisis, sustainability became a paradigm, discourse and policy over the second half of the last century, and has been legitimized by both academic and governmental stakeholders and instances. This vision also takes account of the future and alternative approaches to it; as a model, it is directed and determined by a global framework, which must necessarily be designed to be appropriate to the local context.

---
[1] University of Colima, Mexico.
[2] University of Vechta, Germany.
Email: marco.rieckmann@uni-vechta.de
* Corresponding author: rosthomas@ucol.mx

The world cannot be conceived as it was a few decades ago. The magnitude of the problems we are facing forces us to question the way we organize ourselves, work, study, live and relate to each other. Today more than ever it is necessary to take responsibility for our actions and to do this it is necessary to break paradigms and transform the dynamics of teaching and learning; education must stop being adaptive and become transformative and disruptive. Transformative and disruptive environmental and sustainability education (ESE) does more than view traditional learning spaces as unique, seeing the world as a complex and dynamic system, creating spaces for reflection and negotiation, questioning reality to unearth alternatives, and providing students with tools that allow them to bring their knowledge, competences, and values into play (cf. Lotz-Sisitka et al. 2015, Rieckmann 2020, Rodríguez Aboytes and Barth 2020, Sterling 2011). Education must be transformative and disruptive if it is to develop change agents who are capable of imagining and building different, sustainable futures.

The journey we have taken through the chapters highlights the need to identify what the trends in ESE will be, and what the needs of environmental and sustainability educators will be, in the near future. While it is true that it is important to have a common theoretical framework for ESE, it is also true that each country has its own local problems to address, and that these may necessitate a different approach from that proposed by international bodies. All the countries emphasize the need to avoid falling into the traps of universalized and decontextualized discourse; to seek spaces for dialogue with the international community, but to go beyond what these can offer. The conclusion is that countries must transcend international discourse as well as take advantage of it to generate collaboration and synergies within their own borders. While some countries refer more strongly to UNESCO's Education for Sustainable Development (ESD) discourse and set their own course within this framework (e.g., Austria, Germany, Japan, Kenya, Sweden, Uganda, UK), ESE discourse in other countries deliberately distinguishes itself from UNESCO. In Brazil, Mexico, and South Africa, for example, there were critical voices about the concept of ESD, and in this connection, the concept of (critical) environmental education was preferred. In Canada, too, there was some resistance to replacing the concept of environmental education with ESD. In some countries, the influence of indigenous perspectives was also evident (e.g., naturalist philosophy in India; *Buen Vivir* in Ecuador). In Germany, Austria and the UK, development education and global learning were described as playing an important role alongside ESD and environmental education, although it was not always clear how they related to ESD.

As a model, sustainability encourages reform and innovation; however, it is crucial that the model be adapted to local circumstances. Sustainability is not a discourse that inherently reflects the needs of all communities and populations; on the contrary, it reflects generic issues that undoubtedly correspond to international progress but do not take local needs into account. This is something we all—educational communities and governments—need to do.

Even though the Sustainable Development Goals (SDGs) and Agenda 2030 officially apply to all countries, they have not been formally incorporated into the government programs of several of the countries mentioned in this book; spaces for dialogue are thus essential. A certain level of persistence is identified as necessary to disseminate the concept of the 2030 Agenda and the SDGs, to the detriment of reflection that could lead to the questioning of the relevance of the development model itself. All this diversity, terminological nuance and international trends must include theoretical contributions from the local level, as these represent the voices that are critical of the notion of development when it is restricted solely to the idea of economic growth.

ESE necessarily implies the addressing of complexity. In each chapter, the political, economic, technological, environmental, social, and of course, cultural context must be taken into account in order to base the discourse of the SDGs and the 2030 Agenda on the conditions pertaining in each country and ensure there is the political will to achieve the agreed targets and deliver the agreed actions. Educating students to understand complexity involves problematizing discourses, and analysis of the triggers and consequences of problems, the beneficiaries and losers with regard to any given development model.

Cultural identity must be incorporated into any agenda that sees itself as promoting the development of individuals and communities throughout the world. Beyond promoting a hegemonic and unique discourse, sustainability inspires us to think about the particular, the specific, to ensure that Agenda 2030 is also an opportunity for every country to take an in-depth look at local cultures and their needs.

All countries highlight the importance of increasing research into ESE and new methodologies, and of greater levels of reflection. However, a truly relevant educational policy is one that is capable of modifying educational environments so as to produce citizens who are forward-looking and can think critically, systemically, and ethically; that is, to foster the competences to promote sustainability.

All countries identify national experiences, progress, achievements, limitations, and shortcomings with regard to the way they approach and implement environmental and sustainability policy. However, more effort is needed to put in place programs and projects that promote knowledge, values, and competences to address environmental and socio-economic problems and to raise awareness of the need for a sustainable future. Although progress is being made with ESE in many countries, the necessary political support and structural integration into the education systems is still lacking in many places. ESE needs to be a more central aspect of teacher education and whole-institution approaches need to be more consistently adopted.

As some of the chapters show, few countries have responded sufficiently when it comes to climate change education (In Austria, however, climate action is a central aspect of ESE). Compliance with the Paris Agreements is already not considered essential for those that have not been able to overcome the institutional and governmental barriers; the remaining stakeholders will be able to do very little on their own. This highlights the urgent need to link theory and practice and emphasize that a multidisciplinary approach is an appropriate way to meet the sustainability challenges of the 21st century.

Finally, it is essential that any sustainability-oriented agenda includes shared governance. Governance, understood as the mechanisms, processes and institutions through which citizens express their interests, exercise their rights, satisfy their obligations, and resolve their differences, should be incorporated into 2030 Agendas worldwide. Governance refers not only to the State or public administrative systems, but also to projects, protected areas, natural resources, companies, communities and families. Governance and sustainability are the two prongs of a strategy that will guarantee the long-term conservation of natural heritage. The way in which societies choose to govern their natural resources has profound consequences, so a better understanding of governance processes for participatory management of our natural heritage is an essential step towards sustainability. All countries are showing the will to promote ESE; however, not all have the same advocates promoting it in the same way. While governments in some countries (e.g., Austria, Germany, Japan, Sweden, the UK) are pushing ESE-related activities relatively strongly, in other countries ESE depends more on grassroots/bottom-up activities (e.g., Australia, Chile, Mexico, New Zealand, USA). Hence, it is important that the institutional framework starts with governments and is supported by other institutions, including higher education institutions, non-governmental organizations, and private initiatives.

References

Lotz-Sisitka, H., A.E.J. Wals, D. Kronlid and D. McGarry. 2015. Transformative, transgressive social learning: rethinking higher education pedagogy in times of systemic global dysfunction. Current Opinion in Environmental Sustainability 16: 73–80. https://doi.org/10.1016/j.cosust.2015.07.018.

Rieckmann, M. 2020. Emancipatory and transformative global citizenship education in formal and informal settings – empowering learners to change structures. Tertium Comparationis: Journal für International und Interkulturell Vergleichende Erziehungswissenschaft 26(2): 174–186.

Rodríguez Aboytes, J.G. and M. Barth. 2020. Transformative learning in the field of sustainability: a systematic literature review (1999–2019). International Journal of Sustainability in Higher Education 21(5): 993–1013. https://doi.org/10.1108/IJSHE-05-2019-0168.

Sterling, S. 2011. Transformative learning and sustainability: sketching the conceptual ground. Learning and Teaching in Higher Education 5(2010-11): 17–33.

For Product Safety Concerns and Information please contact our EU
representative GPSR@taylorandfrancis.com
Taylor & Francis Verlag GmbH, Kaufingerstraße 24, 80331 München, Germany

www.ingramcontent.com/pod-product-compliance
Lightning Source LLC
Chambersburg PA
CBHW060804220326
41598CB00022B/2535

9 780367 702434